Reaction Engineering in Direct Coal Liquefaction

ENERGY SCIENCE AND TECHNOLOGY

A Series of Graduate Textbooks, Monographs, Research Papers

Series Editors: C.Y. WEN and E. STANLEY LEE

Consulting Editor: Richard Bellman

No. 1 RICHARD C. BAILIE
Energy Conversion Engineering, 1978

No. 2 C.Y. WEN and E. STANLEY LEE (Eds.)
Coal Conversion Technology, 1979

No. 3 YATISH T. SHAH
Reaction Engineering in Direct Coal Liquefaction, 1981

Other Numbers in preparation

Coauthors

N. L. CARR Gulf Research and Development Company Pittsburgh, Pennsylvania 15230	pp. 213–284
Y. C. FU Pittsburgh Energy Technology Center U.S. Department of Energy Pittsburgh, Pennsylvania	pp. 1–23
J. A. GRAY Gulf Research and Development Company Pittsburgh, Pennsylvania 15230	pp. 24–121
S. KRISHNAMURTHY* Department of Chemical and Petroleum Engineering University of Pittsburgh Pittsburgh, Pennsylvania 15261	pp. 122–161, 162–212, 285–381
S. V. PANVELKER Department of Chemical and Petroleum Engineering University of Pittsburgh Pittsburgh, Pennsylvania 15261	pp. 285–381
S. J. PARULEKAR Department of Chemical and Petroleum Engineering University of Pittsburgh Pittsburgh, Pennsylvania 15261	pp. 213–284
R. G. RUBERTO Gulf Research and Development Company Pittsburgh, Pennsylvania 15320	pp. 162–212
G. J. STIEGEL Pittsburgh Energy Technology Center U.S. Department of Energy Pittsburgh, Pennsylvania	pp. 285–381

* Present address:
Mobil Oil Research and Development Company
Paulsboro, New Jersey

Reaction Engineering in Direct Coal Liquefaction

Yatish T. Shah

Department of Chemical and Petroleum Engineering
University of Pittsburgh, Pittsburgh, Pennsylvania

1981

Addison-Wesley Publishing Company
Advanced Book Program/World Science Division
Reading, Massachusetts

London · Amsterdam · Don Mills, Ontario · Sydney · Tokyo

Library of Congress Cataloging in Publication Data

Shah, Yatish T.
Reaction engineering in direct coal liquefaction.

(Energy science and technology ; no. 3)
Includes index.
Contents: Processes for direct coal liquefaction by solvent extraction / Y.C. Fu and Y.T. Shah – Structure and properties of coal and the mechanism of coal liquefaction / J.A. Gray and Y.T. Shah – Heteroatom removal during liquefaction / Y.T. Shah and S. Krishnamurthy – [etc.]
1. Coal liquefaction. 2. Chemical reactions.
I. Title. II. Series.
TP352.S45 662'.6622 81-3650
AACR2

ISBN 0-201-07381-1
ABCDEFGHIJ-HA-8987654321

Manufactured in the United States of America.

To my wife, Mary

Contents

Series Editors' Foreword

Accelerated research and development for alternate energy resources to supplement the dwindling supply of petroleum and natural gas is currently under way throughout the world. The energy problem has suddenly created new demands on the effective dissemination of large volumes of information created in several languages as well as on science and engineering education. A serious gap already exists between research publications and the more comprehensive and systematic treatment of subjects found in a monographic text. One of the purposes of this Series is to fill this gap. A second purpose is to provide a vehicle for use as a teaching text. In fact, the magnitude of the energy problem and its solution through the creative application of scientific and engineering principles requires that essentially all scientists and engineers possess some background in the basic concepts of energy production, distribution, utilization, and related economics and social environmental implications. However, most of the currently available books and monographs are neither adequate nor suitable for use as textbooks. Much of the material published in the last several years addresses itself either to the general public or to a group of specialists involved in research and development.

Needed are books that are logically developed to emphasize underlying principles and the applications of these principles to energy and, when appropriate, to provide problems suitable for use by instructors and students in classes.

Volume	Authors	Title
1.	R. C. Bailie	*Energy Conversion Engineering
2.	C. Y. Wen & S. E. Lee	*Coal Conversion Technology
3.	Y. T. Shah	*Reaction Engineering in Direct Coal Liquefaction

The following books are planned:

Mathematics and Modeling of Geothermal Energy
Trace Element and Pollution in Energy Production
Fluidized Bed Combustion
Biomass Conversion Engineering
Water Uses in Energy Production
Chemical Process Energy Analysis and Synthesis

It is the sincere wish of the Editors that this Series will be able to play a key role in filling the existing gaps in information and in education.

Of all the synthetic fuel technologies currently under development, coal liquefaction is potentially the most attractive. Direct and indirect coal liquefaction processes can convert coal to fuel oil, gasoline, gas and chemical feedstocks. However, synthetic fuels made from coal are expensive. The high cost is due in part to low thermal efficiencies and high capital costs for synfuel plants. Direct coal liquefaction processes appear to enjoy higher thermal efficiencies than indirect coal liquefaction processes which first convert coal into a gas that is recombined to form liquids. In order to reduce the high capital cost caused by the many process steps required, the severity of temperature and pressure, and the excessive requirement of expensive hydrogen, many reaction engineering studies in direct coal liquefaction processes have been made.

There is a great need for a more systematic approach to interpret and organize the voluminous amount of published literature. This book, the third volume of the series, addresses this need; it will serve as a convenient guide to researchers working in the field as well as a useful introduction to coal liquefaction reaction engineering for scientists and engineers entering this area of research.

C. Y. WEN
E. STANLEY LEE

Preface

The production of liquid fuels from coal has been a possibility since the beginning of this century. It was in vogue during the Second World War, primarily in Germany, but interest in it waned shortly thereafter. It is ironic that in the closing years of this century we must look back to coal as a source of fuel. Such a circumstance has largely been nurtured by the uncertainty in world petroleum supplies.

Production of liquid fuels from coal largely falls into two main categories – direct and indirect. In the former, fuels are produced by the dissolution of coal in an organic solvent while, in the latter, the coal is fragmented into CO, CO_2, H_2 and CH_4 for suitable recombination at a later stage. In dealing with direct processes we have to date accumulated sufficient knowledge to require the need for a comprehensive review of the state of the art. Although progress has been made toward the understanding of the fundamentals of liquefaction, problems that impede its commercial application still exist. It is hoped that in these areas the book will stimulate the novice. While a number of recent books have treated direct coal liquefaction in general terms, the scope of the present book is kept narrow (i.e., only pertaining to reaction engineering) so that an in depth analysis of the subject is possible.

The book begins with a review in Chapter 1 of the various processes currently of promise. In Chapter 2 there is an overview of the structure and mechanism of coal liquefaction. The chemistry associated with coal liquefaction has been described at great length in a number of publications by Dr. D. D. Whitehurst and his co-workers at Mobil Research and Development Company, Princeton, New Jersey. Here we present only an "engineer's overview" of the subject. Chapter 3 reviews the present state of the art on the removal of heteroatoms (containing sulfur, oxygen, and nitrogen) during liquefaction. Lumped kinetic models, which are essential to reactor modelling, as well as their validity, are discussed in Chapter 4. The use of these models along with other hydrodynamic, mixing and transport parameters required for the design of commercial coal liquefaction reactors are discussed in Chapter 5. Chapter 6 reviews various problem areas and possible solutions in upgrading coal liquids to marketable products.

An effort such as this cannot be successfully completed without help from many friends and colleagues. I should especially like to thank all my coauthors, Drs. S. Krishnamurthy, N. L. Carr, J. A. Gray, and Y. C. Fu, and Mr. G. J. Stiegel, R. Ruberto, S. Panvelker, and S. Parulekar for their dedicated and punctual efforts.

This book presents as much their thoughts as mine. I gratefully acknowledge the help of Dr. C. Y. Wen, one of the editors of this series, in every stage of this project. Many constructive suggestions made by Dr. Wen and his graduate students have been most valuable and are appreciated. The comments and discussion on several chapters by Dr. I. Wender and Dr. D. C. Cronauer were useful. The painstaking effort by Mrs. J. Glasser in typing several versions of this book is, of course, most appreciated. Last, but not least, such a project could never have been successfully completed without love and sacrifice from my family and particularly my wife.

YATISH T. SHAH

1. Processes for Direct Coal Liquefaction by Solvent Extraction

Y.C. Fu and Y.T. Shah

INTRODUCTION

Direct liquefaction refers to the processes in which coal is reacted under high hydrogen pressure at elevated temperature in the presence or absence of hydrogen-donor solvent and catalyst. Three alternatives are pyrolysis, solvent extraction, and catalytic hydrogenation. This book deals with reaction engineering associated with the processes for solvent extraction of coal and for catalytic hydrogenation of coal in the presence of a solvent. The topic is of particular current interest because as described in this chapter, three processes of this type are currently being developed with the joint cooperation of industry and government (Department of Energy).

The ultimate goal of liquefying coal is to increase the ratio of hydrogen to carbon and to remove the heteroatoms (S and N) and ash from coal. With the increasing shortage of liquid fuels, coal liquefaction will be needed for producing boiler fuels, transportation fuels, and home heating oils. The technology for liquefying coal to synthetic fuel has been around for some time. Development of direct coal liquefaction processes began in Germany in the 1920s. The Pott-Broche process [9, 27] (solvent extraction) and the Bergius process [8, 27, 63] (catalytic hydrogenation) were operated on commercial scale during the 2nd World War. In the early 1950s, the U.S. Bureau of Mines constructed a demonstration plant at Louisiana, Missouri, and demonstrated the production of liquid fuels from coal by catalytic hydrogenation [34]. The efforts were resumed in earnest by 1962 by the Office of Coal Research, a predecessor of the Department of Energy (DOE).

Since 1962, major advances have been made in understanding the chemistry of coal liquefaction and several modifications have been developed including coal liquefaction by hydrogen transfer from donor solvents and pyrolysis of coal in the

---

Yatish T. Shah, Reaction Engineering in Direct Coal Liquefaction, ISBN 0-201-07381-1

presence of hydrogen for a very short period of time [4, 16, 26, 29, 30, 33, 36, 53, 54, 60, 62]. Other concepts that have also emerged are catalytic hydrogenation in ebulated-bed [10, 19, 38, 39, 40] and fixed-bed [2, 8, 27, 28, 63, 76] reactors, zinc halide hydrocracking [27, 21, 22, 44–46, 67–69, 74, 78], the reaction of CO-containing synthesis gas and steam with coal [5, 7, 25, 73], supercritical extraction of coal [1, 47, 57, 72], solvolysis of coal [6, 37, 42, 58], and the two-stage liquefaction process which combines solvent extraction and catalytic hydrogenation [11, 23, 24, 50, 65].

In this book, however, we restrict our discussion to reaction engineering associated with three major variations of liquefaction technology that are now undergoing large-scale development through the cooperative efforts of industry and DOE [18]. These are the solvent refined coal (SRC-I and SRC-II) process, the Exxon donor solvent (EDS) process, and the H-coal process. The 6-ton/day SRC pilot plant is operated at Wilsonville, Alabama. The 50-ton/day SRC pilot plant at Fort Lewis, Washington, has been in operation since 1974; and the 250-ton/day EDS pilot plant at Baytown, Texas, and the 600-ton/day H-coal pilot plant at Catlettsburg, Kentucky, have begun operation in 1980. In addition, two large SRC demonstration plants (SRC-I and SRC-II), both at 6000-ton/day capacity, have already been planned for construction. These plants will provide the basis for the establishment of a synthetic fuels industry.

1.1 Solvent Extraction of Coal

Coal may be extracted and dissolved at temperatures of 625–725 K by the action of hydrogen donor solvents with and without catalyst or molecular hydrogen present. During the coal dissolution, hydrogen from the donor solvent reacts directly with the coal molecule to produce hydrocarbon liquids. Other reactions such as depolymerization, hydrocracking, and sulfur removal reaction also occur in the dissolver. Hydrogen-donor compounds exist in various boiling ranges of coal liquids, but may be more concentrated in the higher-boiling-point fractions and the hydrogenated coal-derived solvents.

Several processes of different modes have been developed. In the early Pott-Broche process [9] developed in Germany, pulverized dried coal was mixed with a coal-derived solvent in a 2:1 solvent-to-coal ratio, pressurized to 10–15 MPa and heated at 688–703 K for about one hour. The slurry was filtered, and the filtrate was vacuum distilled to recover solvent and product extract. Two solvent extraction processes that are currently in the most advanced stages are the SRC process and the EDS process.

1.1.1 *Solvent-Refined Coal*

This process was initiated by the Pittsburg and Midway Coal Mining Company in the early 1960s. At present, there are two SRC pilot plants in operation in the United States: a 50-ton/day pilot plant at Ft. Lewis, Washington [51], and a 6-ton/day pilot plant at Wilsonville, Alabama [48].

A. SRC-I PROCESS

In the SRC-I (solid fuel product) operating mode, shown in the flow diagram of Figure 1-1, pulverized coal is mixed with process-derived solvent. After adding hydrogen, the slurry is pumped through a preheater and into a dissolver which operates at 700–740 K and at a pressure of 10–14 MPa. Coal conversions of 92 to 95% are achieved during the residence time of 20 to 60 minutes. The dissolver effluent is separated into gases and slurry in a separator. The raw gas is scrubbed with a dilute caustic solution to remove hydrogen sulfide and carbon dioxide. A portion of the hydrogen stream is purged to prevent accumulation of undesirable byproducts (C_1-C_4 hydrocarbon gases and CO), but most of the gas is recycled into the process. The slurry product, after being cooled to 588 K and depressurized to 0.8 MPa, is passed through a precoated rotary drum or leaf filter to separate undissolved coal and mineral matters from the SRC solution. Other mineral separation methods (Kerr-McGee extraction, etc.) are also used. In a commercial plant, these residues would be sent to a gasifier to produce process hydrogen. The filtrate is vacuum distilled to separate the solvent from the nondistillable product (SRC), which is then solidified by cooling. The SRC product is almost ash free ($<$0.16%) and has a sulfur content of 0.6–0.9% and a heating value of approximately 37.2 MJ/kg.

In the Ft. Lewis pilot-plant experimental program, the effects of dissolver temperature and slurry space rate on product yields were investigated. The material balance runs indicated that increasing slurry space rate resulted in a higher yield of SRC and a lower yield of process solvent per unit of coal feed. Increasing tempera-

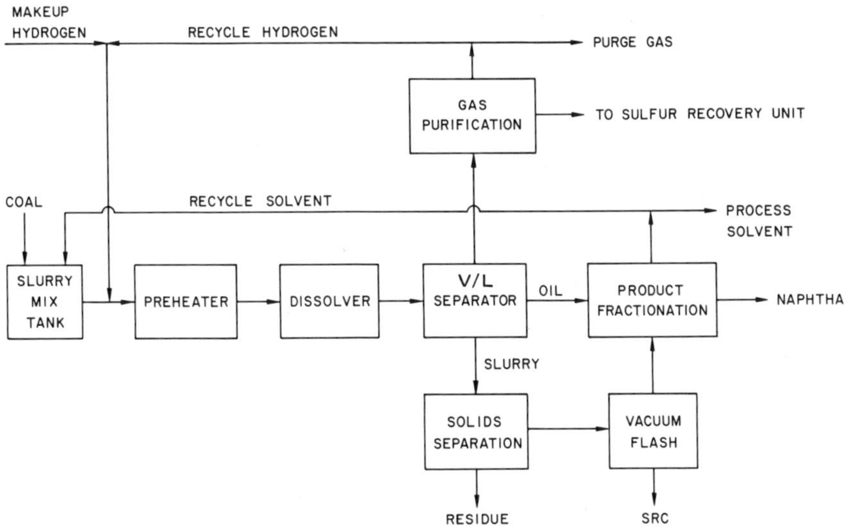

Figure 1-1. SRC-I Process scheme [51].

ture within 727–739 K resulted in decreases in both SRC and process solvent yields. Typical operating conditions and typical product yields are shown in Table 1-1.

The activity of the donor solvent and the solvent-to-coal ratio are also important factors for maintaining good coal conversion. Experimental works in a bench-scale unit [59] and at the Wilsonville pilot plant [12, 48] have indicated that, in general, SRC recycle solvents with higher hydrogen content and higher boiling range are more active. In correlating the Wilsonville performance with the Ft. Lewis operation, it was found initially that Ft. Lewis was able to operate at a lower solvent-to-coal ratio. However, after changing the boiling range of the recycle solvent from 450–700 K to 505–700 K, Wilsonville was able to reduce the solvent-to-coal ratio from 2.2 : 1 to 1.8 : 1.

The process equilibrium to maintain solvent balance depends on operating variables such as hydrogen pressure, residence time, and temperature of the reactor. In the SRC process, the stream of coal slurry and hydrogen-rich feed gas flows upward in a preheater of a long helical coil and then flows upward through a bubble-column dissolver. While the residence time in the dissolver is in the order of 20–60 minutes, the residence time based on effective reaction volume occupied by the slurry in the preheater under plug flow conditions is very short. The coal conversion and the SRC yield are related to the preheater residence time and outlet temperature. Preheater gas bypass runs [14, 15] were conducted at the Wilsonville pilot plant with part or all of the feed gas bypassed around the preheater. By bypassing

Table 1-1

SRC-I Product Yields [51]

Operating Conditions	
Coal	Western Kentucky 9/14
Coal feed rate, maf	1837 kg/hr
Solvent feed rate	2860 kg/hr
Hydrogen purity in feed gas	94 mol%
Dissolver outlet temperature	724 K
Dissolver pressure	10.3 MPa
H_2 consumption	2.4 wt% of maf coal
Product Yield (wt% maf Coal)	
C_1–C_4 hydrocarbon gas	3.7
Light oil (466 K–)	5.1
Wash solvent (466–522 K)	4.0
Process solvent (522–727 K)	4.4
SRC	63.0
Ash	9.6
Unreacted coal	5.4

part of the cold hydrogen gas around the preheater to quench the feed to the dissolver, higher preheater outlet temperatures can be maintained.

An increase in hydrogen partial pressure at the dissolver causes increased reaction rates, along with an increase in hydrogen consumption, and results in a higher coal conversion level [12, 13]. At reduced hydrogen partial pressures, the solvent cracking rate increased, and the gas and light-ends production also increased.

Studies at Wilsonville [12, 13] also indicate that bituminous coals of Kentucky 9 and 14 (Colonial Mine), Illinois No. 6 (Monterey Mine), and Indiana V (Old Ben Mine) are attractive for use in the SRC process, but Wyodak subbituminous coal (Belle Ayr Mine) and Utah bituminous coal (Emery Mine) are not. It appears that certain components of mineral matter in coal, iron compounds in particular, may have a catalytic effect on hydrogen transfer, resulting in an increase in coal conversion. High vitrinite and low oxygen contents of the coal are also desirable to achieve high SRC yield at low hydrogen consumption.

Plants at both Ft. Lewis and Wilsonville have attained operational reliability, but some mechanical problems with the rotary pressure precoat filters and the screens of the pressure-leaf filter were experienced. Efforts are therefore being made to improve the filter press operation and to develop possible alternative solid-liquid separation methods. Solid-liquid separation processes that are currently under development include the Kerr-McGee critical solvent deashing process [15, 57, 64], the C-E Lummus antisolvent deashing process [55, 56], and alternative deashing processes such as the UOP DEMEX process and filtration with a U.S. Filter Corporation vertical-leaf filter [64].

B. SRC-II PROCESS

The SRC-II process [3, 41, 51, 61] is a modified version of the SRC process. The Fort Lewis pilot plant has been modified to permit recycle of a portion of the reactor effluent slurry in place of the distillate solvent of the original SRC-I process. Because of increased severity of operating conditions in the SRC-II process, the primary product is a liquid distillate fuel with a 490–728 K boiling range.

The flow diagram of the SRC-II operating mode is shown in Figure 1-2. The general process is the same for both modes from the slurry mix tank through the dissolver. In the SRC-II mode, however, the dissolver effluent slurry is split into two streams; one is recycled to the slurry mix tank, while the other is passed into a vacuum flash unit. The vacuum flash condensate is fractionated to yield naphtha and middle and heavy distillates. The vacuum residue consists of heavy oil, unreacted coal, and mineral matter.

The increased conversion to light products in the SRC-II process results in a higher hydrogen consumption of about 4–5%, as compared to 2–2.5% for the SRC-I process. This is caused by the following factors:

1. The coal-oil slurry is subjected to reaction conditions of greater severity. The dissolver pressure is 13.2 MPa, compared to 10 MPa in SRC-I, and the super-

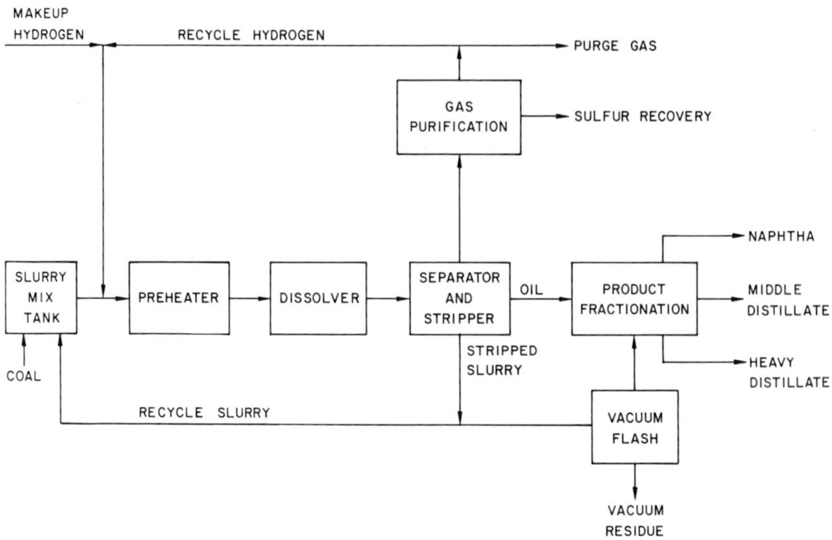

Figure 1-2. SRC-II Process scheme [51].

ficial slurry residence time is two to four times greater in the SRC-II mode owing to the reduced feed slurry rates and higher dissolver volume.

2. Recycle of the coal solution along with catalytic mineral residues allows a greater extent of hydrocracking in the dissolver.

Because the mineral residue is separated as vacuum flash bottoms along with the undissolved coal, the solid-liquid separation step of the SRC-I mode can be eliminated in the SRC-II mode. The elimination of high capital and maintenance costs of facilities for filtration, critical solvent deashing or antisolvent deashing is clearly an advantageous process improvement.

The degree of dissolution of coal and the distribution of products depend greatly on the reactivity of the coal feed. Three high-volatile bituminous coals have been tested to determine the effects of dissolver temperature, coal slurry feed rate, and coal concentration on product yields.

The material balance runs with Western Kentucky coal, Illinois No. 6 coal, and Blacksville No. 2 Pittsburgh Seam coal indicated that higher distillate yield was obtained at decreased coal concentration and increased dissolver residence time. In order to prevent excessive hydrocarbon gas formation and increased hydrogen consumption the dissolver temperature should not be kept too high. The optimum range is 725–735 K. The test results on the three coals are summarized in Table 1-2. The Western Kentucky and Illinois No. 6 coals are more reactive than the Blacksville No. 2 Pittsburgh Seam coal.

Table 1-2

Typical SRC-II Product Yields [51]

Coal:	Western Kentucky 9/14	Illinois No. 6 River King	Blacksville No. 2 Pittsburgh Seam
Operating Conditions			
Dehumidified coal feed rate, kg/hr	900	911	906
Feed slurry composition, wt%			
Deh. Coal	29.5	29.5	30.3
Solvent	33.4	35.6	29.8
SRC	23.9	20.0	24.1
Ash (in recycle slurry)	8.2	11.0	8.3
Undissolved coal (in recycle solvent)	5.0	3.9	7.5
Nominal dissolver residence time, hr	0.98	0.97	1.00
Hydrogen purity in feed gas, mol%	89.8	93.7	91.6
Average dissolver temperature, K	734	730	729
Dissolver pressure, MPa	13.34	13.44	14.09
H_2 consumption, wt% maf coal	4.8	4.7	3.5
Product Yield, wt% maf Coal			
C_1-C_4 hydrocarbon gas	18.4	15.8	13.5
Naphtha (450 K −)	14.2	17.0	11.9
Middle distillate (450–561 K) + heavy distillate (561–727 K)	28.2	30.3	22.3
SRC (727 K +)	26.1	23.0	36.8
Undissolved coal	6.6	5.0	11.9

C. SRC SHORT-CONTACT-TIME (SCT) PROCESS

Recently, there has been considerable interest in developing a two-stage liquefaction concept using short-contact-time coal extraction followed by an upgrading step to produce a distillate fuel oil. Short-contact-time liquefaction (SCTL) has the potential to liquefy coal at lower costs and higher efficiencies than other processes because of lower hydrogen consumption due to lower gas make, higher liquid yield, and higher output per unit reactor volume. This concept is derived from the work of Neavel [52] and Whitehurst et al. [70, 71], who have reported that in a very short time coal can be fragmented thermally into smaller molecules which are soluble in pyridine. This fragmentation produces free radicals that seek stabilization by a hydrogen transfer. This can take place in a hydrogen-donor solvent or nondonating solvent. Very little molecular hydrogen is used.

Mitchell et al. [50] presented data on a two-stage liquefaction coupling SCTL of coal with catalytic upgrading of the dissolved product. The experimental SCTL products were obtained from the Wilsonville pilot plant and laboratory autoclave runs. The SCTL runs at Wilsonville were made by employing preheater only (dissolver bypassed) [14]. It was noted that at 3–4 minutes residence time in the preheater only, the hydrogen consumption and gas make decreased greatly while the SRC yield increased. The SCTL products, in comparison with regular SRC-I product, are more polar and have functional materials containing more oxygen and sulfur. Catalytic hydroprocessing of a blend of a SCTL product with SRC recycle solvent was performed over a Harshaw Ni-Mo catalyst in a fixed bed to produce a distillate fuel.

Although the Consol Synthetic Fuel (CSF) process [23, 24], developed by the Consolidation Coal Company in the 1950s, was a two-stage process, it employed low-pressure solvent extraction in the absence of added hydrogen in the first stage. The operation of a 20-ton/day pilot plant at Cresap, West Virginia, was hampered by many mechanical difficulties. The DOE's coal-liquefaction program now includes a two-stage liquefaction which combines SCTL followed by upgrading using a commercial LC-fining process (expanded-bed hydrotreater) to produce synthetic distillage oils from coal [65]. The LC-fining upgrading process [17] can upgrade SRC extract or SCTL product and maintain continuous operation for extended periods of time as well. Because of the movement of catalyst particles in the liquid-gas medium, deposition of tar and coke on catalyst surface is minimized. A conceptual scheme for the two-stage process is illustrated in Figure 1-3.

The solid-liquid separation for the SCTL reactor effluent streams will have to rely on successful operation of the Lummus antisolvent deashing or the Kerr-McGee critical solvent deashing, which are under development for the SRC-I process. In view of anticipated difficulties in processing SCTL slurry, one must carefully evaluate this shortcoming versus anticipated advantages that can be derived from a newly developed process. A recent evaluation study [11] on a two-stage liquefaction process comprising an SRC-I unit and an LC-fining unit reported that this

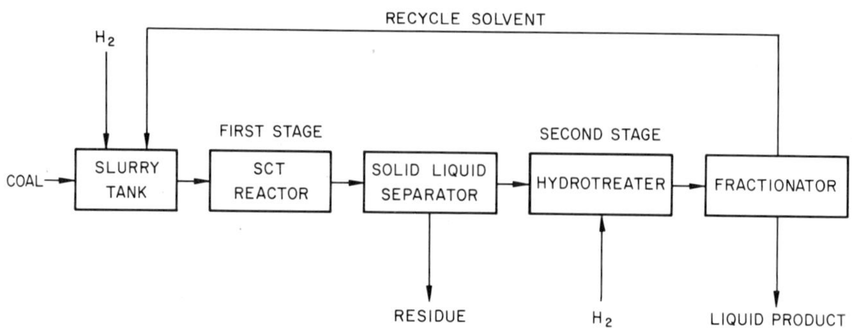

Figure 1-3. Short-contact-time two-stage process scheme.

combination would achieve a significant reduction in hydrogen consumption (as much as 1 wt% based on maf coal) relative to the SRC-II process.

1.1.2 *Exxon Donor-Solvent Process*

The Exxon Research and Engineering Company has been studying coal liquefaction since 1966 and has operated a continuous 1/2-ton/day integrated pilot plant [35]. Since 1976, a 1-ton/day pilot plant at Baytown, Texas, has been in operation under the joint sponsorship of DOE, the Electric Power Research Institute (EPRI), and industry [31]. The 250-ton/day pilot plant has begun operation in 1980.

The EDS process also liquefies coal with a hydrogen-donor solvent under hydrogen pressure. The solvent, however, is a catalytically hydrogenated recycle stream fractionated from the middle boiling range (478–728 K) of the liquid product. The liquefaction system is an upflow, tubular plug-flow reactor that is divided into four sections. The flow scheme of the EDS process is illustrated in Figure 1-4. The hydrogenated recycle solvent is mixed with fresh coal feed and pumped through a preheat furnace into the liquefaction reactor. The reactor operates at 700–744 K and 10.3–13.8 MPa. The reactor effluent is separated by distillation into light hydrocarbon gases, naphtha, distillates, and vacuum bottoms. The vacuum bottom (811 K +) is subjected to Exxon's proprietary Flexicoking process that uses air and steam to produce additional heavy oil and a low-heating-value fuel gas, and separates the mineral matter as ash. The Flexicoking process is a commercial petroleum process that employs integrated coking and gasification units in a fluidized mode of operation. The process hydrogen is produced by steam reforming of C_1–C_4 hydrocarbon gases.

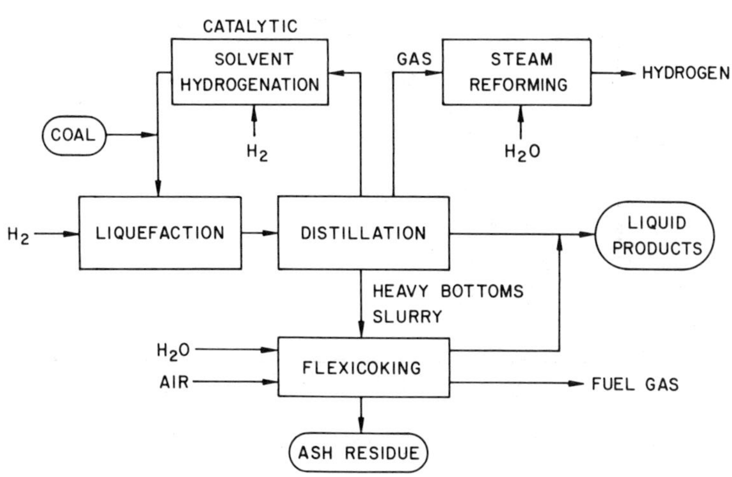

Figure 1-4. Flow scheme of Exxon donor-solvent process.

The hydrogenation of the recycle solvent is conducted in a conventional fixed-bed catalytic reactor using hydrotreating catalysts such as cobalt molybdate or nickel molybdate. Coal conversion and liquid yield are strongly dependent on the molecular composition, boiling-point range, and other properties of the solvent. Exxon uses its proprietary Solvent Quality Index (SQI) as the main criterion of solvent quality and correlates product yields with SQI. The SQI can be controlled by adjusting the hydrotreater temperature, and coal conversion, product yields, and hydrogen consumption decrease significantly when the SQI is less than a certain value. This critical SQI appears to be dependent on the coal type and liquefaction conditions, and the solvent quality requirements are significantly reduced when molecular hydrogen is present in the reactor.

The solvent has two major roles in the liquefaction process: (1) to stabilize the liquid products by donating hydrogen to the coal, and (2) to disperse and transport the coal particles through the liquefaction system. There will be a significant economic advantage if the solvent-to-coal ratio can be reduced. A high concentration of coal could decrease the quantity of recycle solvent required and reduce the size of the equipment needed for a solvent hydrogenation system. The pilot-plant test [35] showed that there is an upper limit of coal concentration, and the use of too low a solvent-to-coal ratio may cause operational problems and result in decreased liquid yield and lower product quality.

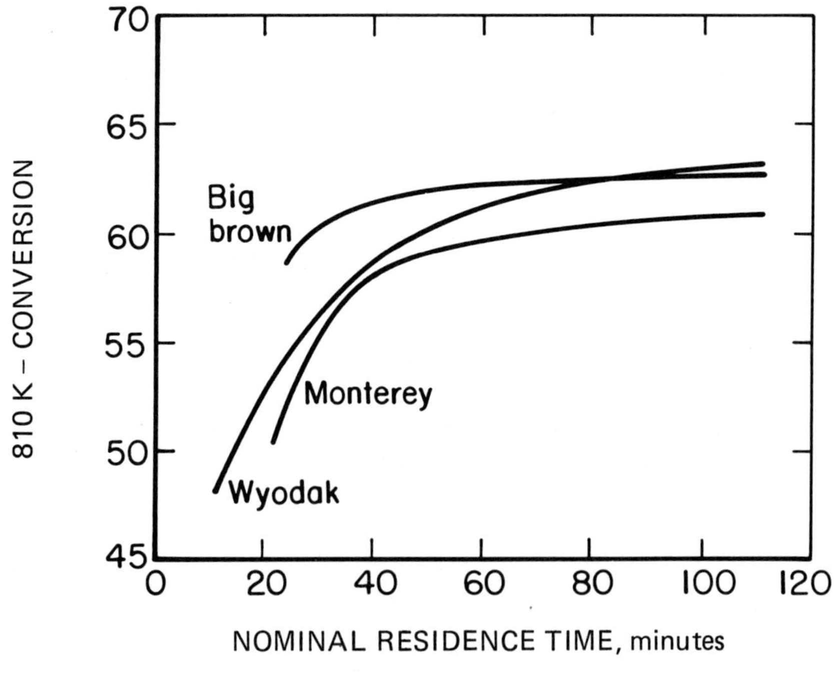

Figure 1-5. Conversion response to liquefaction residence time [49] (722 K, 10.3 MPa).

The specific feature of the EDS process is that the catalyst used in the process is not in contact with the high-molecular-weight asphaltenes, preasphaltenes, or mineral matters, thereby preventing fast deactivation of the catalyst. The use of vacuum distillation and the Flexicoking process is an added feature that permits the separation of solids without encountering mechanical problems.

Several coals, including bituminous, subbituminous, and lignite, have been tested in pilot plants at Baytown [32, 49]. Different coals gave different liquid yield responses to change in operating conditions. Figures 1-5 and 1-6 show the effect of liquefaction residence time on coal conversion and liquid yield, respectively, for three coals. These are Illinois No. 6 bituminous (Monterey No. 1 Mine), Wyoming subbituminous (Wyodak Mine), and Texas lignite (Big Brown Mine). For all three coals, the conversion to products boiling below 810 K increases only slightly after 40 minutes. The low-rank coals gave slightly higher conversion, owing to higher yields of water and carbon oxides. The C_4−811 K liquid yield, on the other hand, is adversely affected at long residence times by cracking of liquids to gases. The optimum residence time is about 40 minutes for Monterey coal, 60 minutes for Wyodak coal, and 25–40 minutes for Big Brown lignite. Table 1-3 shows the liquefaction product yields at these preferred operating conditions. The C_4−811 K liquid yield is highest for Monterey coal; however, substantial liquids are also obtained from the lower-rank coals. The total liquid yields will be higher when

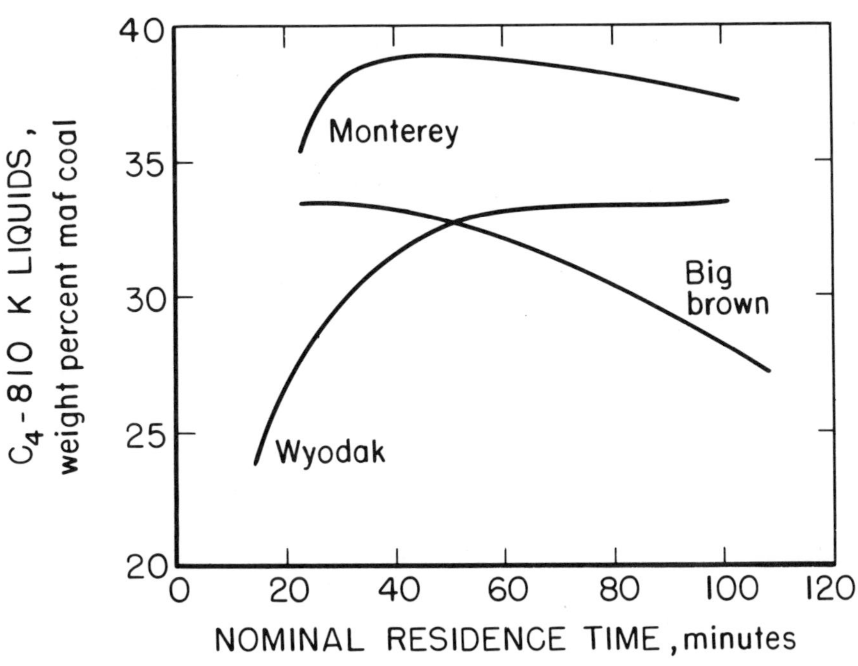

Figure 1-6. Yield response to liquefaction residence time [49] (722 K, 10.3 MPa).

Table 1-3
Product Yields of EDS Process [49]
(10.3 MPa, 722 K)

	Illinois No. 6 bituminous (Monterey)	Wyoming subbituminous (Wyodak)	Texas litnite (Big Brown)
Residence time, min	40	60	25–40
Yields, wt% maf coal:			
H_2	−4.3	−4.6	−3.9
$H_2O + CO_x$	12.2	22.3	21.7
$H_2S + NH_3$	4.2	0.9	1.7
C_1–C_3 gas	7.3	9.3	9.1
C_4–811 K liquid	38.8	33.3	33.3
811 K + bottoms	41.8	38.8	38.1

liquids from the Flexicoking process are combined. The total C_4+ liquid yields from bituminous coals approach 50 wt%, whereas the liquid yields with lignites are about 40 wt% on a moisture- and ash-free (maf) coal basis.

Operations of the pilot plants on the various coals have shown that the lower-rank coals are more difficult to process, primarily because of their high calcium content and high vacuum-bottom viscosity. Calcium carbonate, formed during the liquefaction, deposits on the reactor wall. This problem, however, may be controlled by pretreatment of the coal with sulfur dioxide to convert the calcium to calcium sulfate, which does not decompose to form deposits. The vacuum-bottoms viscosity can also be controlled by increasing liquefaction severity (increasing residence time) or decreasing the severity in the vacuum fractionator. In either case, the bottoms viscosity can be lowered and the bottoms can be pumped more easily to a coking reactor.

1.1.3 *Other Extraction Techniques*

A number of other extraction techniques are also currently being developed. Organic solvents at supercritical temperatures and pressures enhance vaporization of a solid or liquid. This technique was first applied to extracting thermally generated high-boiling coal liquids by the British National Coal Board [72]. The critical solvent deashing process [1, 57] being developed by Kerr-McGee uses solvents at or near critical pressure in solid-liquid separation. Catalytic, Inc. [47] has been evaluating the supercritical extraction (SCE) process since 1976. Compressed toluene solvent gas is used to extract the heated coal at 673–713 K and a pressure of 10 MPa. The vapor mixture containing the solvent and the extract is condensed and depressurized in another vessel. The disappearance of the enhanced solvent properties of toluene causes the extract to precipitate. After flushing toluene from the extract, the extract is recovered from the bottom of a vacuum still.

High-volatile, low-ranking coals are preferred because they give high extract yield without causing agglomeration of coal particles. With low-sulfur subbituminous Wyodak coal, the extract yield was 21 wt% on dry-coal basis. This yield is rather low, but the residual char retains the remaining unreacted coal and is an attractive feedstock for gasification and combustion. Since no molecular hydrogen or donor solvent was used in the SCE process, it would not be an effective desulfurization process to produce a clean fuel. The economic feasibility of this process, therefore, would very much depend on the value of its extract product as a chemical stock.

The government Industrial Research Institute of Kyushu, under the auspices of the Japan's Sunshine Project, has been developing a liquefaction process based on the solvolysis of coal with petroleum vacuum residue [6, 37]. A 1-ton/day pilot plant was completed in November 1978, and is now being tested and operated by Mitsubishi Heavy Industries, Ltd., Japan [42]. The main objective of this process is to utilize resources of coal and petroleum vacuum residue—the feedstock of which Japan has large quantities from refinery plants—to produce a clean fuel, and therefore the process uses a once-through mode.

The solvolysis process uses petroleum asphalt as the solvent to heat-treat and dissolve coal without catalyst under no hydrogen pressure. Petroleum asphalt consists mainly of aliphatic hydrocarbons and does not dissolve coal well. During the course of the SRC process development a petroleum-based solvent was tested in a bench-scale unit as startup solvent [58]. It was reported that the petroleum-based solvent was a relatively poor hydrogen donor and that difficulties in proper operation of the reactor occurred. The Japanese researchers [6, 37], on the other hand, reported that the heat treatment of petroleum asphalt at temperatures above 673 K caused thermal decomposition and polymerization of the asphalt, and resulted in the formation of a pitch rich in aromatic hydrocarbons. This pitch could then dissolve coal at elevated temperatures.

In the solvolysis process, powdered coal is mixed with petroleum asphalt at a coal-to-asphalt ratio of 1:2 and preheated in a furnace to 633–643 K. The slurry is pumped to a reactor and heat-treated at 673–683 K and 0.02–0.05 MPa. The average residence time is 0.8–1.5 hour. A gas stream containing mainly methane is discharged from the reactor, and the slurry of cracked oil (mainly n-paraffins), pitch, and residue is sent to a solid separation unit. Here, the slurry is mixed with a solvent to separate mineral matter and undissolved coal. The extract is sent to a distillation tower to recover middle oil, heavy oil and "refined solvolysis pitch." This refined solvolysis pitch is a uniform mixture of solubilized coal and petroleum asphalt, and can be used as a heavy fuel oil after desulfurization and viscosity adjustment.

Although no pilot-plant data are available, the laboratory data [6] indicated that Pittsburgh and Miike (Japan) bituminous coals were most suitable (among more than a dozen coals tested) when a vacuum residue of Khafji crude oil was used as the solvent. The yields for the Miike coal, for example, were: solvolysis pitch (quinoline solubles) 81.0 wt%, oil 5.8 wt%, gas 4.4 wt%, and residue (quinoline insolubles + ash) 8.8 wt% on feed basis.

1.2 Catalytic Hydrogenation of Coal in the Presence of a Solvent

At elevated temperature and under high hydrogen pressure, coal undergoes thermal reactions to produce free radicals that combine with hydrogen to yield hydrocarbon gases and liquids. The hydrogenation is aided by a catalyst and a solvent that acts as a hydrogen transfer agent. The hydrogen molecules diffuse into and react with the solvent, which in turn donates the hydrogen to the coal particles. The liquid product is formed as the H/C atomic ratio of coal increases from 0.8 to an approximate value of 1.0–1.1. Lighter liquid products could be formed with additional hydrogen uptake. A catalyst would serve to promote hydrogenation and heteroatom (S, N, and O) removal reaction processes during liquefaction. The required hydrogen is usually 4–6 wt% of the feed coal, depending on the severity of the processing conditions.

The earliest catalytic hydrogenation process is the Bergius process [8, 27, 63], developed in Germany and operated industrially during World War II. The process involves a catalytic hydrogenation of a coal-oil slurry at 755 K and 70 MPa in the first stage and a vapor-phase catalytic hydrogenation of the middle fraction product under relatively mild conditions of 672 K and 30 MPa in the second stage. Tin oxalate or iron oxide (2 wt% or more based on coal) was used as the catalyst in the liquid-phase hydrogenation, and tungsten sulfide was used in the vapor-phase hydrogenation [28]. The process is uneconomical mainly because of the extreme operating conditions required and the large amounts of hydrogen consumed (1.95 m^3/liter of middle oil).

1.2.1 *H-Coal Process*

This process has been under development by Hydrocarbon Research, Inc. since 1964. Tests have been carried out on bench-scale units processing 25 kg/day of coal and in a process development unit (PDU) handling about 3 ton/day [10, 19]. Bituminous, subbituminous, lignite, and brown coals have all been evaluated, and sustained H-coal process operation with satisfactory control of the catalyst-bed expansion has been demonstrated in the PDU tests. The H-coal process is now under the joint sponsorship of DOE, EPRI, and industry. The 600-ton/day pilot plant has begun operation in 1980.

The process can be operated in either a fuel-oil mode or a syncrude mode, depending on the type of fuel desired. The flow scheme of the H-coal process is shown in Figure 1-7. Coal, which has been ground and dried, is mixed with recycle oil, then pumped to 20 MPa, and mixed with compressed hydrogen. The mixture is preheated to 644 K and fed to the bottom of the ebullated-bed catalytic reactor. The catalyst is kept in suspension and in a fluidized state in an expanded bed by the

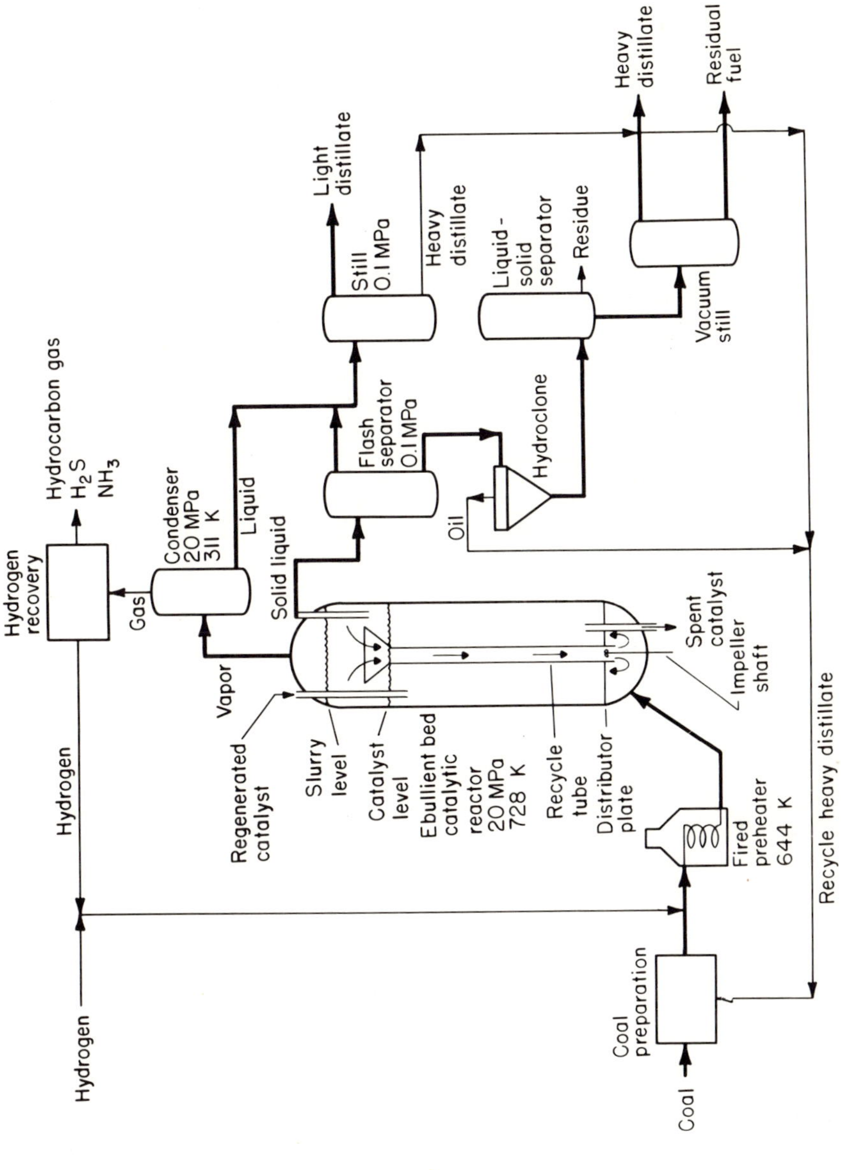

Figure 1-7. H-coal process scheme [18].

upward flow of the slurry and gas through the reactor. Catalyst activity is maintained by the addition and withdrawal of small quantities of catalyst to and from the reactor. The reactor also contains an internal tube for recirculating the reaction mixture through the catalyst bed. The reactor temperature is kept at 727 K. Vapor products leaving the top of the reactor are cooled to separate a liquid condensate. The gas stream is sent to a scrubber to absorb light hydrocarbons, ammonia, and hydrogen sulfide. The remaining hydrogen-rich gas is recycled. The condensate from the reactor vapors is released to atmospheric pressure and fed to an atmospheric still. The liquid-solid product from the reactor is let down into the atmospheric-pressure flash separator. The lighter hydrocarbon liquids that are flashed off are sent to the atmospheric still. The bottoms product from the flash separator is further separated with a hydroclone, a liquid-solid separator, and by vacuum distillation. The atmospheric still yields light and heavy distillates, and the vacuum still yields heavy distillates and residual fuel. A portion of the heavy distillate and hydroclone overflow is recycled as the slurry vehicle.

The severity of operating conditions of the H-coal process affects the type of fuel produced. In the fuel oil mode, a relatively high slurry-space velocity of 1250 $Kg/hr/m^3$ of reactor is used [40]. The reactor-space velocity can be related to the residence time of the slurry phase in the H-coal reactor. The hydrogen partial pressure is lowered to about 12 MPa, and hydrogen consumption is usually less than 4 wt% based on maf coal. In general, at lower hydrogen partial pressures there is less H_2 pickup by the oils, resulting in higher residuum and lower distillate yields.

Table 1-4

Product Yields of H-Coal Process [19, 40]

	Fuel-oil mode	Syncrude mode
Coal feed rate, $Mg/hr\ m^3$	1.25	0.53
Recycle oil/coal ratio	2.1	2.1
Reactor temperature, K	726	726
Hydrogen partial pressure, MPa	12.1	12.6
Yields, wt% of dry coal:		
C_1–C_3	7.71	9.97
C_4–478 K	16.90	23.66
478–797 K	18.28	23.21
797 K +	32.45	19.25
Unconverted coal	6.75	5.68
Ash	10.95	11.67
H_2O	6.67	7.37
NH_3	0.53	0.84
H_2S	2.55	2.65
$CO + CO_2$	1.01	0.95
Total	103.80	105.25
H_2 consumption, wt% of dry coal	3.80	5.25
Sulfur in 478 K + oil, wt%	0.49	0.26

Table 1-4 [19, 40] shows the product yields obtained with the PDU using an Illinois No. 6 bituminous coal. The catalyst was American Cyanamid HDS-1442A, cobalt molybdate on an alumina support, and the operating parameters were equal to those stipulated for the 600-ton/day pilot plant. A 0.49% sulfur fuel oil (478 K +) was produced at 3.8 wt% hydrogen consumption.

Table 1-4 also shows the results obtained in the syncrude mode operating at a space velocity of 530 $Kg/hr/m^3$. In general, with all other variables held constant, the decrease of the space velocity decreases the residuum yield but increases the naphtha yield. The hydrogen consumption is higher in the syncrude mode, at 5.25 wt% based on dry coal. The total liquid yield was 73 wt% and 76 wt% on an maf coal basis for the syncrude and fuel-oil mode respectively.

As with the SRC processes, one of the major problems has been solids separation. The H-coal process uses hydroclones to remove about two-thirds of the solids from the recycled liquid used as the vehicle. In the syncrude mode, the separation of unreacted coal and ash from the liquid can be accomplished by vacuum distillation. In the fuel-oil mode, however, an efficient solid-liquid separation method is needed. The antisolvent precipitation method has been selected, and a C-E Lummus antisolvent deashing unit is under construction at the 600-ton/day pilot plant [56].

Recently, the first demonstration of continuous catalyst addition and withdrawal in an actual PDU operation was undertaken to simulate syncrude mode operations in the pilot plant [38]. Using Illinois No. 6 coal and a catalyst replacement rate of 0.5 kg per metric ton of coal, thirty days of sustained operation was achieved with good stability. A similar thirty-day run simulating the pilot-plant syncrude-mode operation was also conducted using Wyodak coal [39]. Successful operation of the catalyst addition-withdrawal system to maintain equilibrium catalyst conditions was also attained with a catalyst replacement rate of 0.55 kg per metric ton of coal. Even with this high-moisture and high-calcium coal, no significant foaming or calcium carbonate deposits were found in the system. It was reported that Wyodak coal gave lower coal conversion and higher naphtha yield than Illinois No. 6 coal under the same syncrude-mode operating conditions.

Efforts are underway to develop improved catalysts for the H-coal process. Recent catalyst evaluation tests [20] in a bench-scale unit indicated that cobalt molybdate or nickel molybdate on alumina supports with optimum pore size distribution appears to be promising. Improvements in conversion of residual oil to lighter products and in denitrogenation activities have been reported. The catalyst usage is still very high and needs to be reduced.

1.2.2 *Other Catalytic Processes*

A number of other catalytic processes have been investigated in the past which have not yet been carried through on a large scale for a number of reasons. It has been known for some time that metal halides are effective catalysts for the hydrocracking of coal and coal tar [27, 75]. The Conoco Coal Development Company has done considerable research and development work on the use of large amounts of molten zinc chloride ($ZnCl_2$/coal ratio = 1–3) as a catalyst since 1964 [21, 22, 44–46, 77–79]. On the other hand, University of Utah has studied the hydrocracking of coal using small amounts (∿5 wt%) of metal-halide catalysts since 1969 [67–69, 74]. The Department of Energy carried out a considerable research effort on the Synthoil process which used a fixed-bed catalytic reactor, and Gulf Oil Company developed a segmented bed reactor for the Gulf catalytic

Table 1-5

Comparison of Major Coal Liquefaction Processes

	SRC-I	SRC-II	EDS	H-coal
Operating conditions:				
Pressure, MPa	10.3	13.4	10.3	12.3
Temperature, K	724	730	722	726
Residence time, min	40	60	40	30–70
Liquid yield, wt% maf coal	76	70	50*	74
Coal conversion, wt% maf coal	95	95	70	94
H_2 Consumption, wt% maf coal	2.4	4.7	4.3	3.8–5.3
Recycle system	Recycle of process solvent	Recycle of reactor slurry	Recycle of hydrogenated solvent	Recycle of heavy distillate
Reactor	Upflow column reactor	Upflow column reactor	Upflow, plug-flow tubular reactor	Ebullated-bed catalytic reactor
Catalyst		Mineral matter	Ni-Mo for solvent hydrogenation	Co-Mo or Ni-Mo

*C_4–811-K distillate.

coal liquefaction process. Dow Chemical Company is currently developing a novel catalytic process which uses the concepts of homogeneous catalysis.

1.3 Summary

Coal liquefaction technology has been improved substantially during the past 15 years by the development of several modifications of the coal liquefaction process. With the use of hydrogen-donor solvent in coal-oil slurry and an improved catalytic reactor system, coal liquefaction processes can now be operated under less severe conditions to achieve greater efficiencies with less hydrogen consumption. The comparison of the major processes described above and the analyses of typical coal liquid products are shown in Tables 1-5 and 1-6, respectively. The severity of the processing conditions for these solvent extraction and catalytic hydrogenation processes can be tailored to the reactivity of the coal feed and the quality of the product desired. Because of increased severity of operating conditions in the SRC-II process compared to the SRC-I process, the primary SRC-II product is a liquid distillate. Depending on the space velocity or the residence time of coal-oil slurry in the reactor, the H-coal process can be operated in either a fuel-oil or a syn-crude mode. In the EDS process, the optimum residence time can be adjusted for different ranks of coal to obtain the maximum liquid yields.

The use of extruded catalyst granules in the H-coal process eliminates the need for recovering dispersed catalyst powders from the solid residues, but frequent regeneration or replacement of the catalyst is required to sustain the continuous

Table 1-6

Analyses of Typical Coal Liquids

	SRC-I*	SRC-II*	H-coal† Fuel-oil mode	EDS†
Boiling range, K	>600	455–755	455–758	473–811
Sp. gravity, 289/289 K	1.23	1.00	1.10	1.08
Elemental analysis, wt%:				
C	87.24	88.98	86.50	89.40
H	5.77	8.70	7.69	7.70
O	3.93	1.16	3.19	1.83
N	2.15	0.84	0.81	0.66
S	0.75	0.28	0.37	0.41
H/C ratio	0.79	1.17	1.07	1.03

*Kentucky No. 9 and 14.
†Illinois No. 6.

operation. The separation of unreacted coal and mineral matter from heavy oil is the common problem for the H-coal, SRC-1, and other slurry-phase processes.

The major emphasis in the future development of coal liquefaction technology will be on process modifications and catalyst development to improve the process economics. Cost analysis [43] of a typical coal liquefaction process reveals that as much as 30% of the total cost is related to hydrogen production. One approach, therefore, is to use syngas (H_2 +CO) and steam instead of hydrogen as feed gas to the reactor [5, 7, 25, 73]. Efforts are being made in several research laboratories to develop low-cost disposable catalysts for coal liquefaction [64]. The role of coal minerals, pyrite, pyrrhotite, and various iron compounds in coal liquefaction is being investigated. This supportive research may lead to substantial improvements in the process economics of the SRC process.

REFERENCES

1. Adams, R. M., Knebel, A. H., and Rhodes, D. E., *Chem. Eng. Progr.*, **75** (6), 44 (1979).
2. Akhtar, S., Friedman, S., and Yavorsky, P. M., *AIChE Symp. Ser.*, **70** (137), 101, 106 (1974).
3. Anderson, R. P., *Preprints, Div. Fuel Chem., Am. Chem. Soc.*, Vol. 22, No. 6, 1977, p. 132.
4. Anthony, D. B., and Howard, J. B., *AIChE J.*, **22**, 625 (1976).
5. Appell, H. R., Miller, R. D., and Wender, I., "On the Mechanism of Lignite Liquefaction with Carbon Monoxide and Water," presented at the Division of Fuel Chemistry, 163rd National Meeting of the American Chemical Society, Boston, MA, April 10–14, 1972.
6. Arita, S., Osafune, K. O., Yamada, Y., Kakiyma, H., Honda, H., Tagawa, N., and Abe, S., "Solvolysis Liquefaction of Coal Using Petroleum Vacuum Residue," Government Industrial Research Institute Report, Kyushu, **19**, 1060, 1977.
7. Batchelder, R. F., and Fu, Y. C., *Ind. Eng. Chem., Process Des. Dev.*, **18**, 594 (1979).
8. Bergius, F., German Patent 303272 (1922).
9. Broche, H. and Reinmuth, W., *Ulmanns Enzyklopädie der Technischen Chemie*, 3rd ed., Vol. 10, Urban and Schwarzenberg, München, 1958.
10. Calderon, M., *Proceedings of EPRI Contractors Meeting on Coal Liquefaction*, Palo Alto, CA, May 1978.
11. Carson, T. C., Hastings, K. E., Potts, J. D., and Caspers, J., paper presented at the Annual Fall Conference of the Canadian Society for Chemical Engineering, Sarnia, Ontario, September 30–October 3, 1979.
12. Catalytic, Inc., "Operation of Solvent Refined Coal Pilot Plant at Wilsonville, AL," 1976 Annual Report FE-2270-15, June 1977.
13. Catalytic, Inc., "Operation of Solvent Refined Coal Pilot Plant at Wilsonville, AL," 1977 Annual Report, FE-2270-31, May 1978.
14. Catalytic, Inc., "Operation of Solvent Refined Coal Pilot Plant at Wilsonville, AL," January–March 1978 Quarterly Report, FE-2270-34, September 1978.

15. Catalytic, Inc., "Operation of Solvent Refined Coal Pilot Plant at Wilsonville, AL," July–September 1978 Quarterly Report, FE-2270-39, March 1979.
16. Che S. C., Duari-Swamy, K., Knell, E. W., and Zahradnik, R., paper presented at EPRI Contractors' Conference, Palo Alto, CA, May 9, 1979.
17. Chilligsworth, R. S., Hastings, K. E., Potts, J. D., and Unger, H., paper presented at AIChE Meeting, Miami, November 12–16, 1978.
18. "Coal Conversion," 1978 Technical Report, U.S. Department of Energy, September 1979.
19. Comolli, A. G., Battista, C. A., Johanson, E. S., and Laird, C. P., *Preprints, Div. Petrol. Chem., Amer. Chem. Soc.,* **23** (4), 1297 (1978).
20. Comolli, A. G., Johanson, E. S., and Huibers, D. T. A., paper presented at EPRI Contractors' Conference, Palo Alto, CA, May 9–10, 1979.
21. Conoco Coal Development Company, "Zinc Halide Hydrocracking Process for Distillate Fuels from Coal," Annual Technical Progress Report for 1976, Including the Quarter Ending January 31, 1977, February 1977.
22. Conoco Coal Development Company, "Zinc Halide Hydrocracking Process for Distillate Fuels from Coal," Annual Technical Progress Report for February 1, 1978–January 31, 1979, May 1979.
23. Consolidation Coal Company, "Phase I Research on CSF Process Pre-Pilot Plant," R&D Report No. 39, Interim Report No. 1, U.S. Department of Interior, Office of Coal Research, 1971.
24. Consolidation Coal Company, "Pilot Scale Development of the CSF Process," R&D Report No. 39, Interim Report No. 6, U.S. Department of Interior, Office of Coal Research, 1971.
25. DelBel, E., Friedman, S., Yavorsky, P. M., and Wender, I., "The Liquefaction of Lignite by the CO-Steam-Process," AIChE National Meeting, Houston, TX, March 16–20, 1975.
26. Dobner, S., Graff, R. A., and Squires, A. M., *Fuel,* **55**, 114 (1976).
27. Donath, E. E., Chapter 22 in *Chemistry of Coal Utilization,* Supplementary Vol., Lowry, H. H., Wiley, New York, 1963.
28. Donath, E. E., Fuel Processing Technology, **I**, 3 (1977).
29. Duncan, D. A., Beeson, J. L., and Oberle, R. D., "Research and Development of Rapid Hydrogenation for Coal Conversion to Synthetic Motor Fuels," Quarterly Report, April 1–June 30, 1978, FE-2307-38, September 1978.
30. Duncan, D. A., Beeson, J. L., and Oberle, R. D., "Research and Development of Rapid Hydrogenation for Coal Conversion to Synthetic Motor Fuels, "Annual Report, April 1, 1977–March 31, 1978, FE-2307-34, August 1978.
31. Exxon Research and Engineering Company, "EDS Coal Liquefaction Process Development," Phase III A, Annual Technical Progress Report, January 1–December 31, 1976, FE-2353-9, December 1977.
32. Exxon Research and Engineering Company, "EDS Coal Liquefaction Process Development," Phase III B/IV, Annual Technical Progress Report, July 1, 1977–June 30, 1978, FE-2893-17, September 1978.
33. Fallon, P. and Steinberg, M., paper presented at 173rd ACS National Meeting, New Orleans, LA, March 20–25, 1977.
34. Fieldner, A. C., Storch, H. H., and Hirst, L. L., *U.S. Bureau of Mines Rept. Invest.* No. 4651 (1949), 4770 (1950), 4856 (1951), 4942 (1952), 5043 (1953), 5118 (1954), 5236 (1955).
35. Furlong, L. E., Effron, E., Vernon, L. W., and Wilson, E. L., *Chem. Eng. Progr.,* **72** (8), 69 (1976).

36. Greene, M. I., "Research Studies for Producing Liquids by Short Residence Coal Hydropyrolysis," Semi-Annual Technical Progress Report, July 1977–January 1978, FE-2574-7, March 1978.
37. Honda, H. and Yamada, U., *J. Japan Petrol. Inst.*, **17**, 853 (1974).
38. Hydrocarbon Research, Inc., "Laboratory Program to Support H-Coal Pilot Plant Operations," PDU Run 5, FE-2547-19, June 1978.
39. Hydrocarbon Research, Inc., "Laboratory Program to Support H-Coal Pilot Plant Operations," PDU Run 6, FE-2547-26, November 1978.
40. Hydrocarbon Research, Inc., "H-Coal Integrated Pilot Plant," Phase I – Final Report, December 1973–June 1976, Vol. 1, HCP/T-1544/1, November 1977.
41. Jackson, D. M. and Schmid, B. K., paper presented at the Fifth Annual International Conference on Coal Gasification and Liquefaction, University of Pittsburgh, August 1–3, 1978.
42. Japan Ministry of International Trade and Industry, "Summary of Coal Gasification and Liquefaction," Japan's Sunshine Project, Vol. III, August 1979.
43. Kattel, S., "SYNTHOIL Process – Liquid Fuel from Coal Plant," Report No. 75-20, Bureau of Mines, Morgantown, WV, January 1975.
44. Kiovsky, T. E. and Wald, M. M., U.S. Patent 3,725,239 (1973).
45. Loth, R. A., U.S. Patent 3,790,468 (1974).
46. Loth, R. A., and Wald, M. M., U.S. Patent 3,790,469 (1974).
47. Maddocks, R. R., Gibson, Jr., and Williams, D. F., *Chem. Eng. Progr.*, **75** (6), 49 (1979).
48. Maddocks, R. R., paper presented before the Division of Petroleum Chemistry, American Chemical Society, Miami Beach, FL, September 10–15, 1978.
49. Mitchell, W. N., Trachte, K. L., and Zaczepinski, S., *Ind. Eng. Chem., Prod. Res. Dev.*, **18** (4), 311 (1979).
50. Mitchell, T. O., Whitehurst, D. D., and Heck, R. H., paper presented at AIChE Meeting, Miami, November 12–16, 1978.
51. Moschitto, R. D., 13th Intersociety Energy Conversion Engineering Conference, San Diego, August 1978.
52. Neavel, R. C., *Fuel*, **55**, 237 (1976).
53. Oberg, C. L., Combs, L. P., and Silverman, J., Proc. 13th Intersociety of Energy Conversion Engineering Conference, San Diego, p. 402, August 20–25, 1978.
54. O'Hara, F. M., "Synthetic Fuels Process Research Digest," ORNL/FE-1, Oak Ridge National Laboratory, November 1977.
55. Peluso, M. and Ogren, D. F., *Chem. Eng. Progr.*, **75** (6), 41 (1979).
56. Peluso, M., "The C-E Lummus Antisolvent Deashing Process," DOE Project Review Meeting, Pittsburgh, December 5, 1979.
57. Perlaky, C., Weber, W. H., and Rhodes, D. E., "Kerr-McGee Critical Solvent Deashing Process," DOE Project Review Meeting, Pittsburgh, December 1979.
58. Pittsburg & Midway Coal Mining Company, "Development of a Process for Producing an Ashless, Low-Sulfur Fuel from Coal," R&D Report No. 53, Interim Report No. 8, ERDA, FE-496-TI, May 1975.
59. Pittsburg & Midway Coal Mining Co., "Development of a Process for Producing an Ashless, Low-Sulfur Fuel from Coal," R&D Report No. 53, Interim Report No. 7, ERDA, FE-496-T4, September 1975.
60. Rockwell International Corporation, "Partial Liquefaction of Coal by Direct Hydrogenation," Annual Technical Progress Report, August 1977–July 1978, FE-2044-37, February 1979.

61. Schmid, B. K., and Jackson, D. M., paper presented at the Third Annual International Conference on Coal Gasification and Liquefaction, University of Pittsburgh, August 3–5, 1976.
62. Steinberg, M. and Fallon, P., *Chem. Eng. Prog.* **75** (6), 63 (1979).
63. Storch, H. H., Chapter 38 in *Chemistry of Coal Utilization,* Vol. 2, Lowry, H. H., Ed., Wiley, New York, 1945.
64. U.S. Department of Energy, "Solid-Liquid Separation in Coal Liquefaction Processes," DOE Project Review Meeting, Pittsburgh, December 5, 1979.
65. U.S. Department of Energy, "Refining of Syncrudes," DOE Contractors' Conference, Richmond, CA, May 7–8, 1979.
66. U.S. Department of Energy, "Disposable Catalysts in Coal Liquefaction," DOE Project Review Meeting, Albuquerque, NM, June 5–6, 1979.
67. University of Utah, "Hydrogenation of Western Coal in Dilute Phase Under Medium Pressure for Production of Liquids and Gases," Quarterly Reports, October 1970–April 1975.
68. University of Utah, "Applied Research and Evaluation of Process Concepts for Liquefaction and Gasification of Western Coals," U.S. ERDA Report, FE-1200-1, 1975.
69. University of Utah, "Applied Research and Evaluation of Process Concepts for Liquefaction and Gasification of Western Coals," U.S. ERDA Report FE-2006-2, FE-2006-3 (1976).
70. Whitehurst, D. D., and Mitchell, T. O., "Liquid Fuels from Coal," Ellington, R. T., Ed., p. 153, Academic Press, New York (1977).
71. Whitehurst, D. D., Farcasiu, M., Mitchell, T. O., and Dickert, J. J., "The Nature and Origin of Asphaltenes in Processed Coals," EPRI AF-480, July, 1977.
72. Whitehead, J. C., and Williams, D. F., *J. Inst. Fuel,* **49**, 182 (1975).
73. Willson, G. R., Knudson, C. L., Baker, G. G., Owens, T. C., and Severson, D. E., *Ind. Eng. Chem., Prod. Res. Dev.,* **18**, 297 (1979).
74. Wood, R. E., and Wiser, W. H., *Ind. Eng. Chem., Process Des. Dev.,* **15**, 144 (1976).
75. Wu, W. R. K. and Storch, H. H., *Bureau of Mines Bull.* 633, U.S. Department of the Interior, 1968.
76. Yavorsky, P. M., Akhtar, S., Lacey, J., Weintraub, M., and Reznick, A., *Chem. Eng. Progr.,* **71** (4), 79 (1975).
77. Zielke, C. W., Struck, R. T., Evans, J. M., Costanza, C. P., and Gorin, E., *Ind. Eng. Chem., Process Des. Dev.,* **5**, 158 (1966).
78. Zielke, C. W., Klunder, E. B., Maskew, J. T., and Struck, R. T., *Ind. Eng. Chem., Process Des. Dev.,* **19**, 85 (1980).
79. Zielke, C. W., Struck, R. T., Evans, J. M., Costanza, C. P., and Gorin, E., *Ind. Eng. Chem., Process Des. Dev.,* **5**, 151 (1966).

2. Structure and Properties of Coal and the Mechanism of Coal Liquefaction

J.A. Gray and Y.T. Shah

INTRODUCTION

In designing a chemical reactor, it is usually necessary to define the reactants and products in terms of physical properties, thermodynamic properties, chemical composition, and chemical structure. It is also necessary to define the reaction networks, the rates of reaction, and the effect of reaction conditions such as temperature, pressure, composition, and time on the networks and rates. Development of this basic information for the processing of a complex feedstock such as petroleum has been achieved, for the most part, because it is a physical mixture of a large number of readily identifiable types of organic compounds, e.g., straight-chain and branched alkanes, naphthenes, aromatics. Most organic reactions of industrial importance are less complex than those encountered in the processing of petroleum or petroleum derivatives. The design of chemical reactors for coal processing, especially donor-solvent liquefaction, is a unique problem because the primary reactant, in contrast to petroleum, is relatively undefined in its organic structure and reaction chemistry. After a voluminous amount of research over many decades on the origin, chemical structure, and reaction chemistry of various coals, we are far from definitive fundamental descriptions of coal structure. Paleobotanical and petrographic studies of coal clearly illustrate the reasons for this.

Coal is a heterogeneous rock, interspersed with minerals, that evolved from natural organic substances originating millions of years ago. The organic portion contains the elements carbon, hydrogen, oxygen, sulfur, and nitrogen. Coal is not a chemical mixture in the sense of petroleum, but is a mixture of rock types and minerals. The rock types are composed of various macerals that are representative of the source organic materials. It is at this level that one can begin to consider

Yatish T. Shah, Reaction Engineering in Direct Coal Liquefaction, ISBN 0-201-07381-1

chemical structure. Although many chemical structures are likely present in varying proportions within a maceral, certain consistent functional groups can be identified.

The structure and properties of coal play a very significant role on the mechanism of coal liquefaction in a donor solvent. In this chapter we first discuss the origin and physical characteristics of coal, and what we know about chemical structure of the important macerals. This is followed by some discussion on mechanisms of coal liquefaction.

2.1 The Origin and Character of Coal [60, 132, 133, 209]

Coal is believed to have originated from plant debris that was deposited in peat swamps, or in some cases transported and deposited in open bodies of water. In these moist environments, the plant matter underwent a variable biochemical process of partial decomposition, principally by the action of anaerobic microorganisms to form humic acids and then peat. The type of plant material deposited was a function of the geological period and the environment of the particular geographic location. Coal has been identified in every geological period from the Devonian (about 400 million years) to the Quarternary (about 1 million years), with most important deposits originating in the Carboniferous age (about 320 million years) and the Cretaceous and early Tertiary eras (60–130 million years). The character and principal varieties of vegetation that formed coal are different for each of the periods and lead to different types of coals. For example, coal deposits in the United States can be divided into seven major coal provinces as shown in Table 2-1 [75]. Most of this coal is of Carboniferous, Cretaceous, or Tertiary age, and about 45% of the reserves are in the Great Plains Province, consisting mostly of lignite and subbituminous coal [194].

The peat-forming or biochemical stage was interrupted when the peat became covered with sediments, and was compacted and subjected to the pressure of overburden or crustal movement and increased temperature over variable, but very long, periods of time. This second phase of coalification is called the geochemical or metamorphic stage. During the geochemical stage the deposit matured, i.e., underwent metamorphism, progressing in the series peat, brown coal, lignite, subbituminous, bituminous, and ultimately anthracite. The degree of maturity or coalification is called the rank, and is a function of the heat and pressure on the bed as well as the length of time at various conditions. In some cases, coalification was accelerated by extraordinary processes such as rock movement in crustal disturbances, intrusion of coal beds by magmatic rock, and exposure to surrounding radioactive minerals.

It has been shown extensively that for the series from peat to anthracite, a continuous transition exists in the microstructure and chemical and physical properties. Nearly all of these properties or combinations of them can be used to quantify the rank of a coal. These properties include reflectance of polished surface specimens; volatile-matter, moisture, carbon, hydrogen, and oxygen contents; and calorific value. Reflectance and carbon content increase with rank, while volatile-matter, and oxygen contents decrease. The ASTM rank classification used in the United States is based on volatile-matter (or fixed-carbon) content and calorific

Table 2-1

Distribution of Coals in the United States*

	Province						
	Appalachian	Interior	North Great Plains	Rocky Mountain	Pacific	Gulf	Alaska
States	PA, OH, VA, WV, E. KY, AL, TN	(a) East Coal Region: IN, IL, W. KY (b) West Coal Region: OK, MO, KS, AR	ND, SD, MT, N.E. WY	Many basins of different geol. history in WY, CO, UT, NM, AZ	WA, OR, CA	Parts of AR, TX, LA, MS, W. AL	AK
Age	Carboniferous (300 M yr)	Carboniferous	Some Cretaceous, mostly early Tertiary (100–50 M yr)	Mostly Cretaceous some early Tertiary (130–60 M yr)	Tertiary (60–15 M yr)	Tertiary (70–30 M yr)	Cretaceous and early Tertiary (100–50 M yr)
ASTM rank:							
Lign.	□	□	■	□	■	■	□
Subbit.	□	□	■	■	■	■	■
hvC	□	■	□	■	■	□	■
hvB	■	■	□	■	■	□	■
hvA	■	■	□	■	■	□	■
mv	■	□	□	⊠	⊠	□	□
lv	■	⊠	□	□	□	□	□
Anth.	■	⊠	□	⊠	□	□	⊠

*From Given [75]. ■ Principal deposits; ⊠ minor or sporadic occurrences; □ absent.

value [32, 41]. No one parameter is fully satisfactory in quantifying rank over the entire range. Sulfur and nitrogen are independent of rank. Proximate and ultimate analyses, calorific value, petrographic data, rank data, and other properties have been compiled for a large number of selected coals [32, 195]. The elemental analyses of the organic portion of typical coals of varying rank are shown in Table 2-2 [214]. Comparison of these data with similar analyses for the hydrocarbons petroleum, gasoline, and toluene show the relative deficiency of hydrogen and abundance of nitrogen and sulfur in coal.

The minerals that are typically present in coal are shown in Table 2-3. The possible sources of mineral matter include the mineral constituents originally in the plants, detrital matter carried into the peat deposit, deposits from the water in contact with the decaying plants and sludge during the early stages of coal formation, crystalline deposits from water streams percolating through cracks and faults in the coal seams, and products of decomposition or interaction with the coal. For example, pyritic sulfur in Illinois coal seams probably originated from submergence of peat-forming deposits under seawater. Sulfur released from the plant sludge in the form of hydrogen sulfide reacted with soluble iron compounds, precipitating iron sulfide. The ash contents of wood, leaves, and algae are typically $<1\%$, 8–12%, and 10–40%, respectively, making them important sources of minerals.

2.1.1 *Maceral Classification*

In viewing core samples of coal seams or hand specimens of coal, one can see a highly laminated structure, parallel to the bedding plane, consisting of layers varying in luster, texture, and size. Most coals from lignite to anthracite have this characteristic, but it is most outstanding in bituminous coals. Coal has been classified into rock types or lithotypes according to its megascopic properties [32, 42 194]. These classifications are useful for coarse distinction of coals, but are of little value in understanding the organic structure present. When a thin section of coal is examined under a microscope, a wide variety of fossil plant matter can be recognized. Phyterals identified include remnant wood and bark cell tissue; unorganized resins, waxes, and tannins; fungi (sclerotia); spore exines; cuticles, algae, and jellified tissue. Thus, another classification system has developed based on macerals (analogous to minerals in inorganic rocks) that describe the organic source materials. Each plant tissue, organ, or secretion is given a name. For example, spore material would be called sporinite and resinous material resinite. Table 2-4 [32, 42, 194] summarizes the classification of coal macerals, their characteristics, and the source material. Further subclasses of these macerals have been defined (for example, vitrinite is further subdivided on the basis of the remnant cell-wall structure that is visible) but have only been useful in geological studies. European petrographers have developed submaceral classifications specific for lignite and brown coals [31], but these have not gained wide use as yet.

The maceral content of coal is determined by measuring the reflectance of vertically incident polarized light on a polished coal surface using a sensitive photomultiplier cell attached to a microscope [43–45, 195]. A cylindrical briquet is

Table 2-2*

Chemical Constitution of Coals, Petroleum, and some Hydrocarbons†

	Anthracite, Pennsyl-vania	Low volatile bituminous, Pocahontas #3 Bed, West Virginia	High volatile A bituminous Pittsburgh Bed, Penn-sylvania	High volatile A bituminous, Kentucky seams #9, 11, 12 and 13‡	High volatile B bituminous, Illinois #6	Subbitu-minous A, Mammoth Bed Wyoming	Lignite, Beulah-Zap Bed, North Dakota	Petro-leum crude	Gaso-line	Toluene
C	93.7	90.7	84.2	79.1	76.7	76.7	72.6	83-87	86	91.3
H	2.4	4.6	5.6	6.0	5.2	5.6	4.9	11-14	14	8.7
O	2.4	2.8	6.9	7.6	12.8	15.5	20.2			
N	0.9	1.3	1.6	1.6	1.4	1.3	1.1	0.2		
S	0.6	0.6	1.7	5.7	3.9	0.9	1.2	1		
H/C atom ratio	0.31	0.61	0.80	0.91	0.81	0.88	0.81	1.76	1.94	1.14

* Reprinted from Wender [214], p. 103, by courtesy of Marcel Dekker, Inc.

† Coal analyses on moisture- and ash-free basis. Ash content of these coals range from 3 to 12%.

‡ All mined together and blended.

Table 2-3 [60, 132, 133, 194, 195, 209]

The Minerals in Coal

I. Silicates
Montmorillonite $4SiO_2 \cdot Al_2O_3(1+x)H_2O$
Illite $(OH)_4K_y(Si_{8-y} \cdot Al_y)(Al_4 \cdot Fe_4 \cdot Mg_4)O_{40}$
Kaolinite $2H_2O \cdot Al_2O_3 \cdot 2SiO_2$
Muscovite $[K_2]^{XII}[Al_2Si_6]^{IV}[Al_4]^{VI}O_{20}(OH)_4$
Chlorite $H_8(Mg, Fe)_5Al_2Si_3O_{18}$
II. Oxides
Quartz SiO_2
Hematite Fe_2O_3
III. Sulfides
Pyrite FeS_2
Marcasite FeS_2
IV. Sulfates
Coquimbite $Fe_2(SO_4)_3 \cdot 9H_2O$
Szmolnokite $FeSO_4 \cdot H_2O$
Bassanite $CaSO_4 \cdot \frac{1}{2} H_2O$
Anhydrite $CaSO_4$
Gypsum $CaSO_4 \cdot 2H_2O$
Jarosite $K_2Fe_6(OH)_{12}(SO_4)_4$
V. Carbonates
Ankerite $2CaCO_3 \cdot MgCO_3 \cdot FeCO_3$
Calcite $CaCO_3$
Siderite $FeCO_3$
Dolomite $(Ca,Mg)CO_3$

made of crushed coal and epoxy resin, is sectioned by grinding and the sample polished [43]. At a magnification of at least 400X and resolution of 1–2 μm, a point-count procedure for 1000 positions on each of two samples is used to estimate the volume percentage of the various macerals [43]. In another procedure [44] 50 readings on each of two samples are obtained on different particles of vitrinite (or another maceral) to determine the mean maximum reflectance of the vitrinite (or another maceral) and the distribution of maximum reflectance measurements.

Table 2-4 [32, 42, 194]

Classification of Coal Macerals

Maceral Group	Maceral	Origin	Characteristics: Thin section, transmitted light	Characteristics: Polished, reflected light	Characteristics: Coking properties
Vitrinite	Vitrinite (collinite–structureless: telinite–some cell structure)	Plant tissue, mostly wood and bark, either completely or partially humified. Megascopically occurs as vitrain, microscopically mixed with other macerals. Protein, lignin, cellulose.	Cellular structure may or may not be well preserved. Buff/cream color to yellow-tan and pale orange in low rank, tan/orange-red to red-brown and deep red in intermediate rank, and opaque in high rank.	Dominates with grey-white appearance. Reflectance intermediate to other maceral groups. Less than 0.5% for low rank, 0.5–2.5% for intermediate rank, and 2.6–6.0% for high-rank vitrinites.	Chief agglutinating constituent of bituminous coal
Exinite (or liptinite)	Sporinite Resinite Cutinite	Spore and pollen coats. Resins and waxes. Cuticles (waxy cell secretions, waxy coats of leaves).	High translucency, light yellow color in low intermediate rank. Become indistinguishable from vitrinite as rank increases beyond medium volatile bituminous. Distinguished from one another by structural form indicative of their origin. Differentiation becomes more difficult at higher rank.	Dark grey to black appearance. Distinguished from one another by structural form (e.g. resins are typically oval in form and dark) indicative of their origin. Reflectance less than vitrinite.	Most reactive group, first to fuse or volatilize
	Alginite	Algae.			
Inertinite	Fusinite	Same as vitrinite. Has undergone rapid alteration and carbonization forming a natural charcoal. Has residual cell-wall structure and is main constituent of fusain.	Remnant cell-wall structure. Opaque, cell cavities either empty or filled with mineral matter. Varies little with rank.	White, cellular. Reflectance highest of all macerals and independent of rank.	Inert or only agglutinate slightly
	Semifusinite	Same as fusinite.	Intermediate between fusinite and vitrinite.	Intermediate between fusinite and vitrinite.	
	Schlerotinite	Fungal matter. Rarely found in coals.	Structure recognized as belonging to fungi.	Highly reflective.	
	Macrinite	Structureless particles of unknown origin, > 10 μm.	In lignites translucent yellowish-brown to brown. Opaque in higher rank	White, less reflective than fusinite but greater than vitrinite.	
	Micrinite	Unknown attrital matter <10 μm. Probably from macerated plant cell-wall material.		Dark grey in lignite, white in higher rank.	

2.1.2 *Rank Classification by Vitrinite Reflectance*

Reflectance is a convenient property for determining rank because it can be measured on individual coal particles, in contrast to bulk rank properties such as volatile-matter and carbon content (Table 2-5) [139, 195]. Since most coals contain 60% or more vitrinite, such measurements are reasonably representative of the rank of the bulk coal. Figures 2-1 and 2-2 show typical correlations of reflectance with other rank parameters [139, 195]. Reflectance measurements and maceral analyses have been used for process control of the coking quality of coal or a mixture of coals. Reflectance measurements vary with the bedding plane of the coal seam due to anisotropy perpendicular to the layering, with negligible differences in low- and medium-rank coals and progressively more significant differences as anthracite is approached. This anisotropic behavior gives some indication of the effect of pressure in causing a uniform orientation of the chemical structure during the coalification process. Reflectance has also been used in investigating coal structure because measurement in two different mediums (air and cedar oil) enables calculation of the refractive index of coal. The refractive index is related to molar refraction, which in turn is a function of the aromaticity and oxygen substitution of the coal. Various relationships can then be used to estimate structural parameters such as the fraction of aromatic carbon and the molecular weight per mean structural unit [133, 209]. A data bank of a large variety of U.S. coals has been prepared by the Pennsylvania State University, and reference [195] tabulates much of this information including petrographic analyses and vitrinite reflectance values.

2.1.3 *Elemental and Proximate Composition of Macerals*

Some of the individual macerals can be recovered in rather concentrated form by hand-separating the lithotypes under a microscope and by careful gravity separation of the macerals taking advantage of the density order fusinite>micrinite>spor-

Table 2-5

Rank Classification by Vitrinite Reflectance [139, 195]

Rank	Reflectance, %
Lignite	< 0.35
Subbituminous	0.30–0.50
High-volatile bituminous	0.50–1.12
Medium-volatile bituminous	1.12–1.51
Low-volatile bituminous	1.51–1.92
Semi-anthracite	1.92–2.50
Anthracite	> 2.50

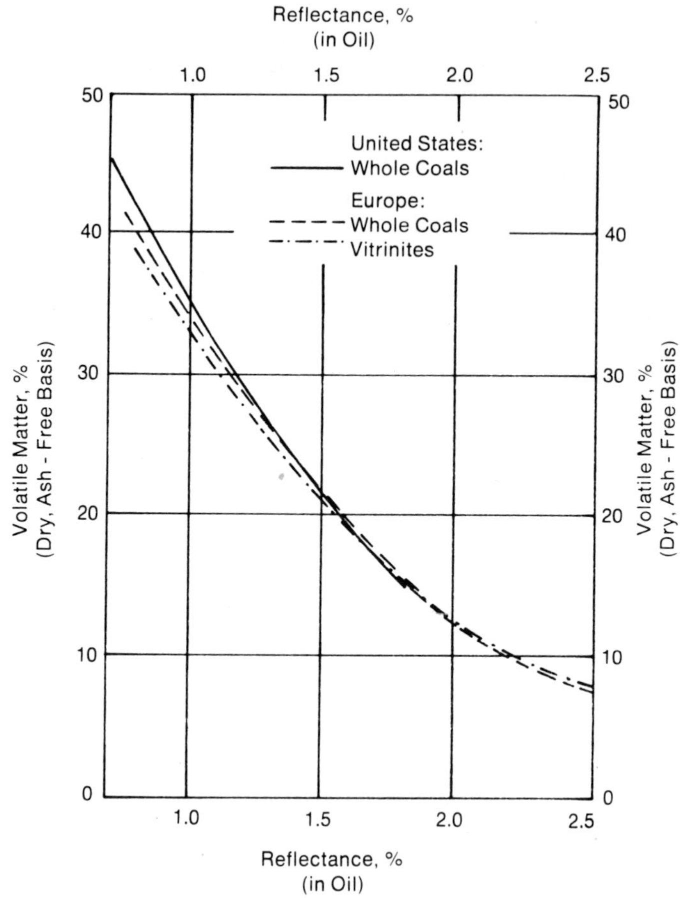

Figure 2-1. Relationship of reflectance of vitrinite to volatile matter in the bituminous-semianthracite range (McCartney and Teichmüller [139]).

inite [52, 133, 167]. Vitrinite is very easily recovered from vitrains, and rather pure forms of fusinite, micrinite, and exinite have been prepared. Many European coals have durain lithotypes containing 60–79% sporinite, while American coals are not known to contain more than 40%. Tables 2-6 [133] and 2-7 [167] show some representative elemental and proximate analyses for various macerals. These data illustrate the following relationships:

Carbon content	$F \gg M > E > V$
Hydrogen content	$E > V > M > F$
Oxygen content	$V > M > E > F$
H/C	$E > V > M > F$

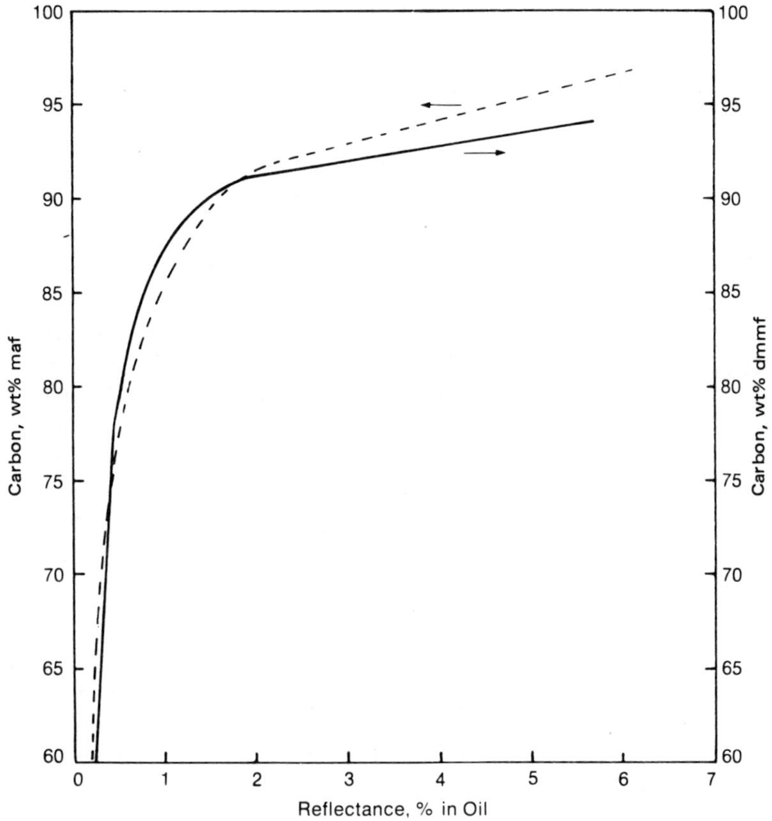

Figure 2-2. Correlation of mean-maximum reflectance of vitrinite with carbon content on dmmf and maf bases [139, 195].

O/C	V>M>E>F
Sulfur content	no trend, similar amount
Nitrogen content	no trend, except lower in fusinite
Volatile matter	E≫V>M>F

Exinite contains the most hydrogen (up to 7.4%) and has a very high volatile-matter yield. Fusinite is low in hydrogen and contains a large amount of carbon compared to the other macerals. The elemental composition of fusinite is invariant with increasing rank, while the exinite changes composition very rapidly in going from the middle to the high end of the bituminous range. At about 89% vitrinite carbon, vitrinite and exinite assume the same composition, and differences from micrinite diminish somewhat, although a significant disparity still exists in hydrogen content and H/C ratio. The differences between micrinite and vitrinite are more pronounced in European coals than in American coals.

Table 2-6*

Analysis of Petrographic Components

% C of the vitrinite	Maceral†	Ultimate analysis, maf						
		C	H	O	N	S	H/C‡	O/C‡
81.5	E	82.2	7.4	8.5	1.3	0.6	1.07	.0776
	V	81.5	5.15	11.7	1.25	0.4	.753	.108
	M	83.6	3.95	10.5	1.35	0.6	.563	.0943
85.0	E	85.7	6.5	5.8	1.4	0.6	.904	.0508
	V	85.0	5.4	8.0	1.2	0.4	.757	.0706
	M	87.2	4.15	6.7	1.35	0.6	.567	.0577
87.0	E	87.7	5.85	4.4	1.45	0.6	.795	.0377
	V	87.0	5.35	5.9	1.25	0.5	.733	.0509
	M	89.1	4.2	4.7	1.4	0.6	.562	.0396
89.0	E	89.6	5.2	3.3	1.3	0.6	.692	.0276
	V	89.0	5.1	4.0	1.3	0.6	.683	.0337
	M	90.8	4.1	3.2	1.3	0.6	.538	.0264
90.0	E	90.4	4.9	2.8	1.3	0.6	.646	.0232
	V	90.0	4.94	3.2	1.35	0.5	.654	.0267
	M	91.5	3.95	2.6	1.35	0.6	.514	.0213
91.2	E	91.5	4.5	2.3	1.2	0.5	.586	.0189
	V	91.2	4.55	2.6	1.15	0.5	.594	.0214
	M	92.2	3.65	2.2	1.35	0.6	.472	.0179

*Adapted from Van Krevelen [133], p. 60
† E - exinite, V = vitrinite, M = micrinite.
‡ Atomic ratio.

An attempt has been made to characterize the maceral/mineral composition of various coal samples by using what is called a microlithotype classification system as shown in Table 2-8 [148, 194]. This system uses 5% concentration levels to separate the various categories, and characterizes the sample according as it consists primarily of one, two, or three maceral combinations, or whether mineral matter is in high concentration. For example, a sample containing more than 95% of the maceral group vitrinite would be monomaceralic and termed a vitrite microlithotype. This classification system has not yet gained as wide a use in coal utilization as has simply characterizing samples by maceral group or maceral analysis. In summary, coal is a heterogeneous substance consisting primarily of various types and concentrations of organic macerals that are physically and chemically different.

The types and characteristics of the organic macerals depend upon the degree of metamorphism and upon the geographic location (environment) and geological

Table 2-7

Analysis of Petrographic Components*

% C of the vitrinite	Maceral†	Ultimate analysis, maf C	H	O	N	S	Ash‡	H/C§	O/C§	Volatile matter**
78.1	V	78.1	5.6	11.7	1.9	2.7	1.0	.848	.113	–
(hvCb coal)	F	90.4	3.5	5.2	0.2	0.7	2.8	.462	.0429	–
	Fusain	89.1	3.4	5.3	0.2	2.0	11.6	.453	.0446	–
83.7	E	85.6	6.7	5.8	1.2	0.7	1.2	.932	.0508	55.7
	V	83.7	5.3	8.4	1.8	0.8	0.6	.749	.0755	31.0
	I	86.7	4.2	7.2	1.2	0.7	6.7	.582	.0626	21.0
83.9	E	84.7	6.4	7.1	1.3	0.4	1.4	.897	.0631	59.8
	V	83.9	5.2	8.6	1.6	0.6	0.6	.742	.0771	32.0
	M	85.9	4.2	8.3	1.2	0.4	5.8	.588	.0727	23.4
84.6	E	84.6	6.5	7.1	1.1	0.7	3.0	.911	.0631	55.3
(hvAb coal)	M	84.6	5.5	7.7	1.6	0.7	2.1	.770	.0685	33.7
	F	83.9	5.1	9.1	1.3	0.6	7.8	.719	.0809	31.4
		90.0	3.4	5.7	0.5	0.4	3.5	.445	.0478	13.4
88.7	Vitrain									
(lvb coal)	Fusain	88.7	4.9	4.1	1.7	0.7	3.3	.655	.0347	33.3
		84.4	4.8	7.3	1.6	2.0	6.3	.673	.0649	24.8

*Adapted from Wender [167], pp. 282–4.
†E = exinite, V = vitrinite, M = micrinite, F = fusinite, I = inertinite.
‡Moisture-free basis.
§Atomic ratio.
**Maf basis, proximate analysis.

period (types of vegetation) in which the source materials were deposited. In studying the chemical structure and liquefaction mechanisms of coals, it is important to understand that each maceral will contain its own set of organic structures and that no one chemical structure can adequately represent the whole coal. However, structural and reaction mechanism studies at the maceral level should yield more meaningful information, although the variation of rank and the heterogeneity within each maceral must be a major consideration. No matter how one tries to separate coal into groups having similar chemical properties, the sample will still be a heterogeneous mixture, not a homogeneous mixture of known chemical compounds such as petroleum. The establishment of a unified approach for characterizing the fundamental chemical and structural properties has not been carried out and is a much debated issue.

Neavel [148] has likened coal to a fruitcake, formed from diverse ingredients and baked to a visibly heterogeneous product. Characterizing the chemical structure

Table 2-8

Microlithotype Classification System*

	Microlithotype	Maceral	Volume percent
1. Monomaceralic	Vitrite	Vitrinite (V)	> 95
	Liptite	Liptinite (L)	> 95
	Inertite	Inertinite (I)	> 95
2. Bimaceralic	Vitrinertite	V + I	> 95
	Clarite	V + L	> 95
	Durite	I + L	> 95
3. Trimaceralic	Duroclarite	V + I + L	each > 5, V > (I + L)
	Clarodurite	V + I + L	each > 5, I > (V + L)
	Vitrinertoliptite	V + I + L	each > 5, L > (V + I)
4. Mineralized	Carbominerite	V, L, I, and mineral matter (MM)	20 < MM < 60

*Neavel [148]; Spackman [194].

of the coal without consideration of its makeup of discrete substances (elements and molecules) is like describing the makeup of fruitcake by grinding it up and analyzing the elemental composition. Given [75] has drawn an analogy between the organic part of coal and a mixture of different copolymers. Each copolymer contains some vinyl alcohol, acrylonitrile, and polystyrene, but with different proportions of the monomer in each. The basic skeleton of each molecule in the mixture would be similar in that they would contain alcohol, nitrile, and phenyl groups, but would be present in different proportions and spatial arrangements. Fortunately, most coals of interest are high in vitrinite, so that the chemical properties are essentially those of the vitrinite component.

2.2 The Chemical Structure of Coal

2.2.1 *Chemical Precursors*

A starting point in developing an understanding of the chemical structure of coal is to consider the chemistry of plants today. Today's plants may be chemically similar to those that formed the source material of various coals. True, it is conceivable that some structural characteristics may have been retained even though much alteration of the chemical structure occurred during the coalification process. It is also important to realize that the plant kingdom underwent much evolution, consequently changing the type of source material available for deposition in the various geologic periods. Thus the basic chemical structures in various eras may be substantially different. Despite these reservations, however, much can be inferred from extrapolation of present-day chemical structure.

The source materials in plants, listed in increasing order of stability during decay, are: [133]

1. Protoplasm (protein)
2. Chlorophyll
3. Oils
4. Carbohydrates
   a. Starches
   b. Cellulose
5. Lignins
6. Epidermis
7. Seed coats
8. Pigments
9. Cuticles
10. Spore and pollen exines
11. Waxes
12. Resins

Some of the chemical structures that make up these components are shown in Figure 2-3 [217]. Only the most stable of these chemical structures would be able to survive the coalification process in any recognizable form. Plant materials above lignin would be expected to decompose rapidly, whereas the last four components would be altered only very slowly. The principal precursors of the vitrinite maceral, usually 70–90 vol% in U.S. coals, are believed to be lignins with some contribution from celluloses, starches, and proteins through interaction of the products of microbial decay [60, 132, 133, 209]. Although the latter components are known to biodegrade rapidly (for example, cellulose is converted to CO_2 and water), the conditions that prevailed in the peat swamps were so varied with respect to aeration and bacterial action that the role of the more easily decayed components is highly questionable and could have varied widely. Since vitrinite contains 10–30 times the amount of nitrogen present in wood, plant and microbial proteins, which are high in nitrogen, must have been important source components and would also account for some organic sulfur in coal. Proteins and associated hydrolysis products as well as amines, amides, and basic heterocyclic nitrogen compounds have been isolated from soil humus, humic acid preparations, peats, and lignites by extraction with mineral acids [132]. Some sort of complexing reactions probably occurred, because the microorganisms would have rapidly consumed protein in free form. Relatively resistant compounds that would also account for nitrogen would include plant alkaloids, porphyrins (chlorophyll to some extent), and chitin [132, 217]. Some of the chemical structures in vitrinite have been shown to be related to decomposition products from lignins and flavinoids, indicating their importance as precursors of vitrinite [18, 100, 161]. This was accomplished by applying sodium amalgam and hot water in a degradation reaction that was developed to study soil humic acids.

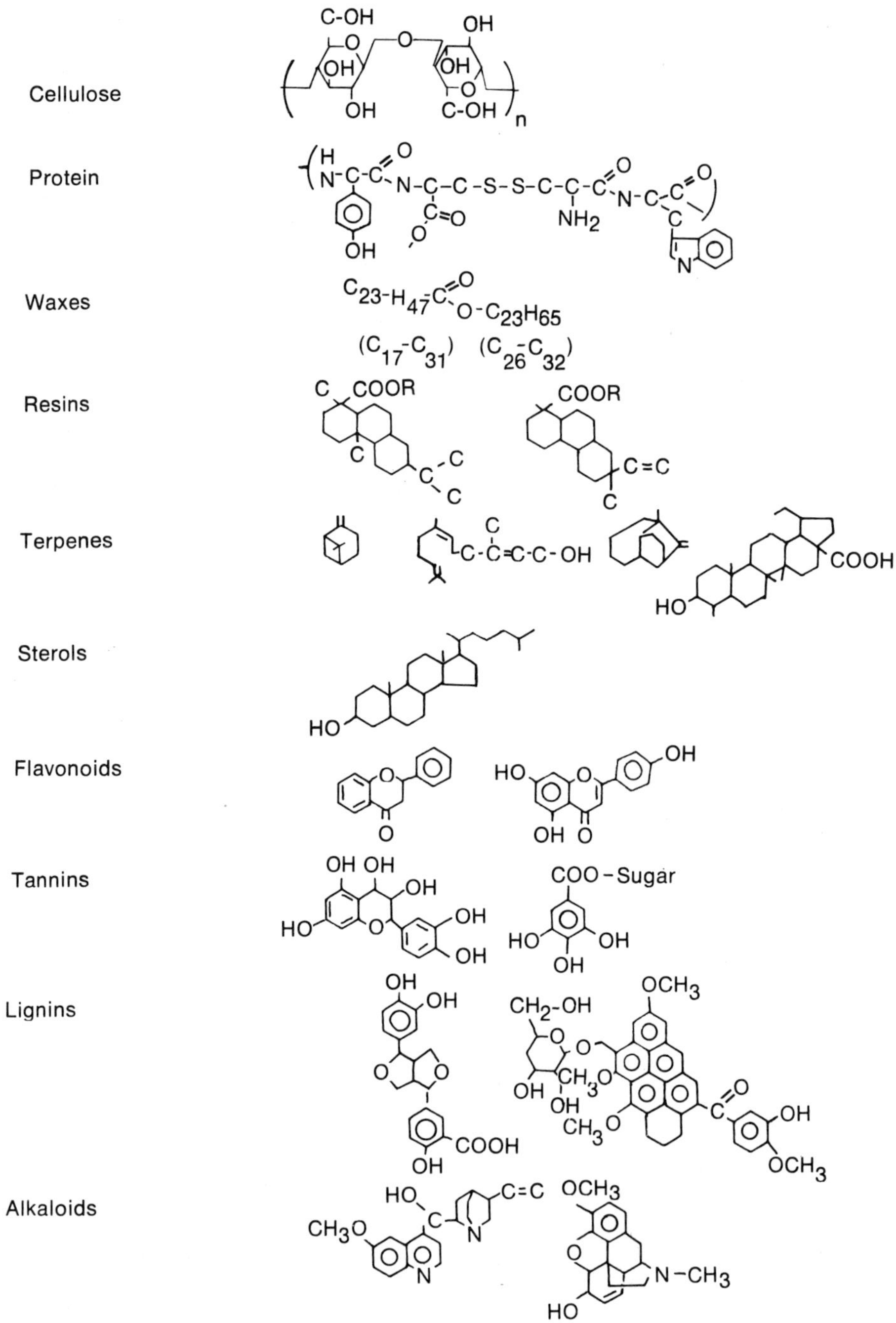

Figure 2-3. Structures of coal precursors (Whitehurst [217]).

Humic acids were extracted from peats and lignites using a sodium hydroxide solution and subjected to the reduction reaction. The products were identified by GC/MS analysis. With bituminous coals, a mild oxidative treatment using aqueous performic acid was needed to yield humic acids for further testing.

In summary, the chemical species that are generally agreed upon as being most resistant to microbial decay and the most likely coal precursors include lignins, waxes, resins, tannins, flavinoids and alkaloids.

2.2.2 *Structural Characteristics*

Many decades of research have been devoted to the elucidation of the chemical structure of coal [60, 75, 132, 133, 210, 219]. Although much has been learned about the arrangement and variation of the five major elemental constituents C, H, O, N, and S in the various possible organic structures of coal, major issues still remain concerning the basic C-H skeleton and the chemical functionalities present. Superimposed on these issues is the variation of structures within each of the maceral classes [148]. Most work has focused on vitrinite because of its predominance in the majority of coals, and a far lesser amount on the other major maceral classes. The gross structural overview that has emerged from this research is that the vitrinite portion of coal is a highly crosslinked amorphous macromolecular network, consisting of a number of stable cluster units or aggregates connected by relatively weak crosslinks (Figure 2-4). Linkages connecting the cluster units include aliphatic carbons [102–104], ethers [196, 199], and weak bonds (possibly hydrogen bonds and charge-transfer complexes) [96, 107, 133, 198, 199]. The weak bonds are evident from the ability to extract up to 40–50% of some bituminous

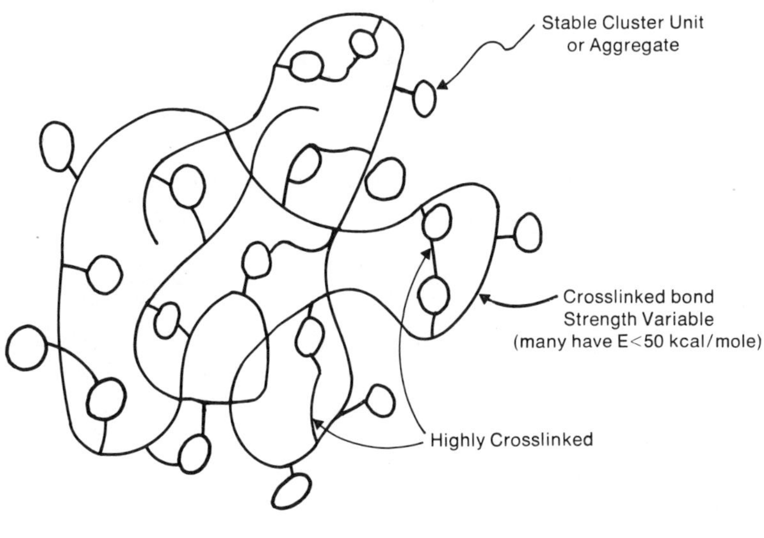

Figure 2-4. The macromolecular network of coal (Whitehurst [217]).

coals using a suitable solvent at temperatures below 523 K, where chemical changes are unlikely during extraction [96, 132, 133, 209]. Of the different macerals, vitrinite yields the most extract and fusinite the least. The extraction yield reaches a maximum near 87 wt% (maf) carbon and becomes negligible above 90%. Solvents containing a nitrogen or oxygen atom with a readily available pair of unshared electrons, such as pyridine, ethylenediamine, and quinoline, are most effective.

A. CARBON SKELETAL ARRANGEMENT AND AROMATICITY

X-ray diffraction studies of coals indicate regions of ordered carbon (not in the normal crystalline sense) and "amorphous" carbon, and attribute the ordered regions to the arrangement of carbon atoms in aromatic rings systems or clusters that are stacked in parallel to varying degrees, with random planar orientation of the lamellae or cluster units [19]. The degree of ordering and the amount of carbon in "amorphous" regions is a function of rank, eventually approaching a graphite-like structure at carbon contents well into the anthracite range. The cluster units are mixed aliphatic-aromatic structures [53]. In a very extensive study of coals and coal macerals varying from lignite to vitrinite of very high rank, the content of "amorphous" carbon was found to correlate well with rank and ranged from 40 to 10 wt% (dmmf) over the range 60 to 94 wt% (dmmf) carbon [26]. This would roughly correspond to a carbon aromaticity of 60 to 90% over this rank range. Up to a rank of about 75 wt% (dmmf) carbon, the average number of carbon atoms per cluster unit is about 8 to 10 and corresponds to crosslinked, single-substituted, benzenoid structures. Between 75 and 80% carbon, the average number of carbon atoms per cluster unit increases to about 15 to 16, corresponding to an average of two condensed rings. The average numbers of carbon atoms and condensed rings increase to about 18 and 3 to 4, respectively, at 90% carbon. In the low-rank bituminous coals the clusters contain 1 to 3 condensed rings, while 90%-carbon vitrinites have clusters containing 2 to 5 condensed rings. The cluster units usually occur singly or in stacks of two or three.

The aliphatic carbon and ether groups (and sulfur analogs of oxygen) linking the condensed aromatic clusters form larger, three-dimensional structures, containing alicyclic groups that can become aromatic by loss of hydrogen [26]. Hydrogen-rich coals and maceral components (exinite) contain a significant amount of alicyclic carbon [54]. Although the cluster-unit layer structures appear similar for the various macerals, exinite contains more "amorphous" material, and the cluster units are more buckled because there are more aliphatic and alicyclic groups than in vitrinite, while micrinite is just the opposite, having more uniform cluster units and less "amorphous" material [26]. Fusinite appears similar in structure to a low-rank coal that has been carbonized at 873 to 1073 K. From this information, the order of carbon aromaticity is inferred to be fusinite>vitrinite>exinite. Ethylenediamine extracts of coals give x-ray diffraction curves nearly identical to the original coal, indicating the extracts are similar in structure to parts of the parent coal: the weakly bonded components.

More recently, a statistical structural analysis based on the analyses of 284 coals and previous x-ray data was used to compute the various structural parameters and the molecular weight of the average cluster unit [188]. The molecular weight varied from 300 to 500 and increased rapidly above 90 wt% (dmmf) carbon. The authors, however, reached the unlikely conclusion that bituminous coals contain on the average 4 to 5 condensed aromatic rings in a cluster unit [124].

Various structural parameters have been deduced in an ingenious way from statistical structural analyses based upon various physical-property measurements, elemental analyses, and the valence requirements of the atoms [133, 209]. As shown in Figure 2-5, the fraction of carbon that is aromatic increases from 67 to 100% as the carbon content of the vitrinite increases from 70 to 94 wt% (maf). These aromaticities were calculated by three different methods using refractive-index, density, and velocity-of-sound data. The molecular weight per mean cluster

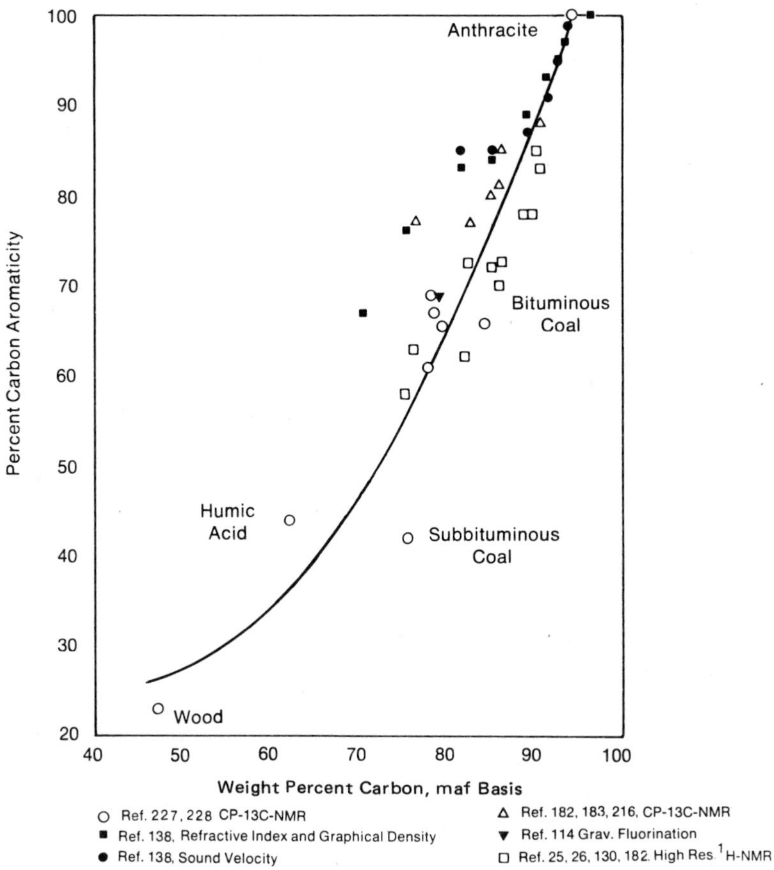

Figure 2-5. Variation of aromatic carbon content with rank.

unit varied from 290 to about 500 below 90 wt% (maf) carbon, and increased to 1060 as vitrinite carbon increased to 96%. The number of condensed rings per mean aromatic cluster was estimated to be 5 to 8 rings below 90% carbon, much too high.

In contrast, some infrared [62] and ultraviolet [63] spectroscopic studies of bituminous vitrinites indicate a considerable aliphatic character, the aromatic units being predominately benzenoid structures rather than polyaromatic structures. Extensive infrared studies [22, 60, 125, 132, 133, 209] of vitrinites ranging in rank from 78 to 94 wt% (dmmf) carbon showed that the structural skeletons were similar. As rank increased, aromatic hydrogen increased, aliphatic hydrogen decreased, and the degree of substitution of aromatic rings decreased. Over the range 84 to 93% carbon, the fraction of aromatic hydrogen increased from about 20 to 50%. Ethylenic double bonds, unconjugated carbonyl groups, and long aliphatic chains were absent, and methyl groups were few compared to methylene groups. Nonaromatic groupings were likely hydroaromatic.

High-resolution ^{1}H-NMR studies [24, 25, 125, 174] of solvent extracts of raw and pyrolyzed coals confirmed the findings of the previous infrared work and yielded carbon aromaticities in the range of 58 to 78% for carbon contents of 78 to 90 wt% (dmmf). These aromaticity values are much lower than those calculated from physical-property measurements as shown in Figure 2-5.

Carbon aromaticities have been determined directly using the recently developed technique of cross-polarization ^{13}C-NMR on samples of wood, humic acid, and subbituminous, bituminous, and anthracite coals [157, 215, 217]. These values are also shown in Figure 2-5, and indicate significantly lower aromaticity and degree of ring condensation than the values calculated from physical-property data. The aromatic-carbon content increases from about 40–50% for subbituminous coal to over 90% for anthracite. Bituminous coals had aromaticities in the range of 60–70%. Other similar bituminous and anthracite coals [206] and coal macerals [175] subjected to CP-^{13}C-NMR analysis have yielded somewhat higher values of aromaticity than in the latter work, e.g. 76% aromatic carbon for a bituminous coal, and 85% aromatic carbon for a bituminous vitrinite containing 85 wt% (maf) carbon. Apparently, aromaticity values may not be uniquely defined by rank and petrography, but may correlate with rank within a wide band that narrows significantly as the carbon content approaches 90 wt% (maf).

Dehydrogenation studies of coal and coal macerals have shown that much of the aliphatic character observed in spectroscopic work is in the form of hydroaromatic structures [156, 163–169]. It has been shown that the hydroaromatic content of coal decreases with increasing rank [166]; that vitrinite and exinite are hydroaromatic, but fusinite is highly aromatic and micrinite contains a nondehydrogenable, nonaromatic-type structure [167]; and that the disparity between fusinite and vitrinite decreases as rank increases [167]. The dehydrogenation was also shown to be quantitatively reversible indicating the hydrogen-donor-hydrogen-acceptor capability of the hydroaromatic structures [165]. It is also likely that the aromaticity and size of the aromatic clusters increases at the expense of hydroaromatic groups during coal metamorphosis [136, 138].

B. METHODS OF COAL DEGRADATION FOR STRUCTURAL ANALYSIS.

Because direct methods of analysis of the chemical structure of coal are few, many approaches have been devised for degradation of the coal macromolecular network into smaller fragments that are more readily soluble and amenable to analysis. These fragments are investigated along with the fragments resulting from model compounds subjected to the same treatment in order to reconstruct what kind of coal molecules would fit the established criteria. The classical techniques that have contributed much to the knowledge of coal structure include: destructive distillation (pyrolysis) of coal to benzene and pyridine derivatives [60, 132, 133 209]; alkyl transfer to phenol and determination of the alkyl phenols [102–104, 106, 108, 128, 129]; liquefaction or solvent refining and examination of products [55, 57, 60, 132, 133, 209, 217, 218]; and oxidation with reagents such as HNO_3, HNO_3-$K_2Cr_2O_7$, $KMnO_4$-OH^-, O_2, and H_2O_2-O_3 to produce benzene carboxylic acids [18, 60, 99, 100, 132, 133].

C. LABILE CLUSTER LINKAGES FROM COAL DEPOLYMERIZATION STUDIES

Work on the depolymerization of coal by alkyl transfer to phenol [106, 128], alkylation and acylation of coal [128], and reductive alkylation of coal [110, 128, 196, 197, 199, 210] indicates that the aromatic units in coal are linked by methylene, ethylene, and ether bridges (diaryl and aryl-alkyl ether). In low-rank coals such as brown coal and lignite, cleavage of aliphatic-ether and benzyl-ether linkages is believed to be a significant contributor to the depolymerization reaction in phenolic solvent. Many aliphatic bridges may also be linked to single phenolic rings. Results with brown coal indicate that methylene, ethylene, and acyl bridges exist between aromatic nuclei, and that tertiary carbon and benzofuran-type structures are present. Aliphatic bridges appear to be the major linkages in high-volatile bituminous coals (80–84 wt% maf carbon) between phenolic rings and condensed rings. Ether oxygen seems to be less important, and most ethers are aromatic. Depolymerization yields are generally poor with high-rank bituminous coals (carbon > 84 wt% maf) because nearly all bridge structures are condensed hydroaromatic rings. The number-average molecular weight of the fragments from depolymerized coals varies from 290 to 1726 but is usually in the range 300–700.

Products from reductive alkylation and acylation have number-average molecular weights of 2000–3000 [196, 197], down to 600–700 [110] depending upon the coal. The reductive alkylation experiments indicate that diaryl and aryl-alkyl ether oxygen are the primary links between the basic structural units in bituminous coals [110, 210], accounting for over 40% of the oxygen [34, 185, 197, 210], and that these coals could be solubilized without breakage of covalent carbon-carbon bonds. Open ether bonds were found to be present in much smaller quantities in low-rank and high-rank (> 90% carbon content, maf basis) coals. A significant portion of the oxygen in the latter coals is in heterocyclic form. Obviously, the depolymerization and alkylation results stress the importance of the structural groups specifically attacked by the respective chemical treatment and are thus not in complete agreement on the importance of ether-oxygen linkages. As will be discussed further on, the distribution of oxygen functional groups is very specific to the coal basin and its geochemistry, so that repetition of the alkylation experiments for coals of the same rank but from different locations may give different results. One can conclude, however, that the important linkages to be broken in reducing the molecular weight of coal to the most basic, stable clusters are: [135, 199]

$$\begin{array}{ll} Ar{-}CH_2{-}Ar & Ar{-}O{-}Ar \\ Ar{-}(CH_2)_n{-}Ar & R{-}O{-}Ar \\ & R{-}O{-}R \end{array}$$

plus analogs with S instead of O.

D. STRUCTURAL INFERENCES FROM PRODUCTS OF COAL OXIDATION

Classical oxidation studies (reviewed in several references [60, 132, 133, 209], using the reagents cited previously, involve either partial oxidation of the coal to humic acid products or further oxidation to benzene carboxylic acids. The former products are difficult to separate and analyze, but the latter products can be used to infer structure or degree of ring condensation, based on some assumptions of the selectivity of the oxidation reaction, the yield of carboxylic acids, and the number of carboxylic acid groups substituted on a benzene ring. The yield of these acids, especially mellitic acid, was observed to increase with rank, while yields of aliphatic acids decreased. The selectivity of the oxidation was generally questionable.

More recent oxidation studies have reexamined the use of the classical reagents [99], and have utilized other reagents thought to be more selective for specific carbon structures [18, 27–31, 47, 48, 99, 100]. Sodium dichromate appears to be selective for oxidation of side chains and alicyclic appendages to aromatic clusters with little degradation of the aromatic rings [99]. The relative abundances of aromatic and heteroaromatic units produced from lignite and bituminous and anthracite coals demonstrated that benzylic structures predominate in lignite and bituminous coal, while three-ring (phenanthrene) condensed structures predominated in the anthracite. The cluster size distribution in lignite (69.4% maf carbon) included a small amount of naphthalene structures, while for the bituminous coals (77.8 and 82.7% maf carbon) they included some naphthalene and phenanthrene units and traces of four-ring units. Three- and four-ring clusters (pyrene, chrysene) dominated the anthracite (91.3% maf carbon) product distribution. The lignite and bituminous coals apparently had an abundance of ring oxygen in ether bridges and furan structures between aromatic nuclei. Nitrogen was found in quinoline, benzoquinoline, carbazole, and acridine structures in the products from bituminous and anthracite coals, and in pyridine units in lignite products. Sulfur was found in the products from bituminous and anthracite coals as benzothiophene, naphthobenzothiophene, and dibenzothiophene.

Photochemical oxidation using ultraviolet irradiation in combination with air and aqueous HCl was not as selective but yielded somewhat similar results. Straight-chain carboxylic acids and methyl-substituted chains were identified in the products from lignite and bituminous coal, indicating the likely presence of these aliphatic units in the coals. Oxidation with a hydrogen peroxide–acetic acid mixture selectively preserved phenolic groups and indicated that the above lignite had about twice the hydroxyl group content of the bituminous coal.

Another oxidation reaction, using sodium hypochlorite, is believed to be selective for tertiary carbon over aromatic carbon and has yielded products from bituminous coals that contained no aromatic compounds other than benzene derivatives [27–31]. It was suggested that at least 50% of the carbon in bituminous coals exists in complex bridged cycloalkanes, i.e., polyamantane or diamondlike structures. The carbon aromaticity was estimated at about 40% over the bituminous range. Further studies [6, 70, 99, 127, 134] of this reaction showed that it is not as selective as once believed and that adamantane structures are stable under coal-liquefaction conditions and would have been identified if present. Fluorination experiments with a high-volatile bituminous coal also give results which do not agree with the polyamantane structure, yielding a carbon aromaticity of 68.6% (77.9% dmmf carbon coal) and 10% carbon in hydroaromatic structures [109]. Although most of the above conclusions are disputed, some tertiary carbon structures have been assigned in more recent work with subbituminous coals [218] as possible structures.

Oxidation studies of bituminous coals and lignite using a trifluoracetic acid–hydrogen peroxide–sulfuric acid reagent that is selective for degradation of aromatic structures while preserving aliphatic structures [47, 48] have shown that ethyl, propyl, and isopropyl groups attached to aryl rings are not likely structures in coal, contrary to prior thinking [16, 104]. None of the bituminous coals appeared to contain arylmethoxy groups, although lignite contained a substantial quantity. In contrast, lignite contained no aryl methyl groups, unlike the bituminous coals. Indane or diarylethane structures are likely present, but benzoate esters and tetralin that is unsubstituted on the aliphatic ring were ruled out as possible structures in coal. This oxidation method is not selective for aromatic structures if polycondensed aliphatic structures are present in significant amounts, as in the case of Wyodak subbituminous coal [46, 47], which gives nonreproducible results and difficult-to-analyze products.

E. STRUCTURAL INFERENCES FROM PRODUCTS OF SHORT-RESIDENCE-TIME COAL DISSOLUTION

Bituminous and subbituminous coals can be nearly completely dissolved in a hydrogen-donor solvent at 698 to 723 K in a matter of several minutes [147, 215]. Careful separation and analysis of the products at these short times as well as at longer reaction times have been very successful in determining what were probably the structural building blocks in the starting coal [55, 57, 61, 183, 184]. The products are usually characterized in terms of solubility classes, although solubility is more related to functionality than to molecular weight or aromatic content. The usual solubility classes are oils (hexane-soluble components), asphaltenes (benzene-soluble, hexane-insoluble components), and preasphaltenes or asphaltols (pyridine-soluble, benzene-insoluble materials). These solubility classes have been subdivided even further using chromatographic separation by sequential elution with specific solvents on silica gel columns, but some overlapping of functionality still occurs. The subclasses are essential in isolating key structural units and constructing models of them [55–57]. The hexane-soluble or oil fraction has little or no functionality and consists, in order of concentration, of hydrocarbons, ethers, thioethers, and nonbasic nitrogen compounds. Saturates are mostly naphthenic, and the molecular weight for the oil class is about 200 to 300. The asphaltene fraction consists mostly of monofunctional compounds such as phenols and basic nitrogen compounds with some ethers also. The molecular weights of compounds in this class range from 300 to 700. Benzene-insoluble materials (preasphaltenes or asphaltols) have multiple functionality and molecular weights ranging from 400 to greater than 2000. Polyphenols with up to five hydroxyl groups per molecule, multiple basic nitrogen groups, and mixed functionalities are present in preasphaltenes.

Model compound work [5, 17, 57, 215, 218] has indicated that at short reaction times hydrogenation of aromatic polycyclic hydrocarbons and destruction or formation of polycyclic saturated structures does not occur, so that any polyaromatic or polycyclic saturated structures present in the starting coal should be present and identifiable in the thermal-liquefaction products. Liquid-chromatography work [61, 218] has shown that up to 40% of the very first products of liquefaction (less than 5 minutes) are of high molecular weight, usually exceeding 2000, or are very highly associated, and that these compounds are then rapidly converted to 300–900-molecular-weight structures. The high-molecular-weight material is found in the highest-solubility subclasses. Apparently the predominant change which occurs to yield the lower-molecular-weight compounds is the loss of functionality by a variety of rather stable nuclei with relatively little disruption. In other words, lower-molecular-weight compounds are formed via a progression into the lower solubility/functionality subclasses. The structures of the compounds in the various solubility subclasses have been identified [218] and are believed to represent the important nuclei in the coal structure for the coals studied. The most probable average structures for one of the chromatographic subclasses for various coals are shown in Figure 2-6 [57]. The most probable structure of one of the highest subclasses for Kentucky coal products is shown in Figure 2-7 [218].

Polarographic [215, 218] and $^{13}C,^{1}H$-NMR [61] analyses of the products at various extended reaction times show that the liquefaction products become much more aromatic due to the conversion of aliphatic carbon into aromatic carbon. Aromatization occurs through dehydrogenation of hydroaromatic structures and rearrangement of polycyclic rings to more stable structures. Thus, if coal were inferred to have the structures as identified in long reaction time or severe condition products without consideration of the initial products, one would think coal is much more aromatic than it really is.

F. FUNCTIONAL GROUPS FOUND IN COAL

The compositions of two typical coals, [57] bituminous and subbituminous, based on 100 atoms are:

Illinois #6, hvAb (Monterey)	$C_{50.8}H_{43}O_{4.8}S_{0.8}N_{0.6}$
Wyodak, subbituminous	$C_{48.4}H_{42.4}O_{8.6}S_{0.07}N_{0.5}$

These compositions show that aside from the carbon and hydrogen building blocks, oxygen is the most abundant heteroatom. The major functional groups identified in

Coal		Kentucky	Hiawatha	Monterey	Wyodak
Composition		$C_{100}H_{82}N_{1.7}O_{9.4}$	$C_{100}H_{84}N_2O_{9.5}$	$C_{100}H_{88}N_{1.6}O_{13.2}S_{1.74}$	$C_{100}H_{86}NO_{18}$
No. polycondensed saturated rings/100C	Coal	4		6	10
	SESC-3	2		4	5
	SCT *				
Structure SESC-3 SCT-SRC					

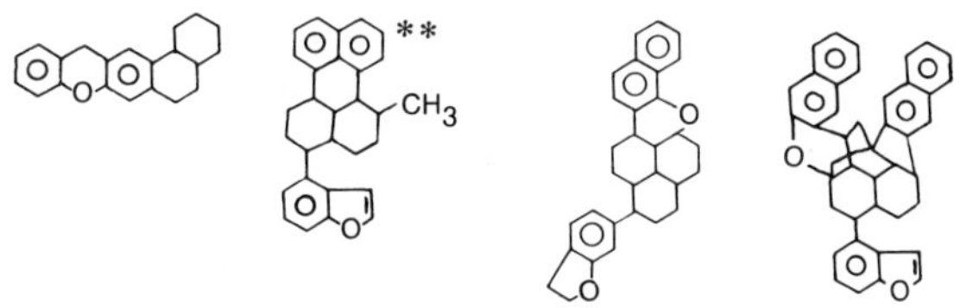

Figure 2-6. Average structures of one solvent subclass for various coals (Farcasiu [57]. *SCT = short contact time <2 min; **Ca. 30-min contact time.

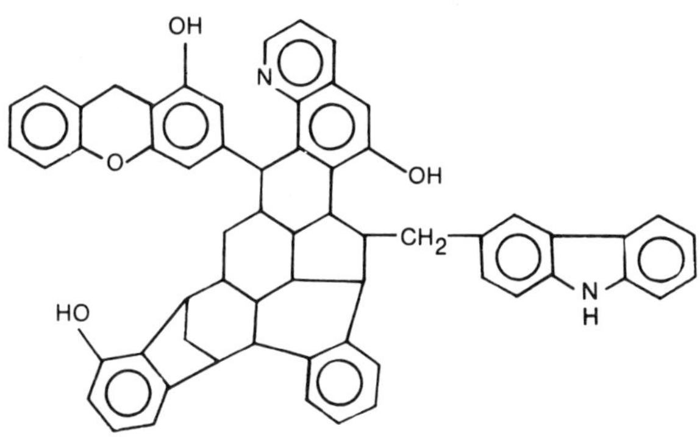

$C_{61}H_{46}N_2O_4$ 73% Aromatic C

900 mol. wt. 60% Aromatic H

Figure 2-7. Formal average structure of Kentucky (Wilsonville) SESC-8 (Whitehurst et al. [218]).

coal [60, 132, 133, 209] are shown in Figure 2-8 [217]. For oxygen the major groups are phenols, ethers, carboxylic acids, and quinones (carbonyl); for sulfur they are thiophenes, sulfides, disulfides, and thiols; and for nitrogen they are pyridines, quinolines, pyrroles, and carbazoles. Metals are incorporated in the organic structure as porphyrins or salts. An example of the latter is the significant ion-exchange capacity due to carboxyl groups in low-rank coals.

The organic oxygen content of coal varies over a wide range starting at about 30 wt% dmmf for brown coals and decreasing to about 1.5 wt% dmmf for anthracites [60, 76, 133, 195, 209]. Various analytical approaches including infrared analysis, polarographic reduction, and chemical methods have been used to identify and quantify the oxygen functional groups for a range of coals [28, 60, 133, 185, 209]. Phenolic hydroxyl is one of the major functional groups that is present in all ranks of coal. Data for a set of 52 American coals covering the range of rank from subbituminous coal (73.9% dmmf carbon) to medium-volatile bituminous coal (90.4% dmmf carbon) have been subjected to various statistical analyses. The per-

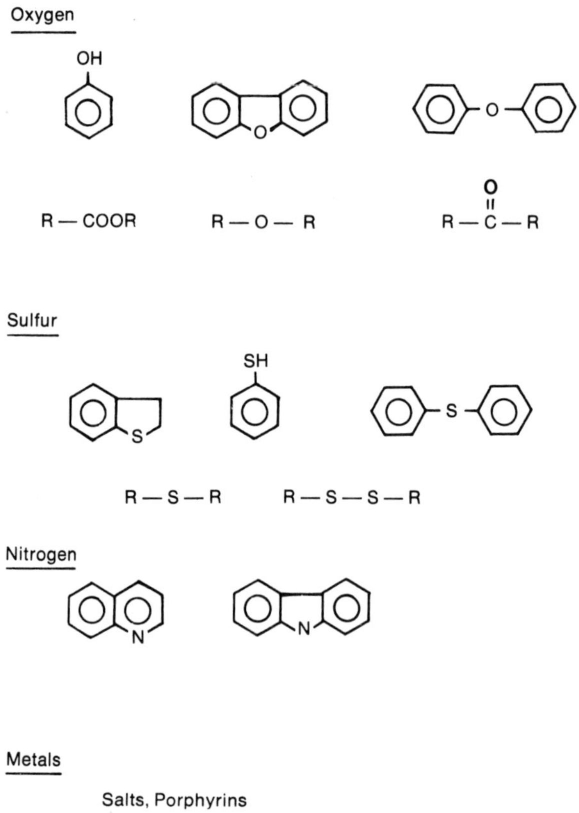

Figure 2-8. Functional groups found in coal (Whitehurst [217], Wiser [223]).

centage of oxygen in dmmf coal has been shown to correlate strongly with carbon content as follows:

$$\text{OH/coal} = 32.67 - 0.341C \qquad \text{(Eastern and Interior Provinces)},$$

$$\text{OH/coal} = 32.08 - 0.341C \qquad \text{(Western Province)},$$

where

C = % organic carbon content, dmmf basis

OH/coal = oxygen as OH, % of dry mineral-matter-free coal.

The hydroxyl oxygen as a percentage of the organic oxygen generally increases with increasing rank but does not correlate well with rank, showing enormous scatter. This indicates that the distribution of organic oxygen into the various functional groups can vary widely for coals within the same rank and depends strongly on the geology and geochemistry of the coal basin. This also explains why results on functional-group distribution published by various researchers were not often in agreement [133].

Functional-group analyses determined by direct chemical means for two coals are shown in Table 2-9 [34, 185]. In the bituminous coal (77.3% maf carbon) ether

Table 2-9

Analyses of Oxygen Functionality in Coals (wt% maf Coal Basis)*

	Burning Star bituminous	Belle Ayr subbituminous
Oxygen type:		
Hydroxyl (Ar–OH)	2.4	5.6
Carboxyl (–COOH)	0.7	4.4
Carbonyl (C = O)	0.4	1.0
Ether (–O–)	2.8	0.9
	6.3	11.9
Oxygen by difference:		
Ash basis	5.9	16.2
Mineral-matter basis	–	16.0

*Cronauer and Ruberto [34, 185].

and hydroxyl groups predominate, while carboxyl and carbonyl groups are less important. At slightly higher carbon contents, carboxylic acid groups are not observed. The predominant oxygen functionalities in the subbituminous coal are hydroxyl and carboxylic acid groups with lesser amounts of ether and carbonyl groups. On an absolute basis, the subbituminous coal contains about 2.5 times as many carbonyl and hydroxyl oxygen groups, and about 3 times as many carboxylic acid groups as the bituminous coal. However, the subbituminous coal contains $\frac{1}{3}$ as many less ether groups as the bituminous coal, and contains a substantial amount of oxygen unaccounted for. Low-rank coals are noted for their content of carboxylic acid and methoxyl groups, and in general for their wide range of oxygen functional groups. It is not known whether the carbonyl groups are present simply as highly associated ketone structures or as polycyclic quinones. The latter is suspected in bituminous coal [60, 76, 133, 209]. As rank increases the variety of oxygen functional groups greatly narrows, eliminating methoxyl, carboxyl, and carbonyl oxygen. Oxygen that is unaccounted for is usually assumed to be present in either linkages or heterocyclic structures [34, 185, 210].

Relatively little work has been published on identification of sulfur functional groups in coal. Several oxidative procedures have indicated that about 60–80% of the organic sulfur in bituminous coals is in the form of dibenzothiophene or similar structures [64]. Reviews of sulfur functional-group analysis in coal [10, 11] and the chemistry, kinetics, and thermodynamics of reactions of sulfur have been published [12]. The results of a reduction-cell method [13] in which various coals were treated with a reducing agent (hydrogen) in the presence of a mixture of solvents and catalyst are shown in Table 2-10 [14]. These data show that in high-rank coals, thiophenic sulfur groups predominate, while thiol, thiophenol, and sulfide groups are most important in the low-rank coals. The amount of organic sulfur in the form of aliphatic sulfides is in the range of 18–25% for all the coals studied, indicating that this may be an intermediate form in the coalification

Table 2-10

Distribution of Organic Sulfur Groups in Five Coals*

Coal	% organic S accounted	–SH	ϕ–SH	R–S–R	Ar–S–Ar	Thiophenes†
Illinois	44	7	15	18	2	58
Kentucky	46.5	18	6	17	4	55
Martinka	81	10	25	25	8.5	21.5
Westland	97.5	30	30	25.5	–	14.5
Texas lignite	99.7	6.5	21	17	24	31.5

*Attar and Dupuis [14].

†Unaccounted sulfur assumed to be thiophenic.

process. Some of the thiol groups are believed present in the form of ionic thiolates, e.g., calcium thiolates. Both reductive and oxidative approaches have indicated that the thiol and sulfide sulfur groups are relatively easy to remove. If the functional-group distribution of sulfur in coal is anything like that of oxygen, it is likely to vary widely for coals within the same rank and will depend strongly on the geology and geochemistry of the coal basin. It is also likely that the thiophenic sulfur content as a percentage of organic sulfur will increase generally with rank, but will not correlate as well with rank as the sulfur content relative to the mineral-matter-free coal.

2.2.3 *Average Chemical Structures*

In order to conveniently visualize the chemical structure of coal and to provide a frame of reference for its chemical behavior, various average structures that satisfy available compositional and spectroscopic data have been proposed. One of the early models for an 82%-carbon vitrinite corresponding to a high-volatile bituminous coal is shown in Figure 2-9 [73]. The basic unit in this model is 9,10-dihydroanthracene in which two benzene rings are linked by two methylene bridges. Another molecular model for a 76%-carbon bituminous coal [223], shown in Figure 2-10, goes a step further by including a distribution of aromatic-unit sizes ranging from one to five rings (a portion of which are hydroaromatic), sulfur and nitrogen functional groups, and connecting linkages between ring clusters consisting

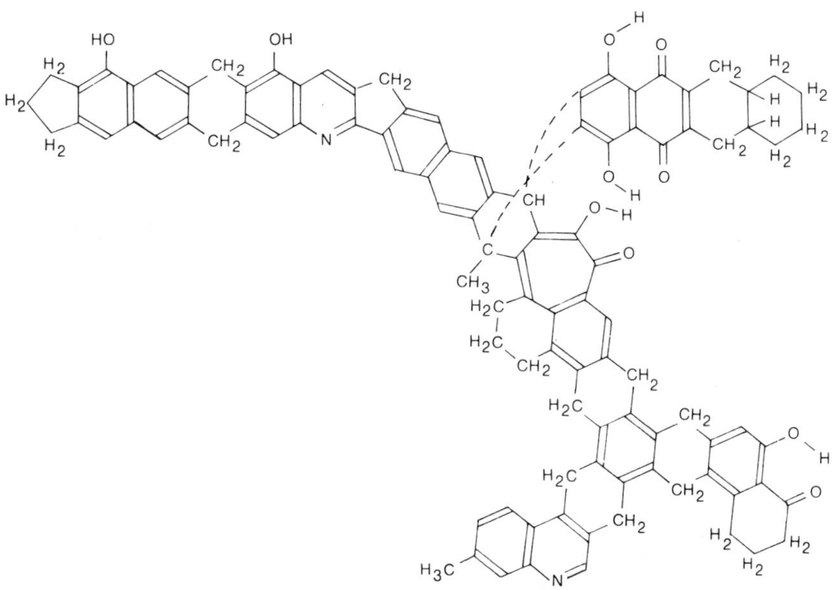

Figure 2-9. Model structure for an 82%-carbon vitrinite (Given [73]).

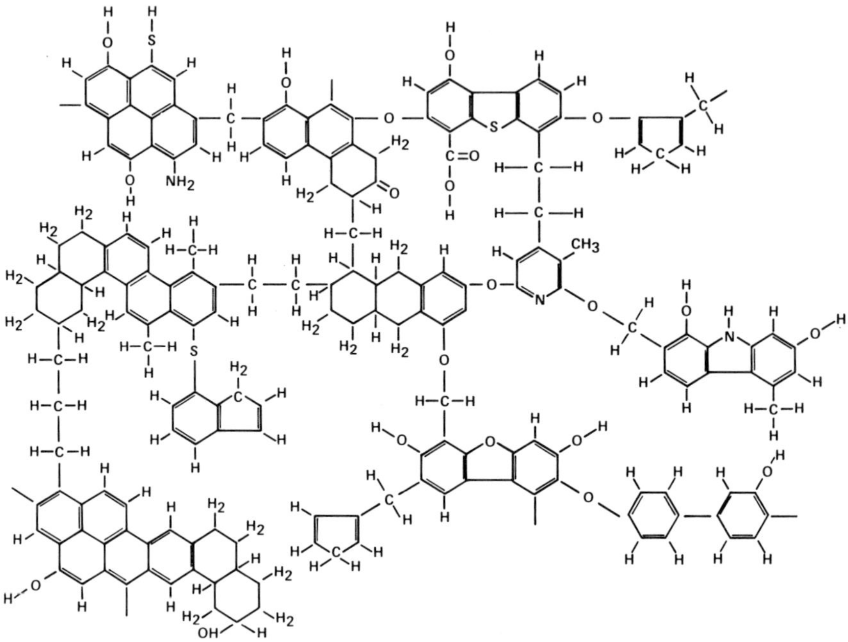

Figure 2-10. Schematic Representation of Structural Groups and Connecting Bridges in Bituminous Coal (Wiser [223]).

of short aliphatic chains (C_1–C_3) and ether bridges. Representations of the basic units of coals of varying rank are shown in Figure 2-11 [213, 214]. These models preserve the basic units that characterize the chemical properties and are not meant to be molecular structures. The average skeletal structure deduced for a molecule in solvent-refined coal produced from Wyodak subbituminous coal is shown in Figure 2-12 [218]. There is a high probability that this structure is the fundamental unit in the initial coal and that the molecular units are linked by weak bonds ($E \leqslant 50$ kcal/mole) such as $ArCH_2$–XAr bonds (X = O, S, C) that are broken in the early states of dissolution [17, 33, 57, 218].

2.3 The Chemistry of Coal Dissolution

2.3.1 *Mechanism of Dissolution*

Based on the chemical and structural information discussed in Section 2.2, the reactive maceral components of coal are believed to be highly crosslinked, amorphous, macromolecular networks consisting of a number of stable cluster units or aggregates connected by relatively weak linkages. The most probable labile bonds under coal-liquefaction conditions include:

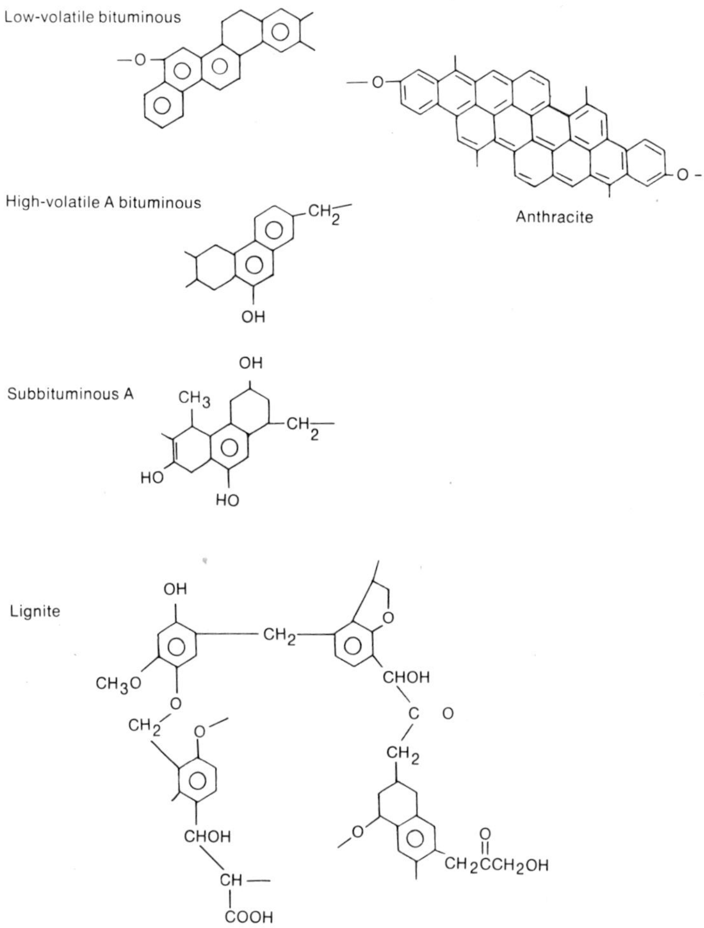

Figure 2-11. Basic structural units for coals of various ranks (Wender [213, 214]).

Ar–CH_2–Ar	Ar–O–Ar	Sulfur analogs of oxygen
Ar–$(CH_2)_n$–Ar	R–O–Ar	Hydrogen bonds
	R–O–R	Charge-transfer complexes

The weak bonds such as hydrogen bonding and charge-transfer complexes can be overcome to some extent at temperatures below 523 K, yielding extracts of up to 40–50% of some bituminous coals [96, 132, 133, 149, 209]. As the temperature is raised above 523 K and enters the temperature regime of donor-solvent coal-

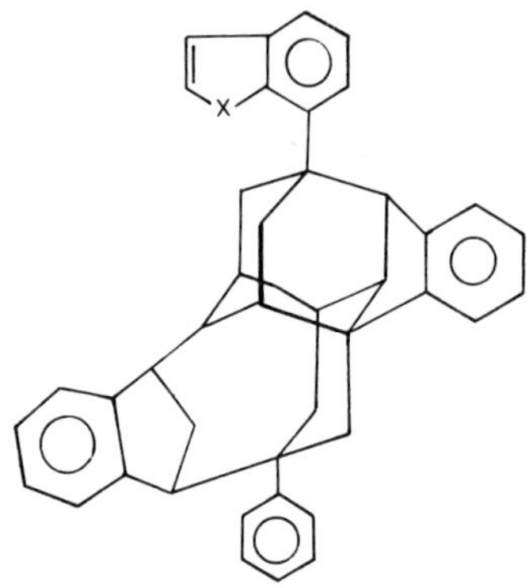

$C_{43}H_{36}NO_2$ 62% AROMATIC C

600 MOL. WT. 52% AROMATIC H

Figure 2-12. Average skeletal formula for Wyodak SRC (Whitehurst et al. [218]).

liquefaction processes (673–733 K), thermal processes rapidly occur whereby the distributed activation energies are overcome for the various labile bonds, yielding reactive fragments of free radicals [38, 39, 57, 98, 147, 216, 218, 221, 222]. If hydrogen is available to a radical from redistribution of hydrogen in the organic matrix (autostabilization) [105] or from donation by the solvent, the radical will be capped, yielding a stable species with molecular weight in the range of 300 to 1000. At conditions where hydrogen is unavailable or the concentration of radicals becomes excessively large, the fragments can recombine or polymerize to form very high-molecular-weight compounds and coke. The autostabilization process involves little in the way of net hydrogen transfer to the thermal fragments from the solvent, but mostly redistribution of hydrogen within the coal matrix, with the solvent acting as net shuttler of hydrogen [38, 39, 147]. Figure 2-13 illustrates what are believed to be the general processes occurring during donor-solvent liquefaction. The intermediate stabilized species are the preasphaltenes (or asphaltols), which are further reduced in molecular weight to asphaltenes (somewhat stable aromatic nuclei), and then to distillable oils and hydrocarbon gases. The latter compounds are also generated at each step of the main reaction path as byproducts [35]. As preasphaltenes and asphaltenes are further decomposed, the aromaticity of the products can increase substantially due to dehydrogenation of hydroaromatic structures [216, 218]. In fact, as asphaltene conversion increases and fewer polar functional groups are present, the remaining byproduct asphaltenes are much more aromatic (condensed) and are hydrogenated at a relatively slow rate [21].

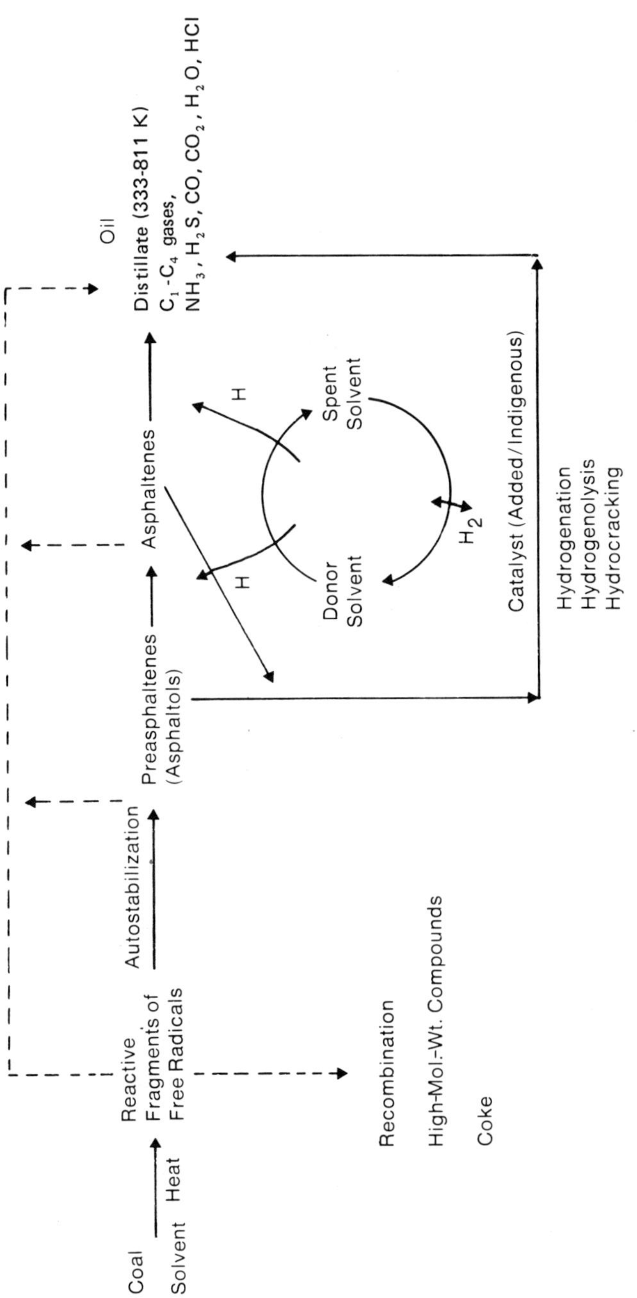

Figure 2-13. General reaction paths occurring during donor-solvent coal liquefaction (Han and Wen [98]).

2.3.2 *Effect of Coal Properties on Coal-Liquefaction Behavior*

A. COAL PARTICLE SIZE

Early studies of the effect of coal particle size on solvent extract yield for a variety of solvents, including anthracene, pyridine, tetralin, benzene and acetone, show that particle size over the normal range of interest (1 to 1200 μm) has only a small effect on the yields, until the size is reduced to about 1 μm [60, 132, 133, 209]. For example, extraction of a bituminous coal in which the largest particle was 1 μm gave a 30.1% benzene extract yield at 533 K compared to 20.6% for an 841–1000-μm sample of coal [132]. Extractions of lignite and brown coal showed virtually no effect of particle size, while extract yields increased only a modest amount with rather large decreases in particle size for bituminous coals. Much of this work was performed at temperatures below 533 K with weak solvents such as benzene. Under more severe conditions of coal liquefaction, these particle-size effects would be expected to disappear completely. This indeed has been confirmed in long-reaction-time rocking-autoclave tests at 713 K for 2.38-mm and 210-μm Elkol subbituminous coal particles in a coal-tar solvent [154], and in short-reaction-time microautoclave tests at 597–713 K for 74–149-μm Ireland Mine (Pittsburgh Seam) bituminous coal particles in tetralin [39].

Very careful short-reaction-time autoclave studies of the particle size ranges < 45, 45–75, 75–425, and 425–600 μm for several bituminous coals and a non-coking subbituminous coal have indicated that particle-size effects, if they exist, are within the variation of the data due to minor differences in parameters such as time, temperature profile, hydrogen pressure, solvent/coal ratio, and experimental error in the product workup [218]. In addition to experimental-parameter differences, detailed examinations of each of the particle-size ranges indicated that the elemental and mineral composition varied with size and could also have caused some of the variations in the conversion and yields. With Kentucky coal, there were only small variations in the overall elemental analyses, and the ash varied by ± 1.5%. However, the 45–75-μm size fraction contained significantly more pyrite than the 425–600-μm size fraction. In the case of an Illinois bituminous coal, carbon, hydrogen, and oxygen increased, and sulfur and ash decreased as particle size increased. The 425–600-μm size contained 11.5% ash, while the <45-μm size contained 16.7% ash, but the pyritic-sulfur contents were nearly identical. Variation in maceral content of the different size fractions was also strongly suspected.

B. COAL SURFACE PROPERTIES

Porosity and surface-area measurements of various coals before and after partial liquefaction have shown that the coal particle pore structure is not conducive to the intraparticle transport of donor-solvent and liquefaction-product molecules and therefore does not play a significant role in the dissolution process [215, 218]. The coal particles in these studies had about 5–10% void (volume), very little micropore structure (i.e. pores smaller than 300 Å), low total pore volume, and low N_2 surface area. Total pore volumes were typically in the range 0.08 to 0.16 cm^3/g. Pore volumes for the <300-Å pores increased in the residue after exhaustive pyridine extraction or liquefaction in a donor solvent, but did not exceed 0.07 cm^3/g. Other studies have also shown for coals from anthracite to lignite that total pore volumes are typically less than 0.16 cm^3/g [68]. About a third of the pore volume is too small to admit organic molecules, while an additional third is capable of diffusion of solvent molecules and low-molecular-weight liquefaction-product molecules [218]. During dissolution the void volume fraction of these two size ranges only increases slightly. The remaining third is accounted for by large cavities and cracks, i.e. the macropore structure. Since much of the initial products of dissolution have molecular weights greater than 1000 and dissolution is essentially complete within two minutes at usual liquefaction conditions, it is very unlikely that the micropores in coal contribute significantly to the diffusion of initial product molecules from the coal particles. It has been proposed that a shell progressive mechanism with simultaneous reaction within the particles may best represent the physical process of dissolution.

The measurement of pore volume and surface area of coal by the BET low-temperature adsorption measurement does not adequately characterize the micropore structure. The reader should consult references [68], [133], and [140], where other methods are reviewed, including activated diffusion with CO_2. Bituminous coals have surface areas in the range 30 to 100 m^2/g, the majority of which is accounted for by ultrafine capillaries of widths below 40 Å and accessible only through pore mouth openings of widths 5–8 Å. Temperatures not too far below ambient temperature must be used to measure this area, and measurements at liquid-nitrogen temperatures are unsatisfactory. For this reason reported nitrogen surface areas are low by a factor of up to 100. For all practical purposes, however, the nitrogen surface area may be the only area truly accessible to organic molecules within the time frame of rapid dissolution.

C. PHYSICAL PROCESSES IN COAL PARTICLE DISINTEGRATION

Several theories have been postulated to describe the mechanisms of coal particle dissolution. One theory considers coal to be a matrix of large strongly linked micelles that contain intimately associated smaller, less strongly bonded micelles that are normally trapped within the matrix unless it is swollen by a solvent [133]. Both types of micelles are continuous series, instead of single discrete classes, and can be progressively dispersed in a suitable solvent by increasing the temperature. The dividing line in the continuous series between the aggregates resistant to disintegration and the extractable ones is determined by the temperature. Thus, dissolution occurs by removal of colloidal-size units directly from the coal matrix. This theory implies that the skeleton of the coal particle should remain intact during dissolution, assuming external forces on the particles are not significant.

In another theory [71], coal particles are visualized as dissolving as a unitary substance similar to gelatin in water. However, retrogressive reactions cause much of the coal to precipitate within a short time. This would be similar to mesophase formation and subsequent solidification into coke particles.

Recent photomicroscopy studies of coal particles at various stages of dissolution have contributed much to understanding the physical processes in coal dissolution [92, 94, 215, 218]. Unstirred autoclave experiments at temperatures up to 723 K have been performed on West Kentucky bituminous coal and Wyodak subbituminous coal in various solvents, and the residues were recovered by Soxhlet extraction and examined under a microscope. The overall size and shape of the particles were not changed despite a conversion to pyridine solubles of 80–85% for the Kentucky coal and about 70% for the Wyodak coal. The highly converted particles had very jagged edges suggestive of substantial chemical attack, while the less converted residue particles looked more like the parent coal. The residue particles from the Kentucky coal were larger than the starting coal particles, indicating substantial swelling; while the residue particles from the Wyodak coal were slightly smaller than the parent coal. The Wyodak residue had a layered-type porous structure, while the Kentucky residue had a cellular-type porous structure similar to that observed in chars and coke from coking bituminous coals. Both residues were very fragile and easily shattered. The coal particles retained their integrity even when temperature, time, and solvent conditions were changed and conversion to pyridine solubles was over 70%. These results clearly show that the physical integrity of the coal particle is not necessarily lost under coal-liquefaction conditions, and that the remaining mineral matter and organic material can retain the skeletal shape of the particle. Results of experiments in which dry coal was injected into the autoclave were identical to those in which the coal was "saturated" with solvent prior to reaction.

If any mechanical stresses, such as stirring, are applied to the particles while they are undergoing dissolution, they will disintegrate very rapidly. However, stirring at reaction conditions will not disintegrate the particles unless dissolution reactions have occurred to such an extent as to weaken the particles severely. This can be viewed as a progressive surface erosion of the organic matrix forming a macropore system that extends into the bulk of the particle. The micropore system of the particle is not significantly altered, nor does it appear to play a significant role in dissolution. Because particle disintegration occurs in less than two minutes in a stirred reactor, it is no surprise that particle-size effects from intraparticle diffusion are negligible. Nonetheless, reactions occurring within the coal particle are probably very different from those occurring in the bulk solution and contribute differently to the product distribution.

The effects of agitation rate, temperature, hydrogen-donor activity of the solvent, and hydrogen pressure on coal-particle disintegration have been investigated. Agitation rates as low as 120 rpm caused complete disintegration of the coal, so that the reactor turbulence encountered in commercial process reactors will probably do the same. At 673 K, disintegration of the coal particles is completed in less than 2 minutes in the presence of a donor solvent. Kentucky coal in a creosote solvent [94] began to fragment at a temperature of 613 K. At 623 K, none of the original coal particles remained intact, and at 633 K, disintegration appeared complete. Very good hydrogen donors, such as tetralin and creosote oil, disintegrated the coal very effectively at 673 K, while paraffin oil had no effect on the particles, and decalin, a weak hydrogen donor, had an intermediate effect. In the absence of a hydrogen-donor solvent and hydrogen, the residue particles appeared not to be physically different from the parent coal particles. When a hydrogen pressure is used at 673 K with a nondonor solvent, particle fragmentation is substantial, but not nearly as complete as the case where a hydrogen-donor solvent is used with a nitrogen atmosphere. Thus, the most effective and rapid hydrogen transfer to the coal comes from the donor solvent. This would imply that once the initial dissolution reactions are complete, the rate-determining step may be the rehydrogenation of the donor solvent by activated hydrogen in solution with gaseous hydrogen as the source.

The particle size distribution in reacted coal-anthracene oil mixtures was studied as a function of the degree of coal conversion to pyridine-soluble material [72]. The particle-size distribution data show that the fraction of particles less than 10 μm in diameter gradually increases with conversion up to about 85% conversion. At higher conversions, the fraction of particles smaller than 10 μm increases very dramatically. Based on these results, it is unlikely that once a coal slurry has been preheated to the dissolver temperature in a donor-solvent liquefaction process, any coal particles larger than 10 μm will remain; therefore, the reactions within the dissolver can be treated as homogeneous except for the possible heterogeneous catalytic contribution of the mineral matter.

D. COAL CHARACTERISTICS

It is a well-known fact in coal-liquefaction technology that coal characteristics, such as rank, petrographic composition, mineral-matter content, and geological history, can significantly affect the performance of a coal in coal-liquefaction processes. In order to adequately control the coal feedstocks in large-scale liquefaction plants so as to keep the process operating conditions and productivity within desired limits, it has become a necessity to develop a fundamental correlation between coal characteristics and liquefaction performance. Early research in this area showed that liquefaction yields varied widely with petrographic composition [59]. This is expected in light of the chemical differences between the various macerals, as discussed in an earlier section of this chapter. The macerals considered reactive include vitrinite, pseudovitrinite, and exinite, while the inertinite macerals (fusinite, semifusinite, micrinite, macrinite) are considered unreactive in coal liquefaction [78].

Researchers at Pennsylvania State University have undertaken in recent years the important task of correlating coal-liquefaction behavior with coal characteristics for United States coals [1, 77–80, 229]. They have selected and evaluated hundreds of coals from most of the seven major coal provinces. Although the earlier portion of this work did not establish predictive correlations, it was shown that conversion varied widely with petrographic composition (the content of reactive macerals) and rank, and that geological history (the coal province) may be an important factor [77, 78]. With regard to the latter characteristics, coals of the Rocky Mountain and Great Plains provinces were formed during the Cretaceous and early Tertiary periods (60–120 millions years ago), whereas coals in the Appalachian and Interior provinces were formed during the Carboniferous age (300 million years ago). Consequently, coals of the same ASTM rank from each of these provinces are not biochemically or geologically the same, since the vegetation in the two eras was quite different, and so were such factors as the rate of burial and geothermal gradient experienced by the coal bed.

Conversion in the early Penn State liquefaction tests was measured as benzene-soluble products yield, and the tests were performed in batch, rocking autoclaves in the presence of a catalyst [77]. In some tests the catalyst was impregnated on the coal and no solvent was used, while in some others, performed by Gulf Research and Development Company in cooperation with Penn State, an anthracene oil solvent was used in the presence of a proprietary catalyst, and filtrate yields were determined in addition to benzene solubles. Temperatures were in the range of 658–698 K, with hydrogen pressures in the range 5.51–15.26 MPa. Heatup generally required several hours, and the hold time was either 1 or 3 hours. Of the nineteen coals studied for rank and geological-history effects, lignites, subbituminous, medium-volatile and low-volatile bituminous coals gave relatively low yields [77], and high-volatile bituminous coals gave the highest conversions. But within this broad range, effects of rank and geological factors could not be distinguished. All of these coal samples contained more than 75% vitrinite, and most contained more than 80%. It was also noted that very high mineral-matter content was coincident with the high conversions observed for two geologically young bituminous coals from the Pacific province, implying mineral-matter effects due to catalysis.

In subsequent experiments maceral concentrates were prepared from a New Mexico subbituminous coal, a Kentucky bituminous coal, and a Montana lignite in which the total content of reactive macerals varied [78]. For both subbituminous and bituminous coals, the liquefaction yield correlated very well with the reactive maceral content. The exinite and micrinite contents of the fractions from the New Mexico coal were small and relatively invariant, while the predominant vitrinite and fusinite contents varied inversely to one another. The filtrate viscosity was lowest for the fractions rich in mineral matter, again implying a catalytic effect. Exinite (mostly sporinite), micrinite, and vitrinite were the major maceral components in the Kentucky coal fractions; there were smaller amounts of fusinite. There is some indication that the exinite (sporinite) was more reactive than vitrinite. This is consistent with the high hydrogen and volatile-matter contents of sporinite as well as its yield of low-viscosity liquid at pyrolysis temperatures below 673 K [74, 209]. On the other hand, micrinite is known to be relatively unchanged during the coking of coal [23, 150, 209]. No obvious correlation existed between conversion and reactive maceral content for the Montana lignite fractions. However, a substantially higher conversion was obtained for the fraction rich in pyrite than for any other fraction even though the reactive maceral content was relatively low for this fraction. One of the most important conclusions that resulted from the preceding work is that the catalytic effect of mineral matter is superimposed on the effects of the other important factors of rank, geological history, and petrography, and makes interpretation and correlation of data very difficult.

Regression Analysis of Coal Characteristics

Having gained the above experience, it was realized that simple correlations between conversion and individual coal characteristics would have very limited success in predicting liquefaction behavior. Therefore, an approach based on a stepwise multiple-regression analysis was adopted and applied to large data and sample bases [1, 79]. Sixty-eight coals from the two eastern and three western provinces were studied under the following standard set of liquefaction conditions using a 125-cm^3 tubing bomb reactor: 40 g tetralin, 10 g coal, no catalyst or gaseous hydrogen, initial nitrogen purge, heatup to 673 K in 5 min, and 3-hr hold time. The vessel was rocked through a 50° angle at a frequency of 85 min^{-1}. Suspicion that the level of agitation was inadequate in these tests was confirmed in later experiments with a modified system using some of the same coals. Nevertheless these first results represent a major advance in developing predictive correlations.

Plots of conversion to benzene-soluble products for all the coals versus carbon content, total reactive macerals, vitrinite reflectance, calorific value, or H/C ratio showed a great deal of scatter, indicating that no simple correlation could be obtained between conversion and any single property. There was also no obvious grouping of coals according to geological province in these plots. However, a plot of conversion against total sulfur content showed that conversion increased continuously and significantly with increasing sulfur content of the subset of 35 Appalachian and Interior province coals. Applying the stepwise regression procedure on all 68 coals for a large number of independent variables gave the following correlating equation:

$$\text{conversion} = 3.03\,S_T + 0.27\,\text{TRM} - 0.82C + 109, \qquad (2\text{-}1)$$

where

conversion = benzene solubles yield, wt% dmmf coal

S_T = total sulfur, wt/ dry coal

TRM = total reactive maceral content, vol% dmmf coal

C = carbon content, wt% dmmf coal

This equation accounted for 78.7% of the variance and gave partial correlation coefficients of $S_T = 0.85$, TRM = 0.57, and $C = -0.69$. Application of the regression procedure to the subset of Appalachian and Interior province coals greatly improved the correlation, indicating that the younger western-province coals may form a separate and distinct population whose liquefaction behavior would be better correlated with rank and petrographic parameters and not sulfur content. The subset correlation was

$$\text{Conversion} = 3.4\ S_T + 0.31\ \text{TRM} - 0.66C + 91.9, \tag{2-2}$$

where 93.8% of the variance was explained, and the partial correlation coefficients were $S_T = 0.93$, TRM = 0.80, and $C = -0.52$.

The high partial correlation coefficient for sulfur in both equations probably is a result of two factors. First, the pyrite in the coal is rapidly reduced to pyrrhotite, which then acts as a catalyst (also, the pyrite itself may be somewhat catalytic). Second, some of the organic sulfur groups are extremely labile and are easily removed. In the process, they form free radicals that accelerate further hydrogenation.

The experimental procedure used in the above study was modified further to eliminate any effect of agitation and to improve the reproducibility of the measurements [79]. It was found that the tubing bomb reactor needed to be agitated up and down at a frequency of 400 min^{-1} for the first two minutes of submersion in the 673-K sand bath, and that the reaction was unaffected during the remainder of the 1-hr hold time as long as the frequency was maintained between 100 and 400 min^{-1}. All coals were run in duplicate and, where necessary, in triplicate, using a tetralin/coal ratio of 2 volumes per unit weight of coal. Conversion was defined in terms of the ethyl acetate–soluble product yield, a change from the earlier benzene extraction. A mean standard deviation for replicate runs of 1.5% was achieved, corresponding to a precision of the mean conversion value from just two runs of 2.1%. Single determinations were performed in the previous study with a precision of 3.4%.

Cluster Analysis of Coal Characteristics

The statistical correlation of coal characteristics with liquefaction conversion has been taken a step further by applying a cluster-analysis procedure that segregates the complex population of coals tested into groups where differences between groups are maximized and differences within groups minimized [80, 81, 230]. A total of 108 coals were selected such that each of the Appalachian, Interior, and Rocky Mountain provinces were about equally represented, and as wide a range as possible was covered—excluding, however, medium-volatile bituminous and higher-rank coals as well as subbituminous and lower rank coals.

The cluster analysis divided the set into homogeneous populations to which regression analysis could be properly applied. Division of the set into three groups as shown in Table 2-11 was found most effective. The mean values for conversion, total sulfur content, and carbon content are greatly different for the three groups, and they can be generally characterized as Group 1 (high rank, medium sulfur), Group 2 (medium rank, high sulfur), and Group 3 (medium to low rank, low sulfur). Coals from the Interior and Rocky Mountain provinces fall almost completely in Groups 2 and 3, respectively, while about two-thirds of the Eastern prov-

Table 2-11

Characteristics of the Groups Separated by Cluster Analysis*

	Group 1	Group 2	Group 3
General character	High rank, med. sulfur	Med. rank, high sulfur	Med. to low rank, low sulfur
No. of coals from:			
Appalachian province	25	11	0
Interior province	3	25	1
Rocky Mountain province	2	0	37
Mean values:			
Conversion, %	52.9	70.6	64.0
VM	38.4	44.1	43.7
R_o	0.82	0.56	0.56
C	85.0	82.2	79.4
H/C	0.777	0.817	0.854
S_T	1.49	4.76	0.69
TRM†	82.4	87.7	89.7

*Given et al. [80].
†Total content of reactive macerals.

ince coals are in Group 1, and the remainder in Group 2. The application of this type of analysis was discriminatory for 102 out of the 104 coal samples, and in effect represents a new classification system specifically applicable to coal liquefaction.

With the properties more normally distributed after group segregation, multiple regression analyses were performed on the three groups and yielded the correlations shown in Table 2-12 [80, 81, 230]. The second equation for Group 1 was obtained when three coals that were outliers in several bivariate plots were excluded from the regression analysis. The correlations for the Group 1 and Group 2 coals required two rank parameters with different loadings, illustrating the complexity in formulating a classification method.

The viability of this classification system was strongly supported when a sample base of 60 coals was studied in a continuous-flow reactor (stirred tank) under conditions similar to an actual process [80]. A 2:1 solvent:coal slurry was fed at a rate of 1000 g/hr with 5 (g H_2)/(100 g slurry) to a well-stirred 500-cm^3 reactor at a pressure of 21.7 MPa and temperatures of 713 or 728 K. The solvent was a partially hydrogenated anthracene oil. The nominal slurry residence time was 30 minutes. Using the cluster analysis procedure, the conversions for the various coals were classified, giving the results in Table 2-13. The same relationship exists between Groups 1 and 2 as observed in the batch experiments using tetralin (see Table 2-11).

Table 2-12

Regression Analyses of Three Cluster Groups of Bituminous Coals*

Group	Best correlating equation	Standard error of estimate	Variance explained, %
1	$Conv = -34.8R_o + 50.7\ H/C + 0.16V + 30.5$	3.30	80.0
1	$Conv = -56.33R_o + 100.43$	2.99	83.7
2	$Conv = 0.86\ VM - 22.8R_o + 1.39S_T + 39.0$	2.95	79.2
3	$Conv = 0.93\ VM + 0.28\ TRM - 1.7$	3.12	47.5

conv = % conversion, wt% dmmf
R_o = mean maximum reflectance of vitrinite
V = content of vitrinite + pseudovitrinite, vol% dmmf
H/C = hydrogen-to-carbon atomic ratio
TRM = total content of reactive macerals, vol% dmmf
VM = volatile matter content, wt% dmmf

*Given et al. [80].

Characteristics of Coals from Various Countries

Some limited studies on Japanese and Australian coals agree in part with the trends of the previous results [37, 50, 109, 142, 228]. It was pointed out that the percentage of volatile matter may not be a good correlating parameter for reactivity because it includes moisture and carbon oxides which vary greatly with rank [142]. By using the percentage of volatile carbon as a correlating parameter instead of

Table 2-13

Mean Values for Conversion of Coals in Different Groups – Continuous-Flow Reactor Experiments*

	Group 1	Group 2
Mean Conversion at 713 K, %	49.1	74.4
Standard deviation	7.4	6.2
Number of data	12	14
Mean Conversion at 728 K, %	59.8	78.9
Standard deviation	6.1	7.2
Number of data	9	6

*Given et al. [80]

total volatile matter, three coal-characteristic studies were shown to be in agreement. There was also general agreement on the effect of reactive maceral content on conversion.

The effect of coal properties on the liquefaction of South African bituminous coals has been studied [91]. These coals are unique in that half or more of the reserves contain between 30 and 35% mineral matter, and they are believed to have formed during the Permian age in deltaic or fluvial environments. The inertinite maceral group is also a major component in South African coals.

In comparison with Carboniferous coals of North America and Europe, the Permian coals of South Africa and Australia tend to have thicker seams, lower vitrinite content, higher semifusinite contents, higher mineral-matter content, and lower sulfur content [50]. Conversion data based on toluene solubility correlated very well with the H/C ratio and the volatile-matter and reactive-maceral contents of the coals [91]. Experiments were performed by two methods, one using an anthracene solvent in a batch autoclave, and the other a semibatch tubular reactor without solvent and an impregnated tin catalyst. Liquid-product yield and distribution also correlated well with these same parameters. It has also been shown for South African and Australian coals that the semifusinite content of the inertinite should be considered as a reactive maceral [91, 145].

E. MINERAL MATTER

In studying the effect of coal characteristics on liquefaction behavior of United States coals, it was realized that the inherent mineral matter in coal had a catalytic effect on liquefaction and that high mineral-matter content generally meant that relatively high liquefaction yields would be obtained [1, 77–80]. The total sulfur content strongly correlated with the liquefaction yield of a coal and, in essence accounted for the combined effects of mineral matter and reactive organic sulfur groups. The latter parameters strongly parallel and correlate with total sulfur content, and consequently, the individual contributions are difficult to distinguish. However, with many coals, the pyrite content correlates nearly as well as the total sulfur content with liquefaction yield and may well be the dominant factor [88, 111, 114]. This comes as no surprise, since the catalytic behavior of iron compounds has been well documented, and various iron compounds were used as disposable catalysts in the former commercial German liquefaction plants [116, 117, 143, 163, 211, 227]. The catalytic influence of mineral matter on liquefaction has also been observed for coals from India [144], South Africa [90], Brazil [155], and Australia [114].

The five major groups of coal minerals are the aluminosilicates (clays), oxides, sulfides, sulfates, and carbonates. Table 2-3 lists many of the compounds in these groups. A low-temperature ashing technique has been developed that allows the selective removal of the organic portion of coal, leaving the minerals behind in an essentially unchanged state [83]. The technique is based on an oxygen plasma generated by a radio frequency field and makes ashing possible at a temperature of about 323 K. A study of 65 coals from the Illinois basin using x-ray diffraction

Table 2-14

Mineral Composition of Illinois Bituminous Coals
Estimated by X-Ray Diffraction and Normative Analysis

Mineral	X-ray diffraction, wt% [162]	Normative analysis, wt% [160]
Kaolinite	13	27 (Kaolinite and expandable clays combined)
Expandable clays	18	
Illite	21	20
Quartz	15	11
Pyrite	23	29
Calcite	9	13

after low-temperature ashing of the sample identified the major mineral constituents as shown in Table 2-14 [162]. This table also shows that estimates can be made based on a knowledge of the major inorganic elements and a normative analysis, with the exception of coals rich in dolomite or iron carbonates [141, 160]. In another study, semiquantitative data were obtained for the major minerals in 57 coal samples from various locations throughout the United States [151]. Kaolinite was the most abundant clay at an average value of 30%, and significant quantities of pyrite, quartz, illite, mixed-layer illite-montmorillonite, and calcite were found.

Trace metals might be of some importance in mineral catalysis. Tables 2-15 and 2-16 show typical concentrations of various trace elements [3, 4, 84, 151, 186].

A variety of experimental approaches have been used to study the catalytic effect of coal minerals on liquefaction and to identify the dominant catalytic species. These approaches include: studying the reactivity of a variety of coals of similar rank and petrographic composition, but of widely varying mineral content; removing mineral matter from the coal by gravity separation or by chemical demineralization techniques and observing the effect on liquefaction; and adding various minerals to modify a coal's liquefaction behavior.

Screening Studies of Coals with Widely Varying Mineral Content and Similar Rank and Petrography

Along the lines of this first approach, six high-volatile bituminous coals from the Appalachian and Interior provinces (Carboniferous age) were selected in which the organic content was relatively constant, but the total mineral matter varied by a factor of 4 and the pyrite content by a factor of 36 [87, 88]. Table 2-17 summarizes the key compositional data for these coals. Their liquefaction behavior was determined in a batch autoclave at the following conditions, using creosote oil as a

Table 2-15

Trace-Element Concentration ($>$ 100 ppm) in Mineral Matter from Coals of Different Rank*

Anthracite		Bituminous						Lig. & Subbit.	
		High Volatile		Medium Volatile		Low Volatile			
Element	ppm	Element	ppm	Element	ppm	Element	ppm	Element	ppm
Sn	795	Sr	1515	Mn	1043	Sr	606	Ba	3248
Ba	715	Ba	956	Ba	652	Ba	548	Sr	3011
Zn	568	B	587	Sr	486	Zr	339	B	659
Cu	335	Zr	313	V	284	Cu	281	Mn	445
Cr	251	Zn	236	Zr	237	Mn	207	Cu	423
Mn	223	Cu	223	Cu	228	V	206	Zr	158
V	205	V	190	Ni	192	Zn	171	Sn	101
Ni	182	Cr	147	B	159	Cr	164		
Sr	146	Pb	140	Zn	142	Co	127		
La	117	Sn	130	Cr	123	Y	113		
		Mn	130	Y	110	Ni	104		
		Ni	117						
% LTA	19.2		10.0		10.9		10.5		14.2
% HTA	15.9		7.6		7.9		7.8		9.2
No. of samples	5		24		7		8		13

*O'Gorman and Walker [151].

solvent: solvent/coal ratio (daf basis) 2.3; temperature 703 K; pressure about 11.73 MPa (6.9 MPa initial hydrogen pressure); hold time 30 min; stirring rate 500 rpm.

The conversion and product distribution were found to be a direct function of the mineral-matter content. Although conversion to pyridine solubles was little affected by mineral-matter content, conversion to benzene solubles increased quite substantially as the content of mineral matter increased. The product distributions in Figure 2-14 show that increasing mineral-matter content resulted in a significant increase in pentane-soluble oil and a decrease in preasphaltene yield. The asphaltenes yield increased only slightly with increasing mineral-matter content.

The hydrocarbon-gas yield, hydrogen consumption, and sulfur conversion data shown in Figure 2-15 indicate that the increased hydrogen consumed with increasing conversion and mineral matter is primarily used to reduce the molecular weight of the preasphaltenes and asphaltenes, and for desulfurization. No increase in hydrocarbon-gas yield was observed.

Table 2-16

Trace-Element Concentrations in Ash*

	Appalachian province		Interior province		Western states	
	Element	ppm	Element	ppm	Element	ppm
	Sr	1052	Zr	825	Ba	1467
	Ba	876	Zn	743	Sr	1456
	Zr	704	B	731	Zr	850
	Li	584	Sr	658	B	529
	V	336	Ba	399	Nd	295
	B	265	V	325	Zn	258
	Mn	260	Mn	325	Ce	238
	Rb	239	Rb	276	Mn	212
	Ce	238	Ni	262	Li	168
	Cr	230	Li	235	V	152
	Zn	230	Cr	224	La	128
	Nd	213	Ce	214		
	Ni	209	Co	193		
	Co	184	Nd	183		
	As	159	La	131		
	La	145	Pb	131		
	Y	142	As	119		
	Cu	128	Y	118		
			Ge	104		
% Ash	9.3		10.5		9.8	
No. of samples	600		123		104	

*Abernathy et al. [3].

A multiple-regression analysis on the conversion to benzene solubles revealed that the pyrite content of the coals studied was the most significant of the variables considered. The other parameters included total mineral content, clays, total sulfur, reactive macerals, and carbon content. The simple linear correlation

$$\text{conversion} = 8.16P + 30.32, \tag{2-3}$$

where P is the pyrite content, gave a correlation coefficient of 0.90, while a similar relationship for the clay content resulted in a correlation coefficient of 0.75. Inclusion of both pyrite and clay minerals in the regression analysis gave the best correlation coefficient of 0.97. This analysis implies that the pyrite is the predominant hydrogenation catalyst and that the clays are playing a minor role as cracking catalysts [144].

Table 2-17

Reactive Macerals, Mineral Composition, and Corrected Carbon-Hydrogen Contents of Coals Used in Screening Experiments*

				% Dry Coal Basis						
Coal sample	Reactive macerals† (%)	Carbon (%, dmmf)‡	Hydrogen (%, dmmf)‡	mm§	Pyrite	Total sulfur	Quartz	Calcite**	Kaolinite	Mixed-layer clays
Kentucky No. 11 Fies Mine	92.8	81.6	5.5	20.90	5.43	6.11	3.34	n.d.	0.21	11.92
Kentucky No. 11 Homestead Mine	92.3	81.4	5.5	15.70	4.55	4.48	2.50	1.10	0.32	7.23
West Virginia, Pittsburgh Seam, Ireland Mine	90.7	84.1	5.9	13.82	3.26	4.44	1.46	n.d.	1.19	7.91
Illinois No. 6 Orient No. 4 Mine	95.3	81.9	5.5	12.95	2.80	3.00	2.20	0.78	0.82	6.35
Pennsylvania Pittsburgh Seam, USBM Bruceton	89.5	86.1	5.6	5.18	0.53	1.05	0.54	n.d.	0.33	3.78
Kentucky, Elkhorn No. 3 Guaranty Mine	89.3	85.1	5.7	5.62	n.d††	0.78	trace	n.d.	2.00	3.50

*Granoff et al. [88].
†Sum of vitrinite + pseudovitrinite + exinite.
‡dmmf = dry mineral-matter-free.
§mm = mineral matter.
**n.d. = not detected by x-ray diffraction techniques.
††Pyrite content estimated from pyritic sulfur = 0.15%.

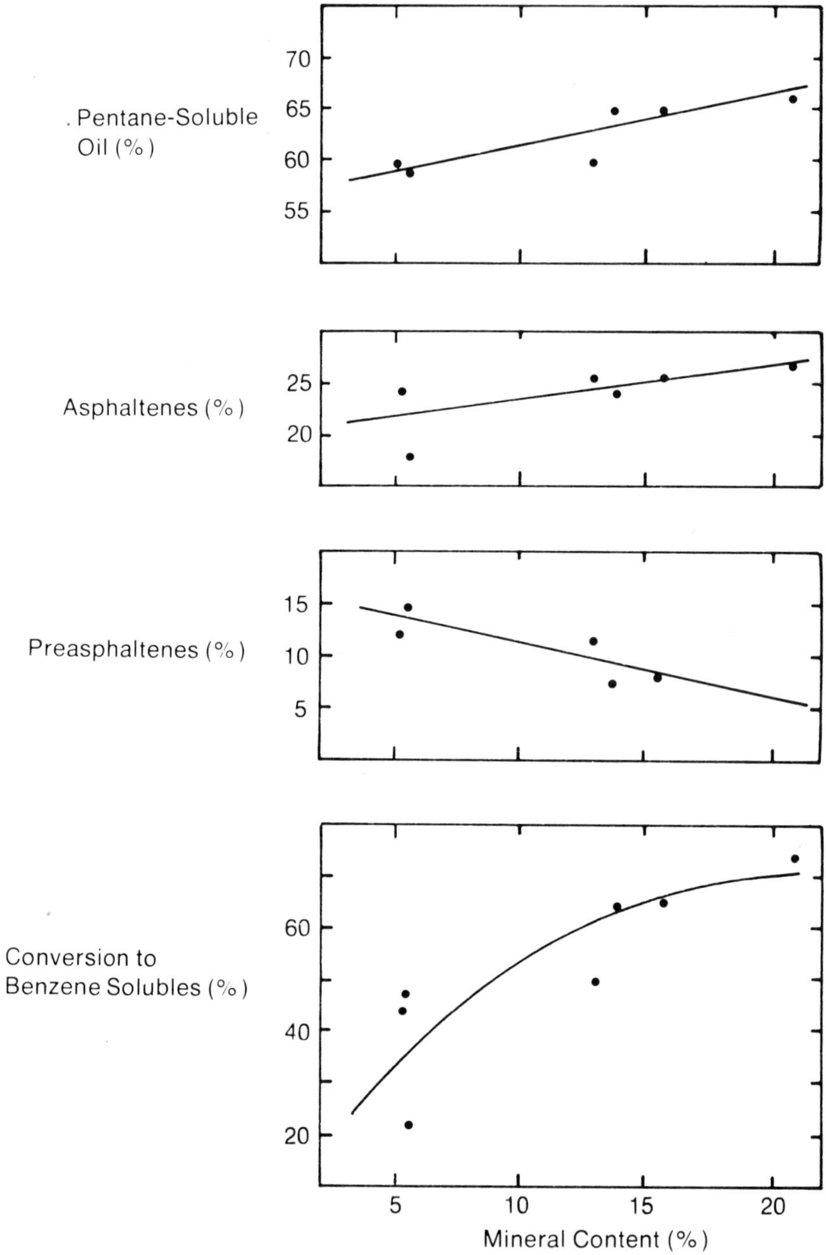

Figure 2-14. Screening series: effect of mineral content on the yield of preasphaltenes, asphaltenes and pentane-soluble oils (Granoff et al. [88]).

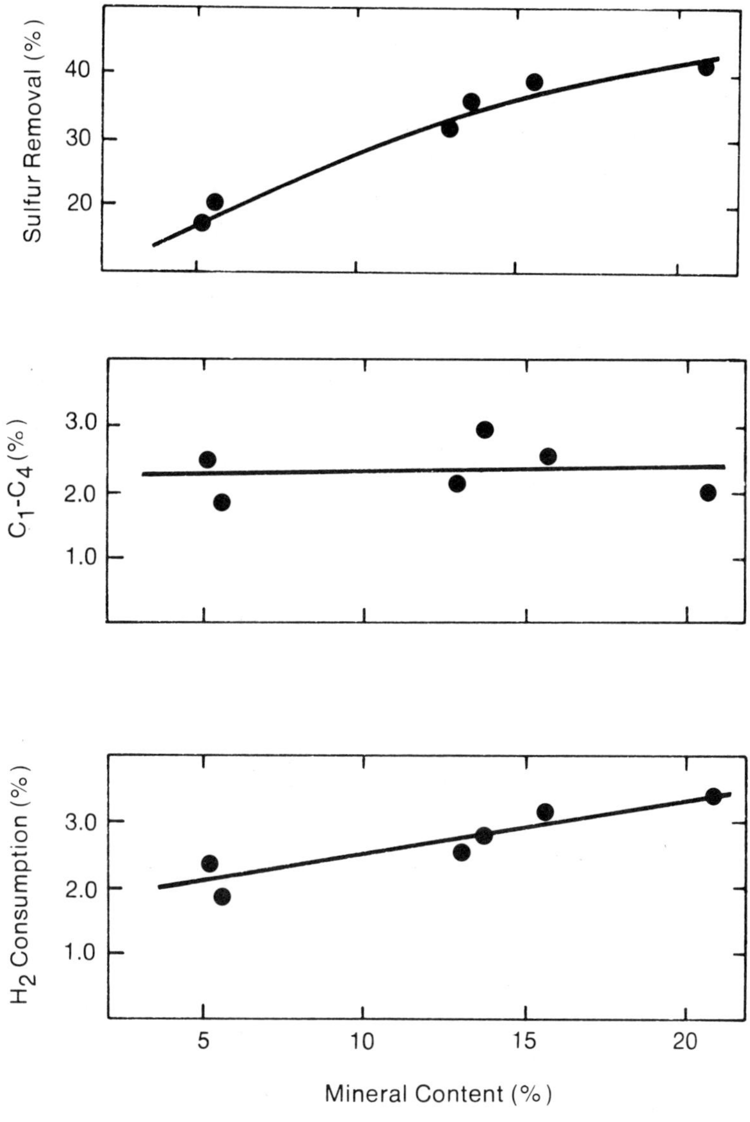

Figure 2-15. Screening series: effect of mineral content on hydrogen consumption, hydrocarbon gas formation, and sulfur removal (Granoff et al. [88]).

Pyrite itself is not the active catalyst but is rapidly reduced to pyrrhotite (FeS_x, $x = 1.0$ to 1.14) under coal-liquefaction conditions [101, 118, 159, 181, 203, 226]. Mossbauer-spectroscopy analysis of iron before and after liquefaction of a

Western Kentucky high-volatile bituminous coal showed that no unreacted FeS_2 or $FeSO_4$ remained in the residue, nor was free iron formed [120].

Not all coals having a large amount of pyrite in the mineral matter are necessarily highly reactive, nor is all of the pyrite reduced during the reaction [80]. Pyrite is probably most effective in coals where it is finely distributed and can readily be reduced to pyrrhotite. If small particles are encased in slowly reacting or unreacting organic matter or, at the other extreme, large particles ($> 10\text{-}20 \mu m$) are reduced on the outside forming a shell, the remaining pyrite may be largely unreduced [80, 187].

It is very unlikely that mineral matter has any effect on the initial coal dissolution reactions, when the free-radical pyrolysis reactions are occurring within a time span of about 2 to 5 minutes at temperatures of 673 K or higher. Although the coal-screening tests discussed above were at reaction times of 30 minutes, the weak correlation of conversion to pyridine solubles with mineral-matter content suggests that this is indeed the case. Some preliminary tubing-bomb reactor data also confirm this belief [82]. In the perspective of a coal liquefaction process, mineral matter probably has a minimal effect, if any, in the preheater, where the reaction time is less than about 5 minutes. The contribution of mineral-matter catalysis likely occurs in the dissolver, where it primarily provides reaction sites for rehydrogenation of the donor solvent, hydrogenation of the pyrolysis products, and dissociation of hydrogen.

Gravity Separation and Chemical Demineralization Effects

Gravity separation and chemical demineralization have been widely used to vary the mineral-matter content in a coal and thereby observe the effect on liquefaction behavior [88, 90, 141, 203]. One of the difficulties with the gravity separation technique is that the macerals tend to separate along with the mineral matter, so that the more inert macerals concentrate in the heavier fractions. Chemical demineralization can alter the organic structure of low-rank coals through ion exchange with carboxyl groups and can cause oxidation if treatment conditions are not carefully controlled.

A North Assam coal was separated into four fractions by specific gravity with the ash content increasing from 1.7 to 64.0 wt% (dry basis) with increasing specific gravity [144]. Total sulfur varied between 4.06 and 7.37 wt%, as shown in Table 2-18. The major minerals present were kaolinite and iron pyrite or oxide (no siderite) in a finely disseminated state. Hydrogenations were performed without solvent or added catalyst in a rocking autoclave at 673 K and 11.87 MPa hydrogen pressure for 3 hours. A direct relationship between mineral-matter content and conversion, based on benzene solubles, was observed. Individual mineral-matter components that correlated with conversion were iron, titanium, kaolinite, pyritic sulfur, total sulfur, and iron/sulfur atomic ratio. The correlation of total sulfur and (total iron)/(total sulfur) atomic ratio with conversion was excellent and suggested that all of the iron, both pyritic and nonpyritic, is converted to a reduced sulfide species that is catalytic, and that one of the roles of organic sulfur is to form hydro-

Table 2-18

Proximate Analysis* and Sulfur Distribution for a North Assam Coal†

Sample	Moisture	Ash	Volatile matter	Fixed carbon	Total sulfur	Sulfur distribution (% of total) Organic	Pyritic	Sulphate
Overall	2.1	4.4	41.5	52.0	4.46	83.8	11.3	4.9
Gravity fractions								
< 1.30 (float)	1.9	1.7	42.7	53.7	4.06	88.4	3.2	8.4
1.30–1.50	1.7	9.5	39.6	49.2	6.23	70.5	13.5	16.0
1.50–1.80	2.5	27.7	32.5	37.3	7.37	40.0	37.6	22.4
> 1.80 (sink)	1.8	64.0	19.9	14.3	5.10	16.1	58.4	25.5

*Wt%, air-dried basis.
†Mukherjee and Chowdhury [144].

gen sulfide, which then converts the nonpyritic iron to pyrrhotite. Addition of elemental sulfur to the sink fraction containing 56% of the total iron in nonpyritic form increased conversion significantly. Although there was notable variation of the maceral composition between the fractions, the mineral-matter effects dominated the results.

In a very similar study using a South African coal, it was shown that transport effects due to agglomeration of the coal particles can greatly influence the results obtained in hydrogenation without a solvent. Increased conversions were observed as increasing amounts of quartz, silica gel, or sand were mixed with the coal in the initial charge, and remained constant once more than 200 g of sand had been added to 30 g of coal. Nevertheless, increasing mineral matter was shown to increase conversion and dominate over any variation in petrography.

A Kentucky bituminous coal was demineralized in a water slurry by high-intensity magnetic separation until the ash content was reduced by 64% and the total sulfur by 25%. As shown in Figure 2-16, the raw coal, a sample simply washed with water, and the demineralized coal were very different in their liquefaction response in a 3:1 creosote-oil solvent:coal slurry. Conversion to benzene solubles increased significantly with increased mineral-matter content, while there was no effect on the removal of organic sulfur. Presumably, the magnetic separation was selective for pyrite removal. Therefore, it seems that variation of the pyrite level has no effect on hydrodesulfurization, and pyrite is a weak hydrodesulfurization catalyst. This has been shown to be the case in hydrodesulfurization of creosote oil [203] and thiophene [97].

An extensive demineralization study was performed where a Kentucky bituminous coal was subjected to gravity separation, yielding five fractions with ash and sulfur reductions as high as 77% and 54%, respectively [88]. This coal and several others were also demineralized by treatment with ferric sulfate, HC1/HF, or combinations of both reagents, resulting in mineral content reductions as high as 71% and

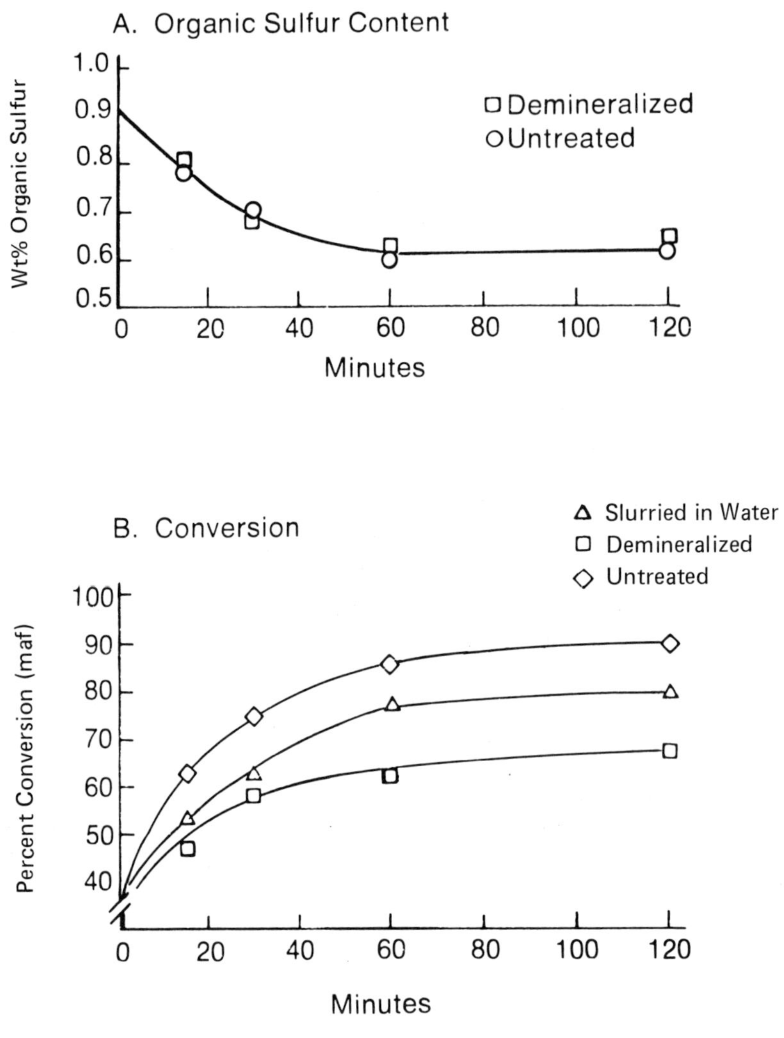

Figure 2-16. Effect of demineralizing coal feed and slurrying coal feed with water on conversion (Tarrer et al. [203]). Reaction conditions: temperature, 683 K; H_2 pressure, 13.8 MPa at 683 K; agitation rate, 1000 rpm; autoclave; 300 cc.

more than 40% pyrite removal. Mineral matter increased with increasing fraction density from 5.81% to 42.87%, while pyrite increased from 1.05% to 14.81%. Mineral matter in the chemically treated coals varied from 6.15 to 18.51%. The liquefaction responses of the raw and treated coals were determined in a batch autoclave at 678–680 K, 30 minutes hold time, 6.9 MPa initial cold-hydrogen pressure, and a creosote oil/coal (daf) ratio of 2.3. Figure 2-17 shows the conversion to

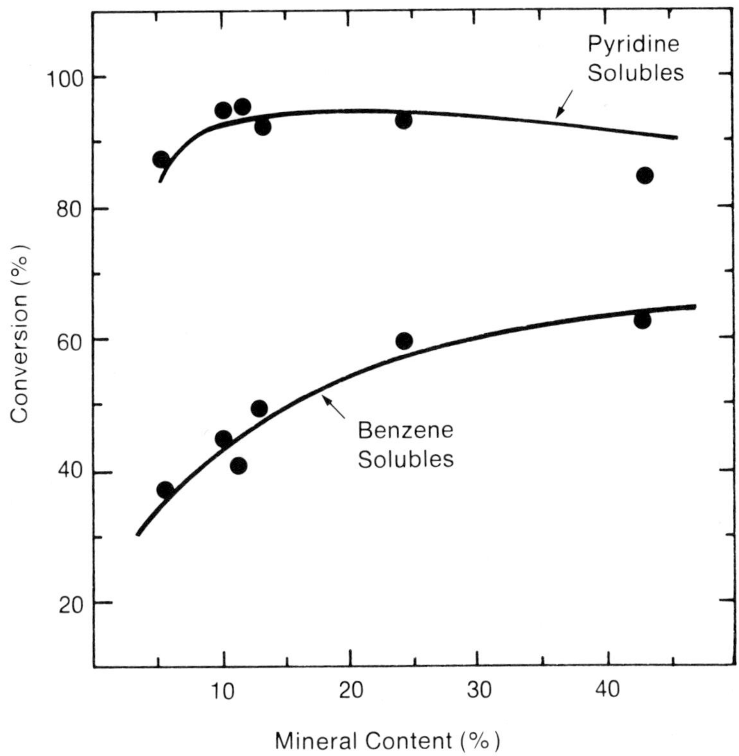

Figure 2-17. Gravity-separated KY11 coal: effect of mineral content on the conversion to pyridine and benzene solubles (Granoff et al. [88]).

pyridine and benzene solubles for the various gravity fractions and indicates that mineral matter has little effect on the initial dissolution but a very strong influence on the hydrogenation of the initial reaction products. The mineral matter is apparently catalyzing the reduction of molecular weight of the components within each solvent class and the conversion of species from a higher-molecular-weight solvent class to a lower one [13]. Typically, for Kentucky bituminous coal, the preasphaltene class would cover the molecular-weight range 650–900, asphaltenes 550–750, and oil ~350–500 [115].

The product distributions for the demineralized samples are shown in Figure 2-18, where the solid curves represent the data for the various density fractions [88]. The results are qualitatively the same as those shown in Figure 2-14 for the coal-screening study except that the yield of asphaltenes is nearly constant instead of increasing slightly. This difference could be a result of the difference in temperature between the two sets of experiments, 703 K versus 678 K. The effect of mineral content on hydrocarbon-gas yield, hydrogen consumption, and whole-product viscosity was essentially the same as observed in the screening studies.

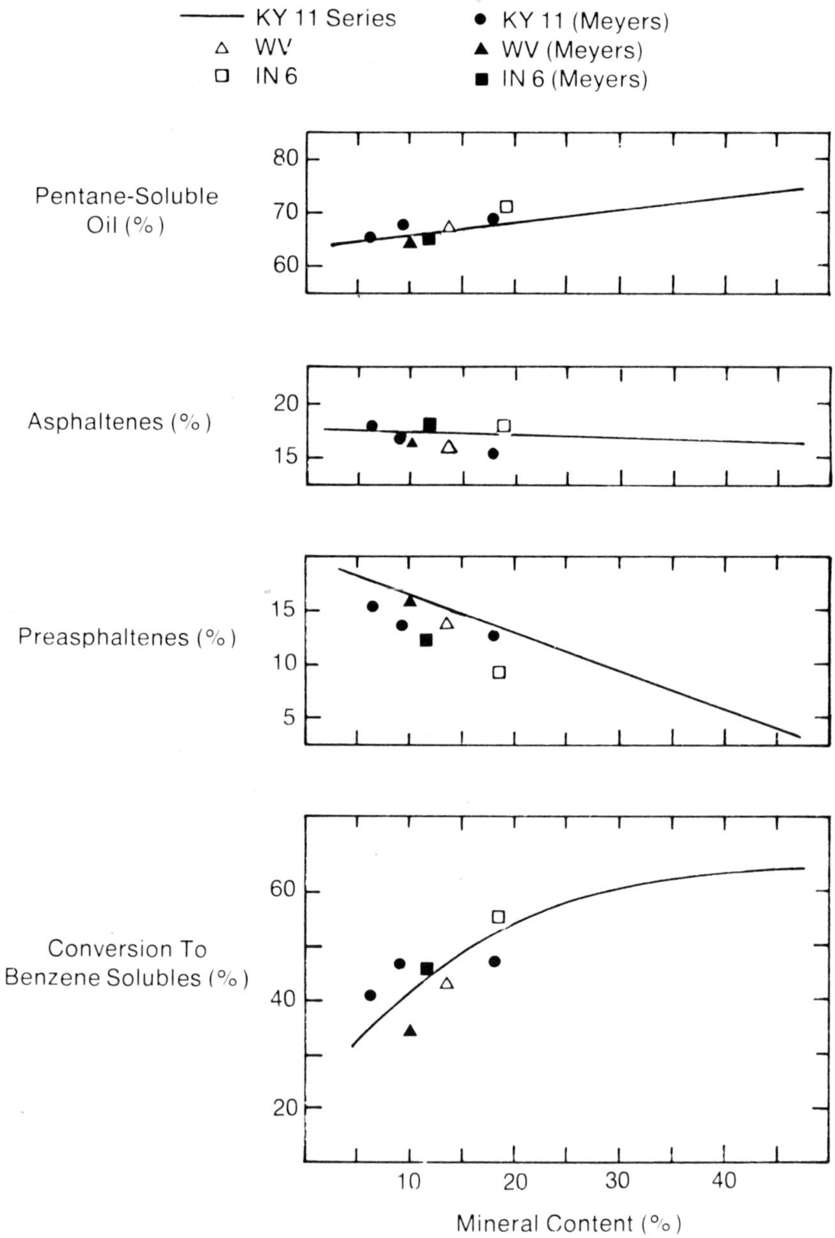

Figure 2-18. Chemically demineralized coal: effect of mineral content on conversion to benzene solubles, and on the yield of preasphaltenes, asphaltenes, and pentane-soluble oils (Granoff et al. [88]).

Mineral-Addition Experiments

Another very fruitful approach used to develop an understanding of the influence of coal mineral matter on coal-liquefaction behavior is to selectively add various mineral constituents to the coal-solvent reactant mix and observe the effect on conversion, product distribution, and sulfur removal [88, 95, 114]. This approach has also been used to study the influence of various mineral constituents and residues on the hydrodesulfurization of thiophene [97], the hydrogenation and hydrodesulfurization of creosote oil [203], and the hydrogenation of anthracene oil [220].

Batch autoclave mineral addition studies were performed with an Illinois No. 6 bituminous coal in a creosote solvent using the additives listed in Table 2-19 [88]. The solvent/coal ratio was 2.5, and 2.5 g of additive were used with 50 g (daf) of coal. The autoclave conditions were 678 K, 6.9 MPa initial cold-hydrogen pressure, and 30 minutes hold time at temperature. Ground cobalt molybdate on alumina catalyst and a blank were used as benchmarks to rank the effects of the other additives. As expected, the cobalt molybdate catalyst gave the largest conversion to benzene-soluble products, while single-crystal pyrite gave no improvement. As shown in Figure 2-19, the other iron sulfides, montmorillonite, and coal-liquefaction residue all effected some improvement in conversion. The pyrite acted as a hydrogenation catalyst and produced the lowest-viscosity product, having the highest hydrogen/carbon ratio of any product except that produced with cobalt molybdate. This is also shown in the product distribution described in Figure 2-20. Pyrite was effective in conversion of asphaltenes to oils but had little effect on the reduction of preasphaltenes. Pyrrhotite and montmorillonite were effective in reduction of both preasphaltenes and asphaltenes in the product. The sulfur content of the products was significantly reduced in comparison with the blank run with pyrrhotite and cobalt molybdate additives, while the pyrite additive exhibited no desulfurization activity. The inactivity of pyrite for desulfurization has also been noted in experiments with thiophene and creosote oil [97, 203]. The anomalous behavior of the single-crystal pyrite compared to observations in coal-screening studies gives support to the belief that the physical form of the pyrite is critical to its catalytic activity. Framboidal pyrite (raspberry-shaped clusters of pyrite) may well be the most active species rather than single-crystal-type pyrite (Table 2-19) [88]. Later work with another form of pyrite derived from coal showed an improvement in conversion to benzene solubles with a West Virginia bituminous coal [89].

Inorganic additives, such as alumina, silica gel, precipated red hematite, heavy magnesium carbonate ($3MgCO_3 \cdot Mg(OH)_2 \cdot 3H_2O$), calcium carbonate and brown coal [114]. A rocking autoclave was charged with 150 g of coal, 15 g of additive, and 150 g of tetralin, and was operated at 704 K, 9.66-MPa initial hydrogen pressure, and a hold time of 1 hour at temperature. None of the additives improved conversion to chloroform solubles except for iron added as hematite, which increased conversion by 20%. It was nearly as active a catalyst as cobalt molybdate.

Table 2-19

Minerals Used in the Spiking Experiments*

Mineral†	Description
Pyrite (FeS_2)	Single crystals from Sinaloa, Mexico, obtained from Ward's Natural Science Establishment. Pulverized to −200 mesh.
Fe_2S_3	Obtained as −200-mesh powder from Atomergic Chemetals Co.‡
Iron sulfide (FeS)	Obtained as −100-mesh powder from Cerac/Pure, Inc. Identified by x-ray diffraction as pyrrhotite, $Fe_{1-x}S$.
Montmorillonite	From Clay Spur, Wyoming. Obtained as −325-mesh powder from Ward's Natural Science Establishment.
ZnS	Obtained as −325-mesh powder from Cerac/Pure, Inc.
Residue	Acetone-washed, vacuum-dried filter cake from an autoclave run with IL6 coal in creosote oil. Used as a minus −100-mesh powder: ash = 48.4%, sulfur = 4.9%.
Co/Mo	Harshaw 0402T Co/Mo on silica-alumina. Catalyst pellets (1/8-in. diam × 1/8-in. length) were crushed and screened to 45 × 100-mesh particle size.

*Granoff et al. [88].
†Each of these was stored in a sealed container in a dry box.
‡It was found that this was probably a mixture of pyrite and pyrrhotite.

The mineral constituents tested for their influence on the hydrogenation and hydrodesulfurization of creosote oil are shown in Table 2-20 [203]. The tests were performed in a batch autoclave, and the decrease in initial hydrogen pressure at temperature was used as an indicator of hydrogenation activity [93, 159]. The screening results are shown in Figure 2-21 and indicate that pyrite and SRC mineral residue are relatively active hydrogenation catalysts. Ground cobalt molybdate catalyst had the highest activity, while the minerals ankerite, calcite, quartz, dolomite, and kaolin gave no improvement over the blank run. The desulfurization activity of each material is shown in Figure 2-22, which indicates that coal ash,

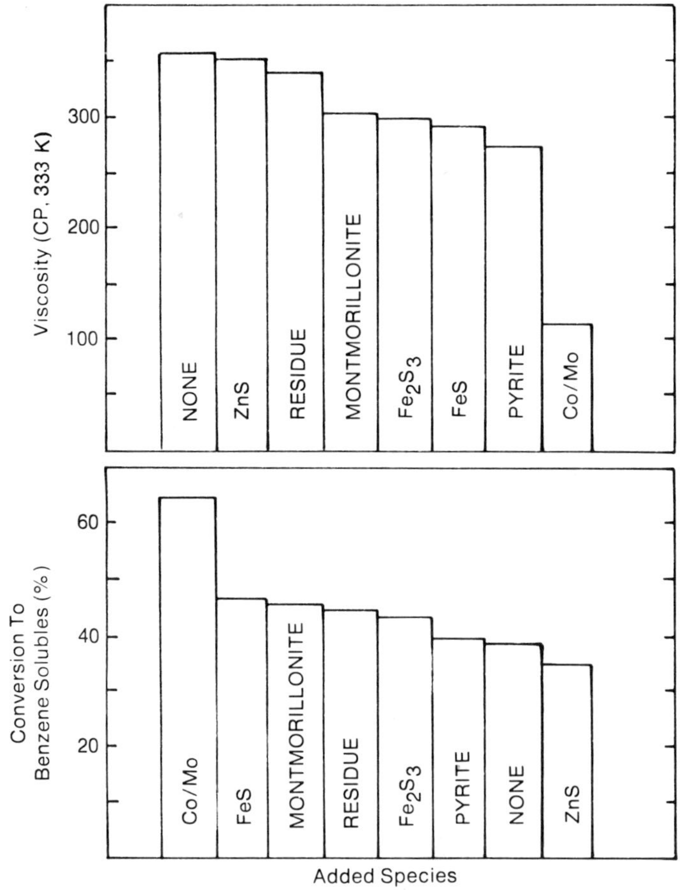

Figure 2-19. Illinois No. 6 coal: effect of added minerals on conversion to benzene solubles and product viscosity (Granoff et al. [88]).

iron, and SRC residue are the most active catalysts after cobalt molybdate. As discussed previously, pyrite has little desulfurization activity.

Catalysis of Solvent Hydrogenation

The catalytic influence of SRC residue is well known and is the basis for the SRC-II process, where mineral residue is recycled from the products to the feed and serves as an added catalyst to hydrogenate the donor solvent [8]. This role of the mineral residue, and in particular FeS, has been confirmed in solvent-hydrogenation studies [203, 226]. In one study, mineral residue from a Kentucky No. 11 hvBb coal having a 15% iron content in the ash was recovered from a coal solution, pyridine washed, rinsed with benzene and acetone, and dried in a vacuum oven [226]. The residue, which contained 40% carbon, 2% hydrogen, and 5.5% sulfur (distri-

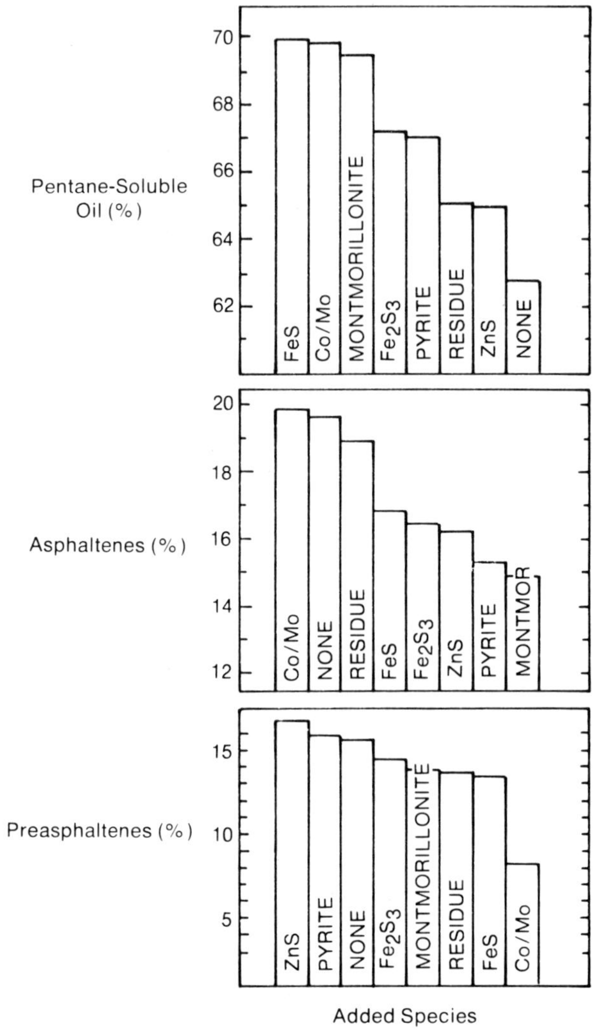

Figure 2-20. Illinois No. 6 coal: effect of added minerals on the yield of preasphaltenes, asphaltenes and pentane-soluble oils (Granoff et al. [88]).

buted as 0.1% pyritic, 0.5% sulfate, 1.9% organic, and 3.0% sulfide), 5.5% iron, and 46.14% ash, was ground to −100 mesh, and 200 g were added to 3800 g of anthracene oil. The anthracene oil was processed in a continuous-upflow tubular reactor at 6.9 MPa hydrogen pressure, 698 K, liquid-space velocity 0.60 hr^{-1}, and gas-space velocity 239 hr^{-1}.

When anthracene oil was fed into the reactor without added residue, the effluent showed no evidence of reaction. When the feed was switched to the anthracene-oil–residue mixture and the transient concentration effects noted, the hydro-

Table 2-20

Descriptions of Coal Minerals or Catalytic Agents Studied*

Species†	Classification	Description
Ankerite (ferriferrous dolomite)	Carbonate	An isomorphous mixture of CaMg $(CO_3)_2$ and $CaFe(CO_3)_2$.
Calcite‡	Carbonate	A crystalline form (hexagonal scalenohedral class of the hexagonal system) of $CaCO_3$. Often to a small extent the Ca is replaced by Fe, Mg, and Mn. Clay, sand, bitumen, and other mechanical impurities may be present.
Dolomite‡	Carbonate	A double salt with equal molecular quantities of $CaCO_3$ and $MgCO_3$, and not an isomorphous mixture of these two compounds. Usually found in a curved rhombohedral form.
Kaolinite‡	Kaolin	A collision type of clay; often found in minute pseudohexagonal (monoclinic) crystals. Chemically, an acid aluminum silicate, $H_4Al_2Si_2O_9$ or $2H_2O \cdot Al_2O_3 \cdot 2Si_2$ (H_2O=14%). Iron is often present in small amounts.
Muscovite‡	Shale	A lamina-type silica substance, having a monoclinic crystal structure; and chemically classified as an acid potassium aluminum orthosilicate, $H_2KAl_3(SiO_4)_3$ or $2H_2O \cdot K_2O \cdot 3Al_2O_3 \cdot 6SiO_2$ (H_2O=4.5%). Often, the potassium is partially replaced by sodium, and some varieties contain an excess of silicon over that indicated above.
Pyrite	Sulfide	A cubic structure of FeS_2, having, in its crystalline structure, a rocksalt-type arrangement of Fe^{2+} and S_2^{2-} ions, with Fe being octahedrally surrounded by S and each S atom having one S and three Fe atoms as neighbors. Uncommonly, Ni, Co, or sometimes both are found substituted for Fe. Obtained from Matheson, Coleman, and Bell (90–95% pure).

Table 2-20 (Continued)

Species†	Classification	Description
Quartz‡	Accessory	A crystalline form of SiO_2; a member of the triangonal trapezohedral class of the hexagonal system.
Siderite‡	Carbonate	A crystalline form of $FeCO_3$, with the brown to gray crystals usually being rhombohedral. Calcium, magnesium, and manganese are usually present in small amounts as replacing elements.
CoMoAl		Commercial catalyst from Laporte Industries, Inc. (Comox 451, 1.5-mm extrudate), surface area 300 m^2/g; pore volume 0.66 ml/g. Chemical analysis: 3.7, 12.8, 0.06, 1.6, and 0.03%. CoO, MoO_3, $Na_2O + K_2O$, SiO_2, and SO_3, respectively.
Iron		Reagent-grade hydrogen reduced iron from Mallinckrodt, Inc.
Reduced pyrite		Solid residue from hydrogenation of creosote oil in presence of 15% by weight of FeS_2 at 678 K, stirrer setting 2000 rpm, and initial hydrogen pressure 20.7 MPa.
SRC residue		Obtained from filter cake from Wilsonville SRC pilot plant. Analysis: 55.2% ash content and 13.6% S for −325-mesh material; and, prior to screening, 30% filter aid, 53% ash, and 2.9% S.
Coal ash		Obtained by burning Kentucky No. 9/14 mixture (7.2% ash) in a muffle furnace at 1000 K. Analysis: 13.7% Fe.
Kaolin		Obtained from W.H. Curtis and Co.

*Tarrer et al. [203].

†All agents were ground to −325 mesh prior to use, except muscovite which was ground only to −80 mesh because of its laminar silica structure, and except as indicated.

‡Minerals obtained from David New, Minerals and Books, Providence, Utah.

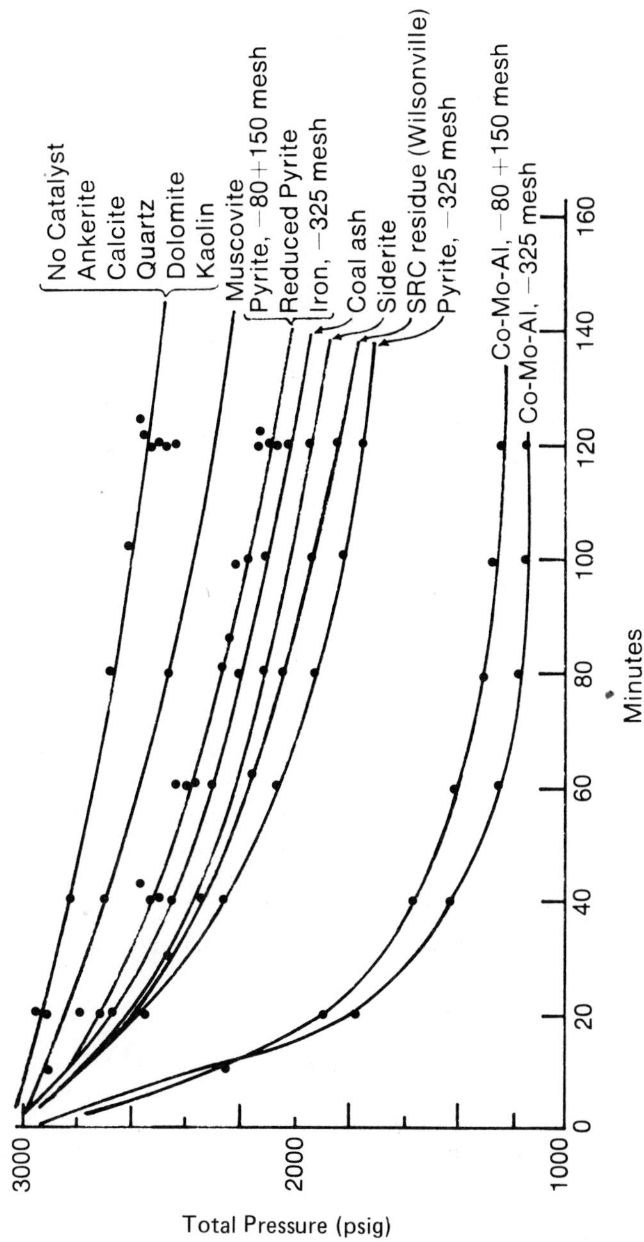

Figure 2-21. Catalytic effect of coal mineral matter as indicated by total pressure (Tarrer et al. [203]). Reaction conditions: temperature 698 K; H_2 pressure 20.7 MPa at 698 K; agitation rate 2000 rpm; autoclave 300 cm^3; charge 100 g creosote oil/15 g catalyst.

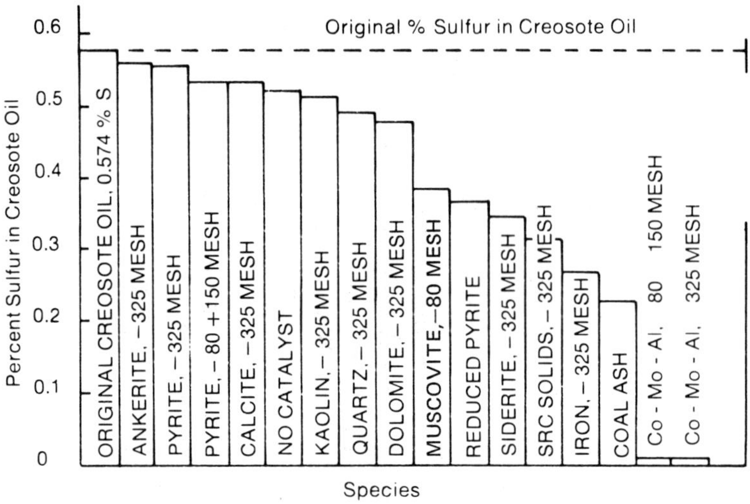

Figure 2-22. Comparison of desulfurization activity of catalyst (Tarrer et al. [203]).

aromatic hydrogen content of the effluent increased dramatically as the concentration of residue in the effluent increased. For a residue concentration buildup equivalent to 0.15% iron in the effluent, hydrogen addition corresponded to 1.4% per percent of iron in the solvent-residue mix. The iron recovered after reaction was in the form of pyrrhotite. Thus, the role of coal-mineral catalysis in coal liquefaction is to accelerate the hydrogenation of the process solvent, thereby indirectly affecting the rate of coal liquefaction. The rate-limiting step in coal liquefaction at temperatures above 643 K has been shown to be the hydrogenation of the process solvent to replenish hydrogen-donor species [93, 95]. This is verified by the increase in rate of liquefaction with increase in the concentration of coal minerals.

Interactions of Iron and Sulfur

In selectivity and performance studies of free-iron and SRC mineral residues in the liquefaction of a Kentucky bituminous coal, Fe was found to be a very effective sulfur scavenger and improved hydrodesulfurization to a level comparable to ground cobalt molybdate [95, 131]. Fe also selectively accelerated certain hydrogenation reactions, with the result that the concentration of pyridine-insolubles was decreased to almost zero and the oil concentration was increased by 27%, while the asphaltene and preasphaltene fractions and the hydrogen consumption were unchanged. SRC mineral residue, on the other hand, was not effective in hydrodesulfurization, left liquid product selectivity relatively unchanged, and increased hydrogen consumption substantially, yielding more H_2S and C_1–C_4 hydrocarbon gases. The residue, however, was more active overall for hydrogenation than the Fe or a fly-ash sample from a coal-fired power plant. Various concentrations of Fe were tested up to 13.3 wt% on coal, and the results show that H_2S formation can be completely eliminated. The reduced iron sulfide species formed catalyzed hydrogenation, resulting in an increased hydroaromatics level in the solvent. It was shown that hydrogen consumption could be substantially reduced by lowering the reactor temperature from 683 K to 653 K and compensating for the reduced temperature by adding 14.3% Fe.

Detailed study of the hydrogenation and hydrodesulfurization reactions have been broadened to include the following coal minerals: SRC residue ash, magnetite (Fe_3O_4), pyrite, hematite (natural Fe_2O_3), reagent-grade Fe_2O_3, commercial Fe_2O_3 catalyst, and reagent-grade reduced iron [69]. In coal liquefaction tests, the Fe_2O_3 additives were more effective for sulfur removal than elemental iron within the range 71–1000% of stoichiometric iron required for sulfur reaction. In either case there was no benefit from concentrations outside this range.

The experiments with Fe_2O_3 consumed more hydrogen than those with elemental iron, so that on a selectivity basis (sulfur removal per unit of hydrogen consumption), elemental iron was more effective. Hydrogen sulfide reacted with Fe_2O_3 so that none was present in the product gases, and little of the Fe_2O_3 was reduced to elemental iron. The degree of desulfurization was essentially the same when the Fe_2O_3 surface area was increased to 60 m^2/g (commercial Fe_2O_3 catalyst) from the previous surface area of 8.9 m^2/g. The observed desulfurization rate for the Fe_2O_3 additive was shown to be independent of Fe_2O_3 particle size and stirring rate, indicating that desulfurization was controlled neither by pore diffusion nor by liquid-solid mass transfer, but was kinetically controlled. Magnetite and residue were found to be less effective for desulfurization than Fe or Fe_2O_3. A model compound study with benzothiophene demonstrated that the additives selective for desulfurization, such as Fe_2O_3 and Fe, gave almost totally an ethylbenzene product, while additives inactive for desulfurization hydrogenated the benzothiophene to dihydrobenzothiophene [69]. Pyrrhotite gave a 2:1 mixture of dihydrobenzothiophene and ethylbenzene.

A mechanism has been proposed for the influence of sulfur and red mud (47.2% Fe_2O_3) on the hydrogenolysis of a Japanese coal in which hydrogen formed from the decomposition of H_2S on the reduced iron sulfide is postulated to be more active than normal gaseous hydrogen [152]. Yields of oil plus asphaltenes increased from 7.9% (dmmf coal basis) to 16.1% by addition of 10% red mud. Addition of sulfur as a promotor with the red mud at levels of 1, 5, and 10% increased conversion to 32.8, 59.8, and 62.9%, respectively. Addition of hydrogen sulfide in the gas without any additives increased conversion to 30.4%. Electron-microprobe sulfur images of hydrogenated coal (pelletized) suggested that H_2S was formed from the S added as promotor and hydrogen, and then reacted with the iron oxide to form iron sulfide. However, the S/Fe mole ratios determined by the x-ray image were 0.34, 0.74, 0.77, and 0.22 at temperatures of 573, 623, 673, and 723 K, respectively. Thus, at 723 K, the data suggest that the iron sulfide reacts with hydrogen to produce H_2S and elemental iron (the formation of elemental iron has never been observed in coal reactions, as discussed previously, and is known to be thermodynamically unfavorable), or H_2S decomposes to H_2 and S in the presence of iron sulfide (also known to be thermodynamically unfavorable). The following mechanism was proposed:

$$H_2 + S \xrightarrow{(Fe_2O_3)} H_2S, \tag{2-4}$$

$$Fe_2O_3 + 2H_2S + H_2 \rightarrow 2FeS + 3H_2O, \tag{2-5}$$

$$FeS + H_2 \rightleftharpoons Fe + H_2S, \tag{2-6}$$

$$H_2S \xrightarrow{(FeS)} H_2{}^* + S, \tag{2-7}$$

$$\begin{aligned} \text{coal} + H_2{}^* \rightarrow \text{gas} + \text{oil} + \text{asphaltene} \\ + \text{unconverted residue} + \text{water}. \end{aligned} \tag{2-8}$$

The second reaction proceeds in two steps as follows:

$$Fe_2O_3 + 3H_2S \rightarrow FeS + FeS_2 + 3H_2O \qquad (473\text{–}513\ \text{K}), \tag{2-9}$$

$$FeS_2 + H_2 \rightarrow FeS + H_2S \qquad (543\text{–}603\ \text{K}). \tag{2-10}$$

FeS is believed to decompose into its elements in the temperature range 673–873 K [235], and the following reactions have been proposed for the reduction of pyrite:

$$2FeS_2 + H_2 \xrightarrow{500\ \text{K}} Fe_2S_3 + H_2S, \tag{2-11}$$

$$Fe_2S_3 + H_2 \xrightarrow{550\ \text{K}} 2FeS + H_2S, \tag{2-12}$$

$$FeS + H_2 \overset{640\ \text{K}}{\rightleftharpoons} Fe + H_2S. \tag{2-13}$$

Several researchers have reported, however, that FeS is not reduced to elemental iron [113, 118, 226]. Unfortunately, the above mechanism provokes more questions than it answers and reflects the present lack of knowledge of the detailed mechanisms of mineral and sulfur catalysis.

2.3.3 *Effect of Solvent Properties*

A. ROLE OF SOLVENT

Coal liquefaction can be characterized as a three-step process involving dissolution, hydrogen transfer, and hydrogenation. The solvent characteristics can affect the rate of reaction and product selectivity in each of these steps. As discussed previously (Section 2.3.1), the reactive maceral components of coal are believed to be highly crosslinked, amorphous, macromolecular networks consisting of a number of stable cluster units or aggregates connected by relatively weak linkages (see Figure 2-23). During the dissolution step, the macromolecular structure frag-

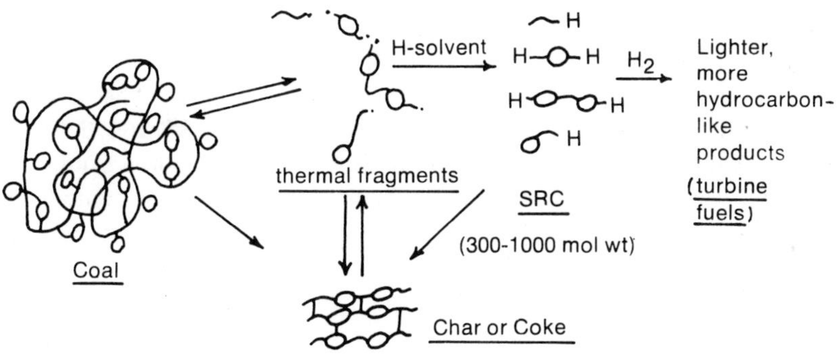

Figure 2-23. Chemistry involved in coal conversion (Whitehurst et al. [219]).

ments into free radicals. If donatable hydrogen is available from the solvent, the free radicals are "capped" and stable species are produced that typically have molecular weights in the range of 300–1000. If a hydrogen-donor solvent is not present, the free radicals will be stabilized by condensation reactions forming compounds of higher aromaticity than the initial cluster units and char or coke. Thus, the solvent dictates the reaction path taken by the intermediate radicals and thereby controls the product selectivity [219].

From a more basic viewpoint, there are at least three pathways available for free radical stabilization: H-abstraction from the solvent or a hydrogen-rich portion of the coal; either elimination of part of the radical molecule with associated group migration, or rearrangement to form another more stable structure, or both; and a combination reaction such as condensation or alkylation with another molecule in the solvent, which could also be a radical. The presence and type of hydrogen donors will determine the preferred reaction path [219].

The physical solvation properties of the solvent largely determine whether or not the converted coal dissolves. For example, SRC solvent contains more phenols and polyaromatics than H-coal and EDS solvents and therefore the former is a better physical solvent for converted coal [219]. Once the coal is in solution, the solvent controls the compositional changes of the intermediate products and the gas formation rate.

It is now generally accepted that the quality of the solvent plays a very important role on the mechanism of coal dissolution and subsequent upgrading of the dissolved products. It has been pointed out that four quantities, namely H-donors, phenols, H-shuttlers, and H-abstractors, play important roles on the determination of the overall effectiveness of the solvent in the liquefaction process [218]. Here we briefly discuss the role of each of these quantities.

Hydrogen Donation from the Solvent

In general, the sources of hydrogen in a liquefaction process are hydrogen gas, solvent, coal, and coal-derived materials such as heavy residue and solvent-refined coal [39, 85, 164, 218]. The demand for hydrogen is a result of the free radicals generated during the dissolution stage. The source of the hydrogen is a function of the demand, and generally follows a hierarchical mechanism with the most active H-donors receiving preference. For example, Figure 2-24 [39, 85, 164, 219] shows that for the conversion of benzaphenone to diphenyl methane in a synthetic solvent, the hydrogen donation occurred first by 9,10-dihydrophenanthrene, then by tetralin, and finally by *p*-cresol [215, 218, 219]. The rate of demand for hydrogen is greatest during the initial stage of liquefaction, when massive thermal fragmentation of the coal structure produces very large numbers of free-radical species, and decays with increasing conversion as the remaining bonds become increasingly more difficult to cleave thermally.

For low-boiling solvents, the concentration of tetralin is a good indicator of the hydroaromatic concentration in the solvent and correlates with the extent of coal conversion in pilot operations [215]. Researchers at Exxon [65, 66] have shown that the conversion of coal to cyclohexane-soluble products (or distillates) can be correlated with a "Solvent Quality Index." This index is an indication of the concentration of hydroaromatics. Figure 2-25 shows the Exxon correlation, and it also shows that for lower-quality solvents elemental hydrogen can help in increasing the degree of coal conversion.

The fact that a high hydroaromatic (or hydrogen-donor) concentration in the solvent gives better conversion, both at short and at long contact times, has been demonstrated in the literature [40, 218] and in other, unpublished reports. The absence of a high hydroaromatic concentration at a long residence time does not

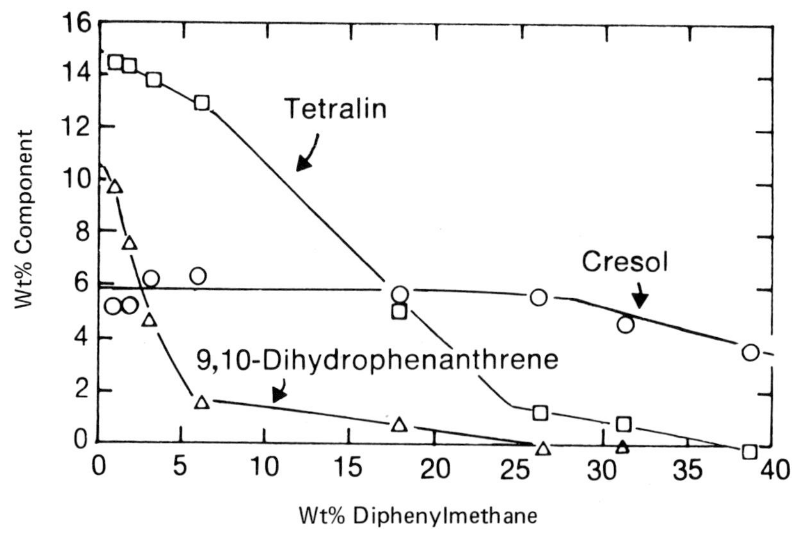

Figure 2-24. Relative activity of H-donors toward benzophenone (Whitehurst et al. [218]).

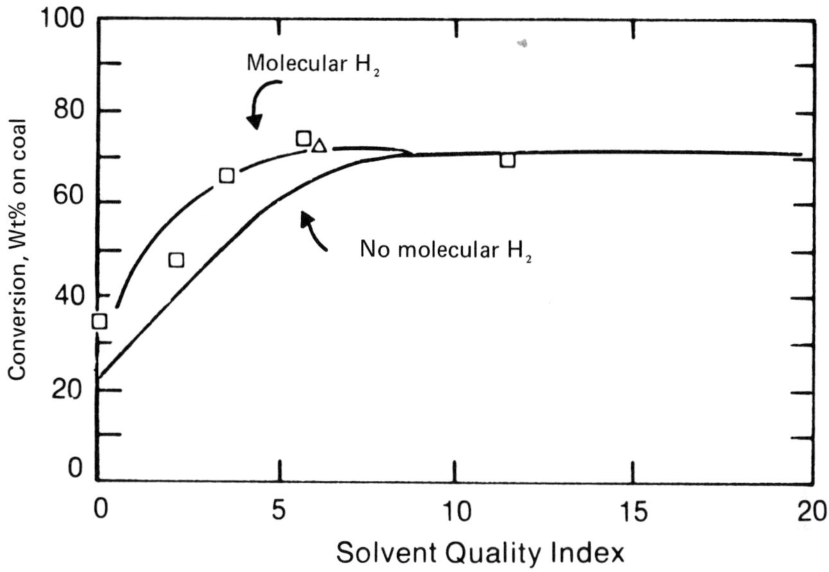

Figure 2-25. Effect of molecular hydrogen on coal conversion (Furlong et al. [65]).

preclude high coal conversion, but the products may contain repolymerized solids (e.g. char). Even at a short contact time, when there is relatively little net hydrogen consumption, the conversion of Burning Star coal was found to be dependent upon the concentration of hydroaromatics in the solvent (see Figure 2-26) [218]. This would indicate that the rate of coal dissolution is only partially controlled by the thermal fragmentation reactions and is strongly affected by the interaction of H-donor with insoluble intermediates by a mechanism such as illustrated in Figure 2-27 [219]. This mechanism assumes that thermal bond rupture, independent of H-donor, must occur first, followed by interaction of H-donor with the radicals that are formed. However, multiple bond ruptures and subsequent H-donor interactions are necessary before the reaction products are soluble. It is not probable that two bond ruptures would occur simultaneously in close proximity, since there would be a strong tendency for bond reformation. Therefore, the first bond rupture and H-donor interaction sequence is completed before the second one, etc., until product solubility is achieved. The kinetics of such a reaction network would explain the dependence of dissolution rate on H-donor concentration.

As mentioned earlier, the demand for hydrogen during the liquefaction process is dictated by the coals and the resulting free radicals. When no hydroaromatics are present, poor solvents, such as *p*-cresol become H-donors. However, when these compounds lose a hydrogen atom, they become reactive themselves and are incorporated into the final product, which is an undesired result. Figure 2-28 classically illustrates how a small amount of hydrogen transferred in the right place can achieve a significant difference in the final product during liquefaction.

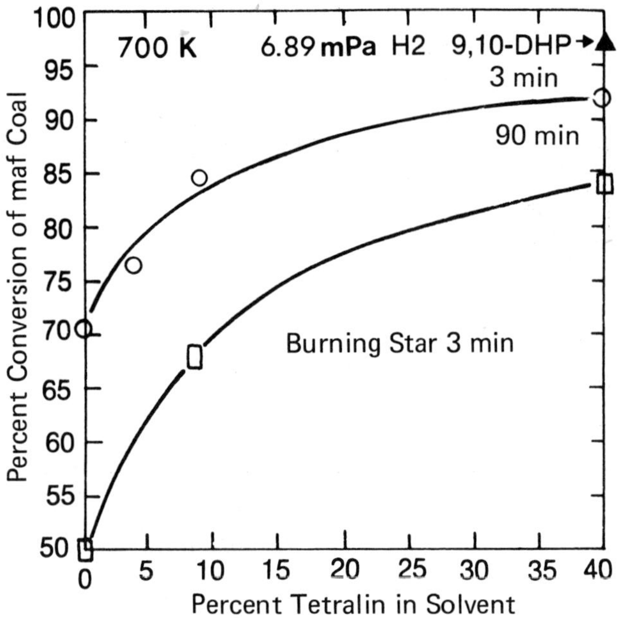

Figure 2-26. Rate of conversion of bituminous coal (Whitehurst et al. [219]).

It is clear from the above discussion that hydrogen donors play a significant role in the mechanism and kinetics of coal dissolution and subsequent upgrading of dissolved coal. Unfortunately our knowledge in this area is still at the development stage and requires more quantitative understanding.

Hydrogen Shuttling

Coal and solvent-refined coal have been reported to be significant sources of hydrogen in themselves during coal liquefaction [39, 46, 85, 164, 219]. The presence of a nondonor, polyaromatic solvent, such as naphthalene, can effectively provide a means of hydrogen redistribution to such an extent that over 80% of the reactive macerals of coal (vitrinite) can be converted to pyridine-soluble products in relatively short reaction times and without the need for hydroaromatic compounds in the solvent or gaseous hydrogen [46, 219]. A proposed mechanism [218], illustrated in Figure 2-29, envisions a H-shuttling phenomenon where hydrogen from relatively hydrogen-rich portions of the coal structure is redistributed to hydrogen-lean species via hydrogenation and dehydrogenation of the nondonor solvent. According to this mechanism there is no net change in the overall H/C ratio or the concentration of the shuttler, but there is a significant increase in the degree of carbon unsaturation or aromaticity in the products relative to the

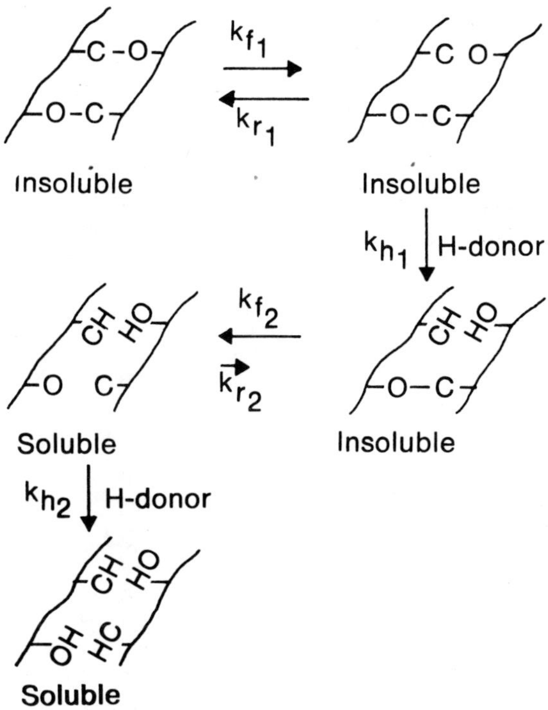

Figure 2-27. Speculative mechanism to explain the dependence of coal dissolution on hydrogen-donor concentration (Whitehurst et al. [219]).

case when external hydrogen is provided. This H-shuttling process has its limitations, however, because the labile hydrogen available within the coal for redistribution is limited in quantity and is very rapidly consumed. Consequently, extensive thermal fragmentation quickly creates a situation where the hydrogen demand far exceeds the labile-hydrogen capacity of the coal. This results in further hydrogen transfer occurring at the expense of char formation.

Not all molecules containing an aromatic nucleus are good H-shuttlers. Pyridine, for example, is very effective in dissolving coal products, so much so that it is used as a means of defining conversion. On the other hand, pyridine cannot promote solubilization of coal, regardless of the reaction time, because the molecule is not labile either for hydrogen acceptance from the coal or for hydrogen abstraction by the coal fragments. The behavior of pyridine is typical for single–ring aromatic compounds in general. Several studies have shown that hydrogen shuttling is more efficient for high-boiling, polycondensed (preferably in an angular orientation) aromatic ring systems [219]. If, however, the solvent is hydrogenated, there is very little difference in the ability of different boiling point cuts to promote conversion.

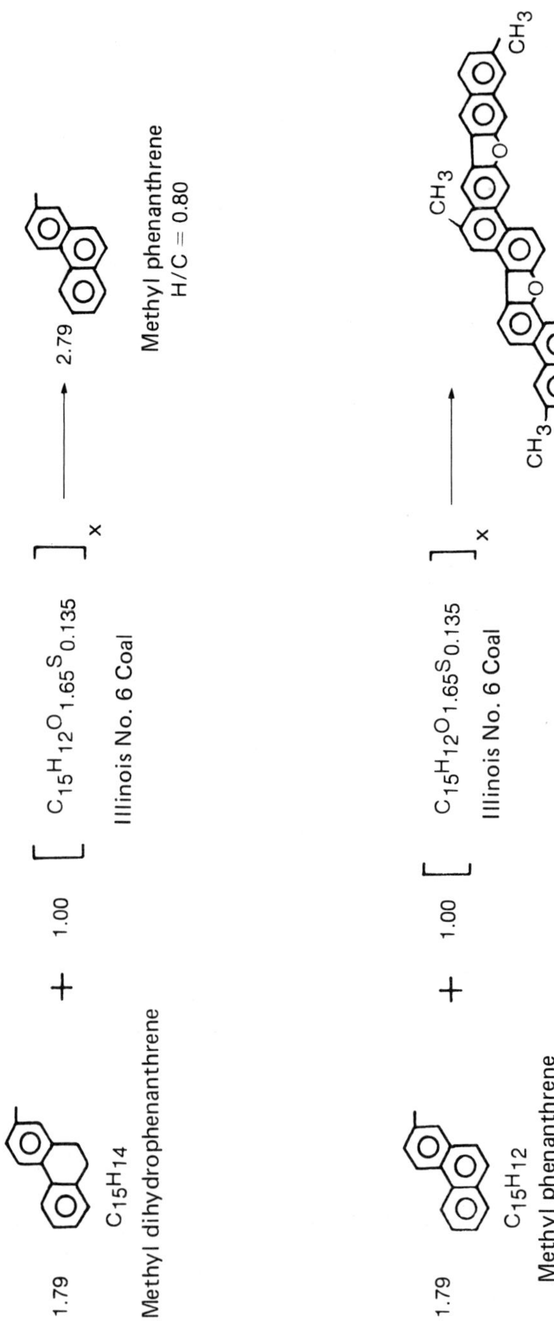

Figure 2-28. Stoichiometry which shows the importance of small quantities of donor hydrogen (Larsen [130]).

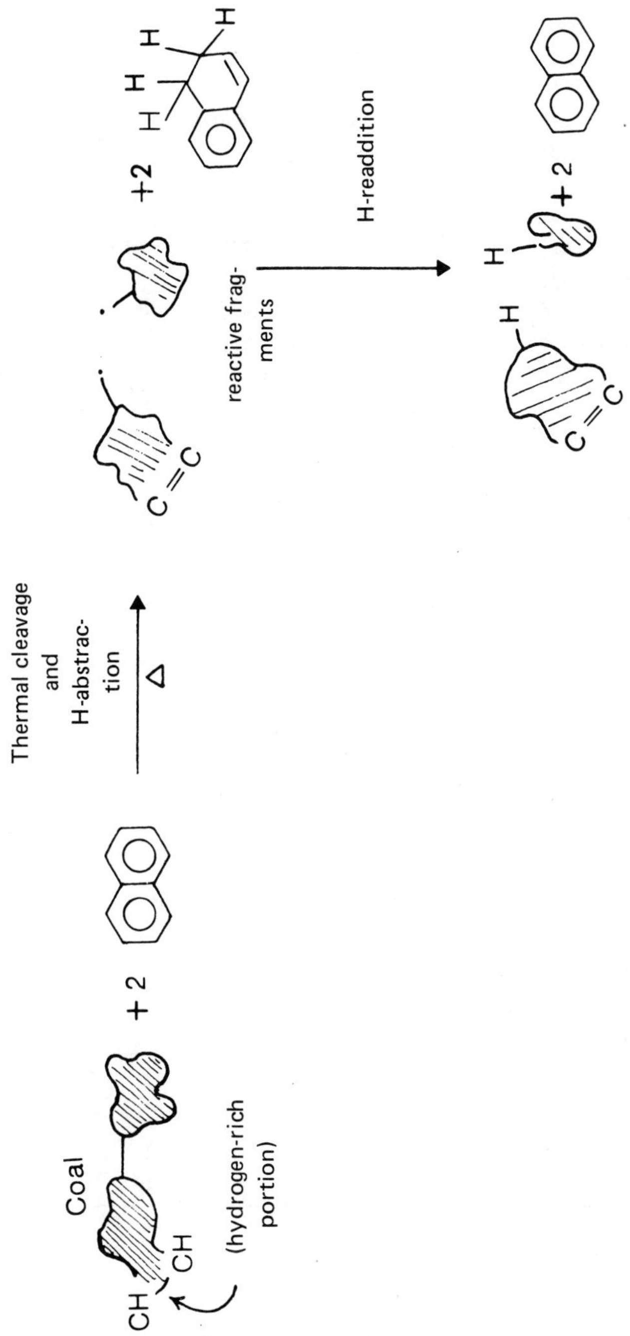

Figure 2-29. A possible mechanism for H-shuttling (Whitehurst et al. [218]).

In situations when the concentration of H-donors in the solvent is limited, H-donors and H-shuttlers can act synergistically to yield an effective solvent. This has been demonstrated using solvents of limited H-donor concentrations and correlating the extent of coal conversion at short reaction times with the total concentration of polyaromatic ring compounds [218]. An example of this is shown in Figure 2-30 for a series of coal-derived solvents having low H-donor capacities [less than 0.5(g H)/(100 g solvent)]. The concentration of H-donors alone did not correlate significantly with the extent of coal conversion to pyridine-soluble products for these solvents.

Both light (< 9 carbon atoms) and heavy (> 9 carbon atoms) phenols contribute substantially to the physical solvation properties of the carrier solvent, and as a result, even very highly polar coal products are soluble under liquefaction conditions. Thus, even a short residence time in the preheater can give as high as 90% conversion to pyridine-soluble materials.

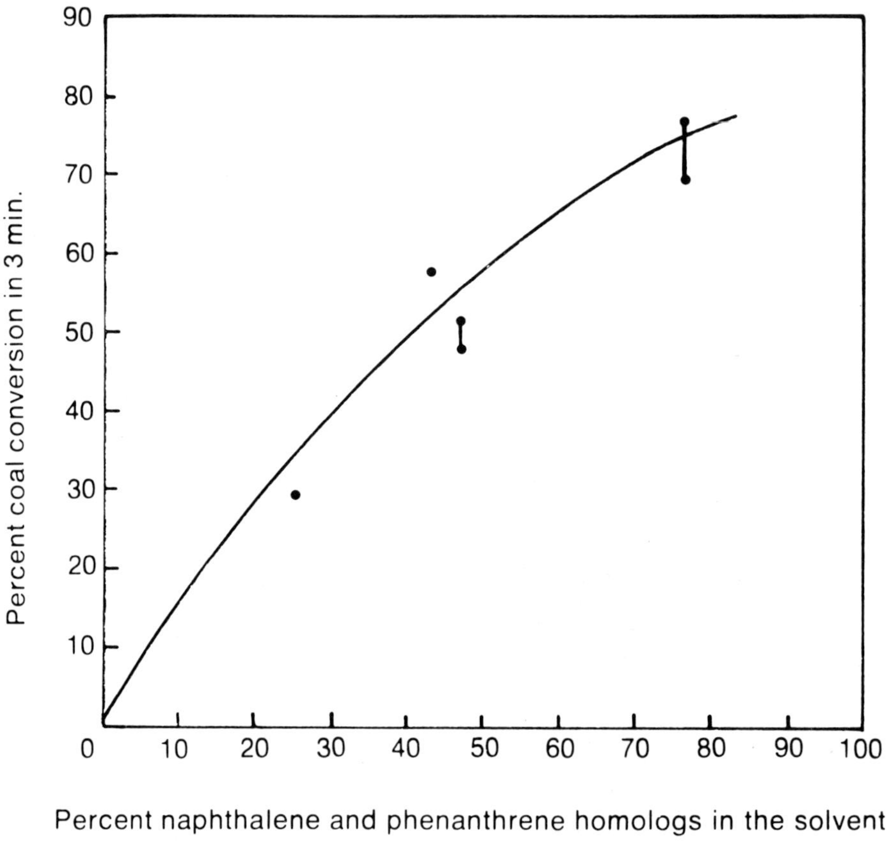

Figure 2-30. Kentucky coal conversion at 4 min (677–678 K) in solvents of limited H-donor capacity (Whitehurst et al. [218]).

The majority of phenolic compounds in recycle solvents are light phenols and usually account for about 2/3 of the total hydroxyl content in the solvent. When there is a high demand for hydrogen and the carrier solvent contains a limited supply of hydroaromatics, the light phenols can donate hydrogen through alkylation reactions with coal fragments. The light phenol components do not contain hydroaromatic structures, and the phenols in general are unusual H-donors in that they can become significantly incorporated into the coal products.

B. REGENERATION OF H-DONOR

In order to achieve a stable steady state in a solvent-recycle coal-liquefaction process, it is necessary to provide a means for replenishing the H-donor capacity of the process solvent. During coal liquefaction the solvent quality degrades, primarily due to the conversion of hydroaromatic components to condensed aromatics. If the hydroaromatic content of the solvent is not adequately replenished, deleterious side reactions will occur such as phenol condensation, solvent dimerization, and char formation. Stable solvent composition can be achieved either by converting coal to provide new solvent or by rehydrogenating the hydrogen-depleted solvent with gaseous hydrogen. Although both methods are used in coal-liquefaction processes, they are slow compared to the rate of coal dissolution. The difficulty is apparent in Figure 2-31, which shows that the yield of solvent-range products does not become significant until after approximately 50% of the oxygen has been re-

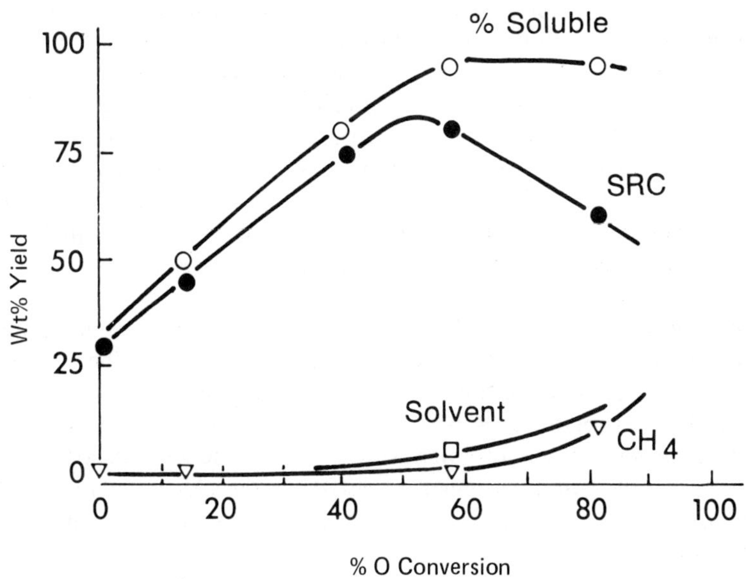

Figure 2-31. Product yields versus percentage of oxygen conversion for West Kentucky coal (Whitehurst et al. [218]).

moved from the coal. Unfortunately, when this point is attained, light hydrocarbons become a major product.

It is generally believed that the rehydrogenation of solvent by hydrogen gas is not thermodynamically constrained, but the reaction is generally slow and requires the presence of a catalyst. Pyrite (which is contained in the mineral matter of coal) catalyzes the hydrogenation reaction [58]. Other fine mineral matter can also help catalyze this reaction [199].

C. ISOMERIZATION AND ADDUCTION OF HYDROGEN DONOR SOLVENTS

Although the primary role of hydrogen-donor solvent is to stabilize H-acceptor free radicals, solvent rearrangement (isomerization), polymerization, and adduction (reaction with free radical acceptors generated from the coal) can also occur. In these side reactions, six-membered hydroaromatic rings contract to five-membered rings and/or adduct to reaction products from the coal. Isomerization and adduction reactions associated with a representative donor-solvent constituent, octahydrophenanthrene, have been investigated, and previous work with other constituents such as tetralin has been reviewed [36].

The rate of isomerization of a hydrogen-donor-solvent component, such as octahydrophenanthrene or tetralin, is first order with respect to the component concentration. The rearrangement rate increases as the number of hydroaromatic rings in the donor molecule increases. Free radicals increase the rate by an order of magnitude. The activation energy for tetralin rearrangement is in the range of 23–32 kcal/mole, depending upon whether or not a source of free radicals is present. In general, most of the usual isomers are formed with little or no effect of free-radical concentration on the distribution of isomers. In addition to forming more stable nondonor molecules from active donor molecules (e.g. tetralin rearrangement to methyl indanes), isomerization causes difficulty in following the extent of coal liquefaction because it is not possible to attribute the difference in transferable hydrogen between feed and product liquids solely to hydrogen donation to radicals. The transferable hydrogen is considered to be the hydrogen on hydroaromatic six-membered rings as measured by ^{13}C-NMR, high-resolution mass spectrometry, or some other technique.

All solvents, both good and poor hydrogen donors, can become irreversibly adducted by acceptor free radicals generated from the coal, and the adduction is enhanced by the presence of sulfur and oxygen functionality or by long reaction times and low temperatures ($<$ 673 K). At temperatures normally used in coal liquefaction preheaters and dissolvers ($>$ 673 K), adduction reactions are most likely to occur only with poor donor solvents. Once the adducts are formed at low temperatures, however, they are not necessarily cleaved by heating to high temperatures.

2.3.4 *Role of Oxygen in the Mechanism of Coal Liquefaction*

A generalized mechanism for solvent processing of coal is illustrated in Figure 2-32. It is generally accepted that the "structure" of the starting coal has an extremely important effect on that of the product yield. The oxygen content and chemical form are important factors controlling the yield of liquid products because there is a strong tendency for small structural units that are potential distillate molecules to polymerize, forming residue via an oxygen-bridging mechanism.

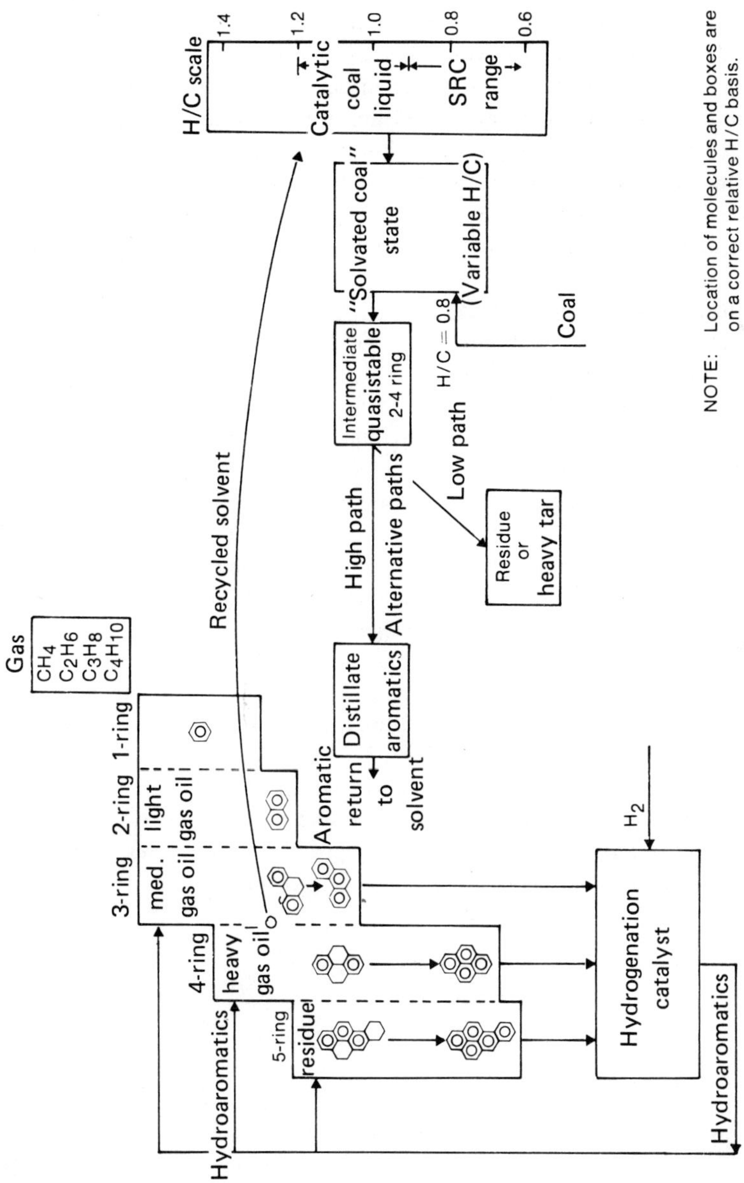

Figure 2-32. A generalized mechanism for solvent processing of coal (Larsen [130]). Locations of molecules and boxes are on a correct relative H/C basis.

The oxygen-bridging reaction is believed to occur very rapidly and to be essentially irreversible at normal coal-liquefaction preheater and reactor operating conditions. During the initial preheating of the coal and solvent, when dissolution progresses rapidly, it is especially important to have sufficient hydroaromatics present to remove oxygen (e.g., hydroxyl oxygen) and prevent residue formation. Thus, one of the major roles of the hydroaromatics in the solvent is to remove oxygen free radicals before retrogressive reactions can occur.

The results of recent work that examined the role of oxygen functional groups in the mechanism of coal liquefaction are shown in Figure 2-33 [201] and indicate that other mechanisms in addition to ether cleavage may be occurring. In these experiments with a high-volatile bituminous coal and tetralin as the solvent, the yields of pyridine-solubles, benzene-solubles, and hexane-solubles plus gas were correlated with the loss of oxygen for various contact times and temperatures. The results show that hydroxyl-oxygen removal strongly correlates with total oxygen removal and apparently has a significant effect on the yield of soluble products. Since little molecular-weight reduction occurs when hydroxyl groups are removed, it is probable that part of the increased yield of soluble products results from the elimination of hydrogen bonding.

Hydroxyl-group blocking experiments using hexamethyldisilazane and the pyridine-soluble fraction showed that replacing the hydroxyl hydrogen with trimethylsilyl indeed increased the benzene solubility, suggesting that hydrogenolysis of hydroxyl groups played an important role in the generation of benzene-solubles during liquefaction [201]. However, the blocking experiments could only account for 40% of the increase in benzene-soluble yield obtained for a differential increase in hydroxyl-group removal in the liquefaction experiments. This would indicate that concurrent mechanisms are occurring.

When the conversions of each oxygen functionality are compared as a function of pyridine-soluble yield as in Figure 2-34 [201], it is apparent that relatively small changes have occurred in the carbonyl and hydroxyl oxygen contents. However, the undetermined oxygen, believed to be primarily ether oxygen, decreases quite dramatically with increased yield of pyridine-solubles plus gas. The amount of ether oxygen cleaved was estimated from the difference between the total oxygen loss and the hydroxyl-plus-carbonyl oxygen loss, and correlated with the percentage of pyridine-solubles plus gas (coal conversion) according to the regression equation

$$Y = 0.0336X - 0.230, \tag{2-14}$$

where Y = ether oxygen cleaved, wt% dmmf coal,

X = coal conversion, wt% dmmf coal.

Based on the slope of this equation and the carbon content of the coal, the number of ether oxygen linkages cleaved per 100 initial carbon atoms was estimated to be 3.0 ± 0.3. This agrees well with the value 2 calculated from the results of reductive alkylation experiments [110].

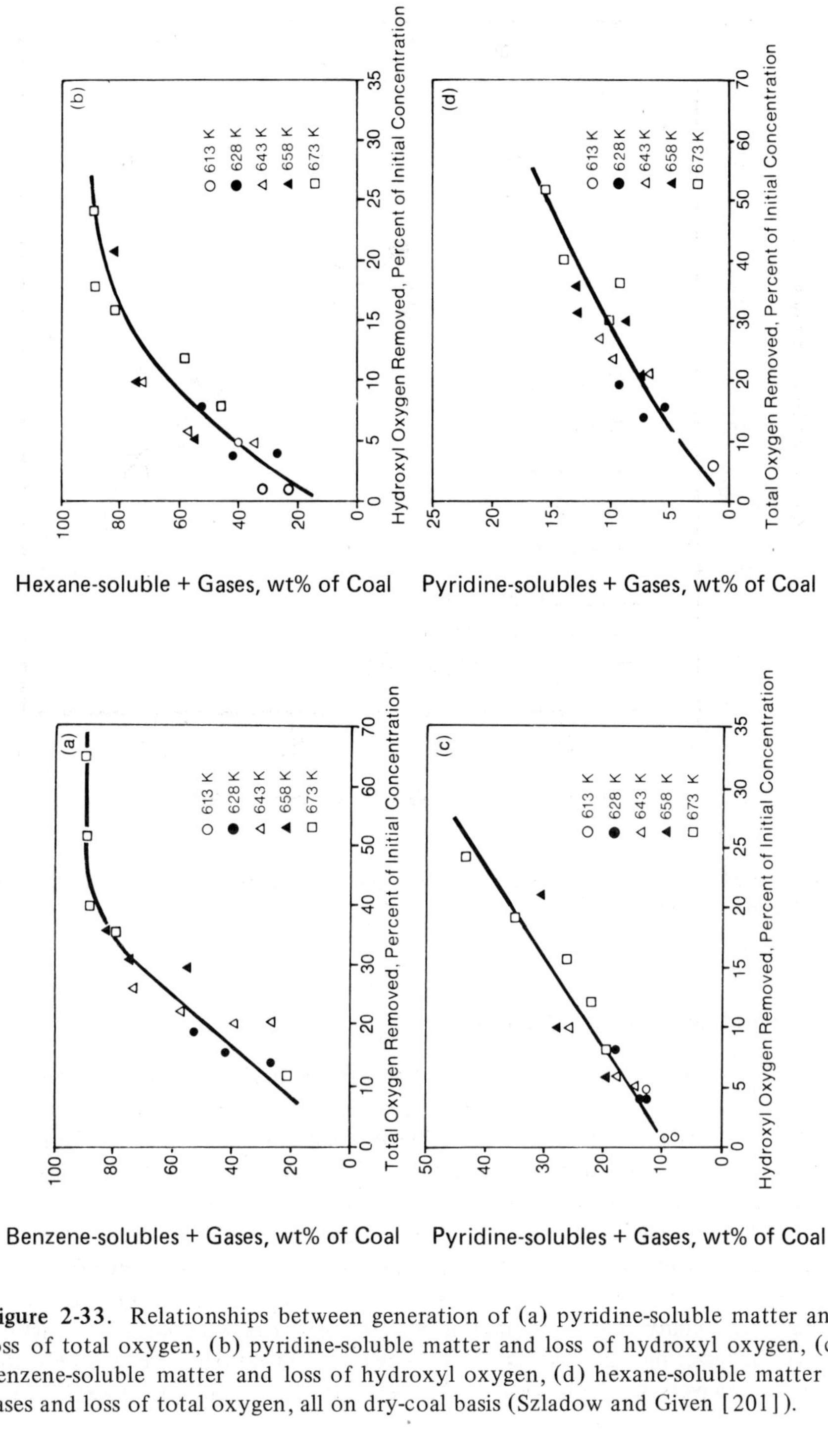

Figure 2-33. Relationships between generation of (a) pyridine-soluble matter and loss of total oxygen, (b) pyridine-soluble matter and loss of hydroxyl oxygen, (c) benzene-soluble matter and loss of hydroxyl oxygen, (d) hexane-soluble matter + gases and loss of total oxygen, all on dry-coal basis (Szladow and Given [201]).

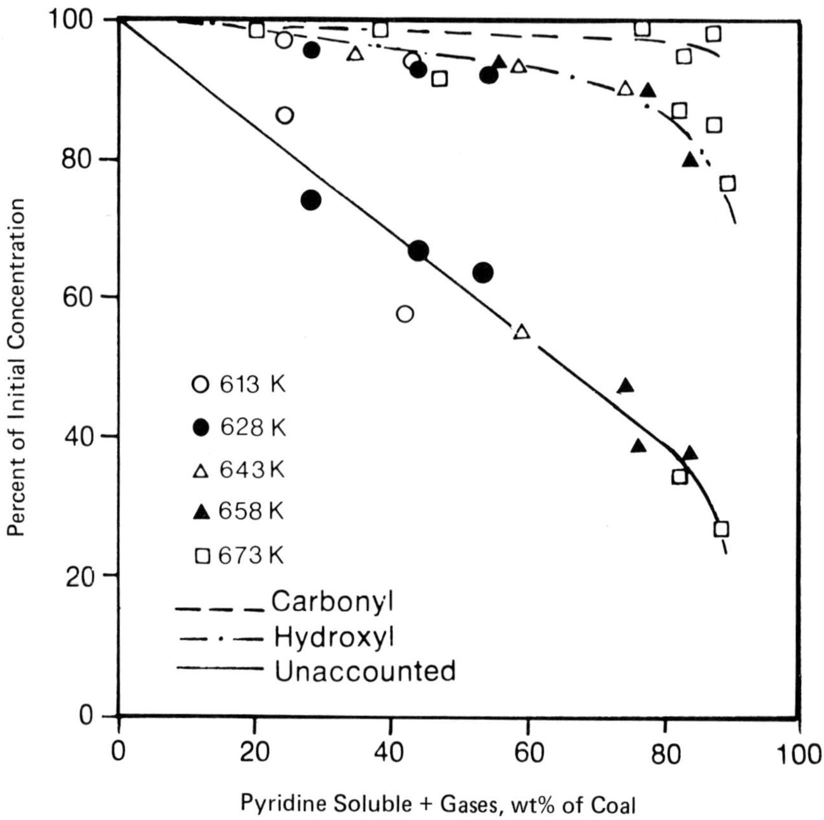

Figure 2-34. Loss of oxygen functional groups versus generation of pyridine-soluble matter, dry-coal basis (Szladow and Given [201]).

2.4 Free Radicals in Coal Liquefaction

In addition to the nature of the solvent, free radicals are generally believed to have a primary role in the behavior of coals during liquefaction processes. Coals have a significant concentration of stable free radicals in their natural form, that were presumably formed in the "coalification" process and probably stabilized by resonance with aromatic clusters. When coal is heated, thermal bond rupture occurs, generating large quantities of free radicals and releasing volatile material, mainly water and light hydrocarbons, in the process. Bituminous coals become fluid upon heating, pass through a temperature of maximum fluidity, and ultimately repolymerize and solidify. Thus, free radicals are also associated with unwanted reactions in coal liquefaction, such as coking in the reactor and retrogressive reactions, that give poor conversion yields. Despite the great significance given to the role of free radicals and their ubiquitous nature, they are not well understood or well characterized, nor has their concentration been measured under reaction conditions.

2.4.1 *Liquefaction as Controlled Pyrolysis*

Coal liquefaction can be viewed as a controlled pyrolysis process in which heating of the coal-solvent mixture causes thermolysis reactions to occur and the solvent controls the allowable retrogressive reactions by donating and shuttling hydrogen that caps the free radicals. This concept in fact has been applied in flash pyrolysis and supercritical gas-extraction processes in which the thermal-reaction products are rapidly quenched or removed from the coal particles before repolymerization occurs.

The common nature of the pyrolysis reaction with coal liquefaction was investigated with a Utah high-volatile bituminous coal in a TGA apparatus in the temperature range of 673–733 K [221]. The pyrolysis yields followed a second-order overall kinetics with an activation energy of 36 kcal/mole. Solvent extraction of the same coal in tetralin was also second order overall, with activation energy of 29 kcal/mole. Catalytic hydrogenation of the same coal, stirred in a coal-derived vehicle oil, with a cobalt molybdate catalyst, was also second order overall with an activation energy of 38 kcal/mole. As shown in Figure 2-35, batch hydrogenation, pyrolysis, and coal dissolution under given conditions approach completion at similar rates, and they are believed to involve similar mechanisms, namely thermal bond rupture to yield free radicals, followed by radical stabilization as the rate-determining step.

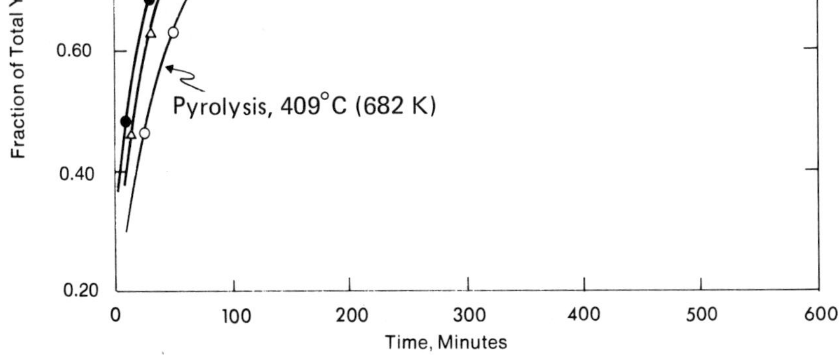

Figure 2-35. Comparison of fraction yield versus time (Wiser et al. [221, 222]).

2.4.2 *Measurement of Free Radicals in Coal and Conversion Products*

A. ELECTRON-SPIN-RESONANCE SPECTROMETRY

Since the discovery of free radicals in coal by Uebersfeld et al. [206] and Ingram et al. [112] using electron-spin-resonance (ESR) spectrometry, ESR has become widely used in the study of the structure of coal and coal products [112, 126, 132, 206, 209]. The technique is attractive because it is a nondestructive instrumental approach that yields quickly and easily the concentration of unpaired electrons and other chemical information. A related technique, electron-nuclear double-resonance (ENDOR) spectrometry, has become very popular recently, because it yields spectra that are much more highly resolved for the same samples than ESR spectra [120, 173, 189]. Published experimental results obtained by the ENDOR technique are very sparse [51, 120], however, so the following discussion will focus on ESR.

The ESR spectral parameters of importance are the *g*-value (gyromagnetic splitting factor) of the radicals, the radical concentration, the peak-to-peak linewidth on the derivative of the absorption curve, and the spectral lineshape [157]. The *g*-value is strongly influenced by the molecular environment of the unpaired electrons. The radical concentration is simply the number of unpaired electrons ("spins") in a given quantity of sample and is estimated by integrating the intensity of the ESR signal. The spectral linewidth of a heterogeneous mixture such as coal is believed to be a function of unresolved proton-electron hyperfine interaction, i.e., interaction when unpaired electrons are in the vicinity of nuclei with magnetic moments. The spectral lineshape is affected by unresolved hyperfine resonance structure and the anisotropy of the sample. Since coal is a heterogeneous mixture of organic compounds the ESR spectral parameters are a composite for all the radicals present. Mineral matter is of such low intensity in the coal spectrum that the ESR signals mainly originate from organic matter [179]. Thus, the position and character of the spectral signal is greatly affected by the relative concentrations of the various types of radicals. Sufficient evidence has also been accumulated to conclude that the ESR spectra of coals and coal derivatives are attributable to free radicals without any significant contribution from charge-transfer interactions [51, 174, 190, 231, 232].

B. ESR SPECTRA OF COALS

The relationship between ESR intensities (free-radical concentration) and coal rank is well known [133] and has been attributed to the formation of radicals through the degradation reactions during the coalification process and subsequent stabilization of the radicals by resonance in the aromatic and heteroatomic structures of the coal. As coal rank increases up to about 94% carbon, the free-radical concentration increases exponentially, as would be expected on the view that resonance stabilization should be greatest for coals containing the largest condensed polynuclear aromatic structures, and such structures are a result of coal metamorphism. The radical concentration for carbon content above 95% falls off rapidly because of electrical-conductivity effects.

There is strong evidence that coals, especially those of low rank that are rich in oxygen, contain heteroatomic free radicals [177, 233]. This is shown by the correlation of heteroatom content with the *g*-values of the coals [233]. In particular, the *g*-values increased with the function $\Sigma \xi_i X_i$, where X_i are the atom fractions of oxygen, nitrogen, and sulfur, and ϵ_i are the spin-orbit coupling constants of the respective atoms. This treatment of the data is based on the generally accepted view that the difference between the *g*-values of the free radicals and the *g*-value of the free electron is greatest when the radicals are unpaired electrons localized to some degree on atoms having strong spin-orbit interactions. Evidence supporting the heteroatom hypothesis is shown in Figure 2-36, where the *g*-value is correlated with the content of oxygen and the content of oxygen plus sulfur [177]. Two of the coals having unusually high sulfur contents were statistical outliers until organic sulfur was included in the correlation. Inclusion of nitrogen in the statistical treatment gave inconclusive results. It is also interesting to note that the ESR linewidth can be linearly correlated with hydrogen contents for coal vitrains [166, 193].

The relationship between the *g*-values of coals and coal rank is reasonably well understood [133, 174] and is illustrated in Figure 2-37. The high *g*-values ob-

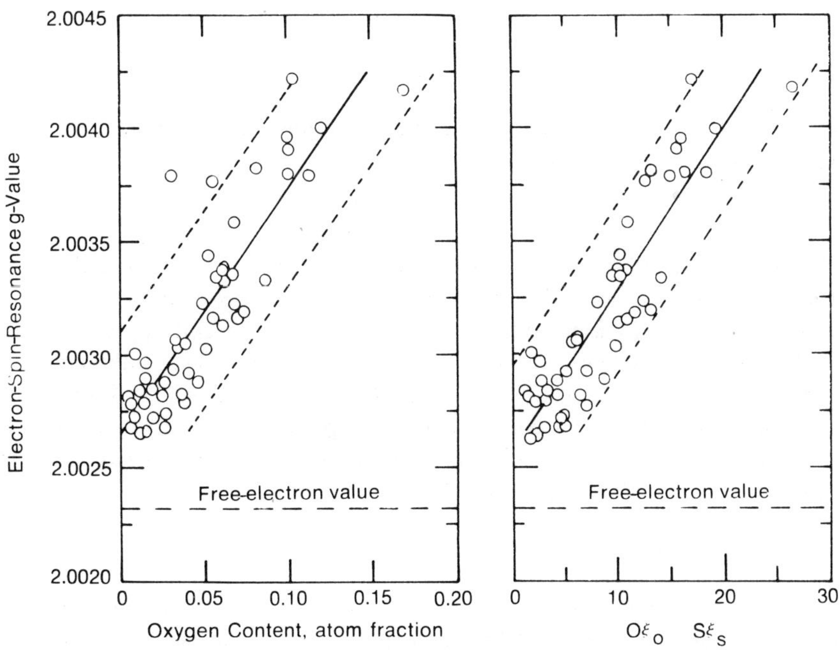

Figure 2-36. Functional dependences of the electron-spin-resonance g-values of vitrains from selected coals (Retcofsky et al. [177]).

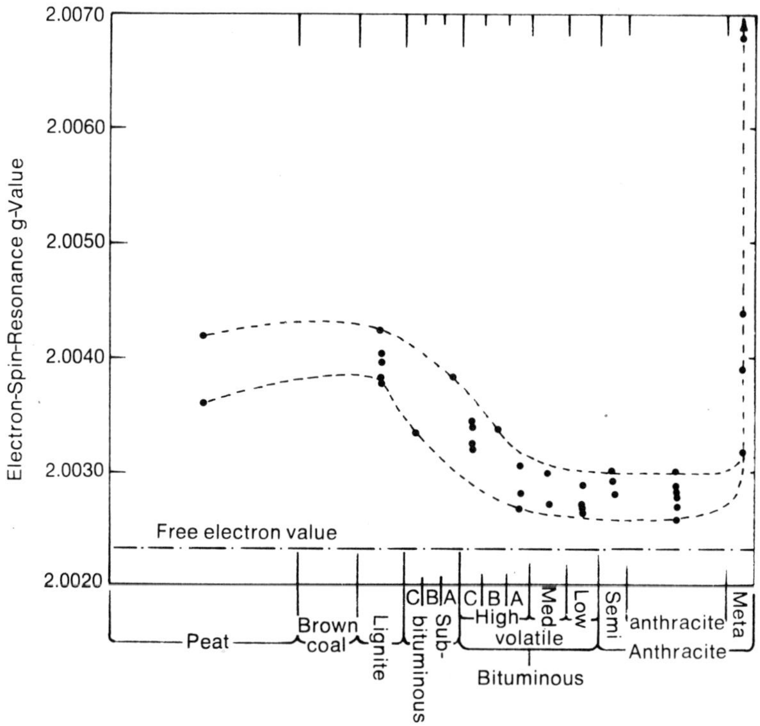

Figure 2-37. Electron-spin-resonance g-values as a function of coal rank for vitrains from selected coals (Retcofsky et al. [174]).

served for peats and lignites are attributed to the presence of aromatic radicals with some partial localization of the unpaired electrons on the heteroatoms, particularly oxygen. As coalification progresses, the *g*-values decrease and approach known values for aromatic hydrocarbon radicals for bituminous and young anthracite coals. In the late stages of coalification the *g*-values become very large, as expected, due to the high aromaticity and condensation into graphite-like structures. ESR line-widths, as a function of rank, for vitrains from various coals have the same qualitative variation with rank as the *g*-values [174, 179].

C. ESR SPECTRA OF PYROLYZED COALS

Coal liquefaction is essentially a controlled form of coal pyrolysis, since the first step in both processes is the thermal fragmentation of the labile structures into free radicals. Consequently, ESR studies of heat-treated coals should yield mechanistic information applicable to coal liquefaction. Several investigators have performed such studies [126, 157, 205, 124, 209], but more recent work [172, 177,

179] indicates that these results may be in error because special precautions necessary to minimize electrical contact between the semiconducting, heat-treated coal particles were unknowingly omitted. It is necessary to grind and dilute the samples of heat-treated coals with an electrically nonconducting medium to avoid determination of grossly inaccurate *g*-values. As an example of this problem, two groups of investigators [157, 205] have reported *g*-values less than that of the free electron for several coals that were heat-treated at typical coal-liquefaction temperatures. The free-electron *g*-value is the theoretically minimum value that can be obtained. Although the low *g*-values are reproducible, they represent apparent *g*-values that reflect the physical properties of the sample rather than the nature of the unpaired electrons.

The effect of heat-treatment temperature on *g*-values for several coals is examined in Figure 2-38 [157]. Two of the coals, Pocahontas #3 low-volatile bituminous coal and Illinois #6 high-volatile bituminous coal, exhibit a slight decrease in *g*-values with increasing temperature to 873 K, suggesting that very minor changes are occurring in the radical centers. This is attributed to the tendency of the radicals to become more hydrocarbon-like by a combination of molecular rearrangement and loss of oxygen as volatile matter. In the case of Wyodak subbituminous coal a relatively large decrease in *g*-value occurs below 773 K. At temperatures up

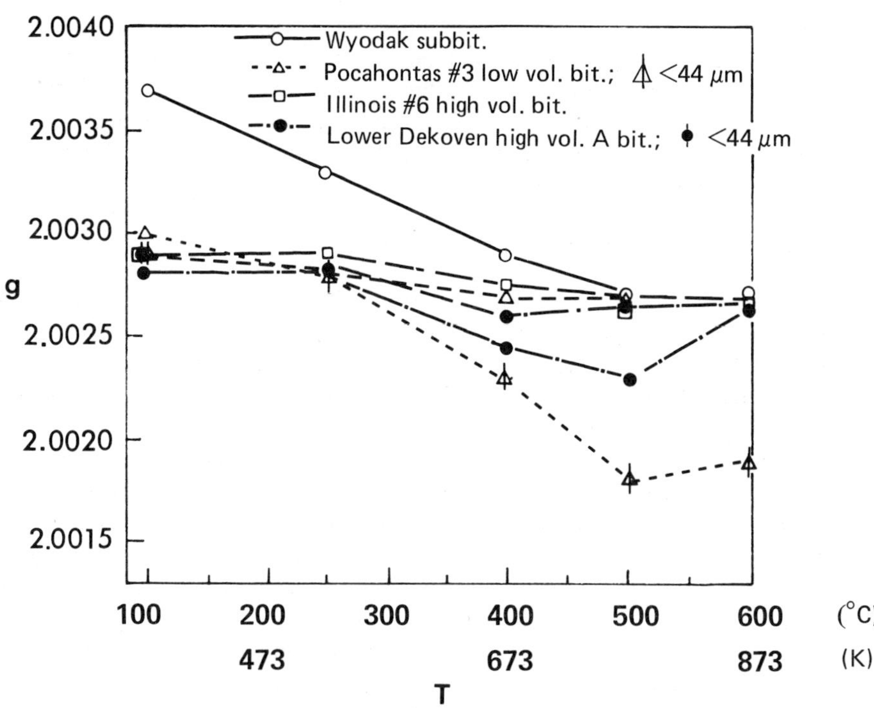

Figure 2-38. *g*-values versus temperature (Petrakis and Grandy [157]). Reprinted with permission from *Anal. Chem.* **50** (2), 307 (1978). ©1978, American Chemical Society.

to 723 K, the destructive vacuum distillation of a Wyoming subbituminous coal similar to the Wyodak coal yields primarily water, carbon dioxide, and carbon monoxide, with the carbon oxides accounting for approximately 40% of the oxygen present in the coal on a moisture-free basis [132]. Yet the radical concentration for the Wyodak coal changes very little at temperatures up to 673 K. Thus, the implication is that the oxygen removal via gaseous products must have taken place in preexisting radical centers or in parts of the coal structure where the unpaired-electron density was large. One bituminous coal, Lower Dekoven, Illinois, gave unusually low g-values between 523 and 773 K. This behavior was interpreted to be a result of the formation of sigma-type radicals similar to those formed by bond breakage during hydrogen abstraction. The radical formed by irradiation of benzaldehyde is an example:

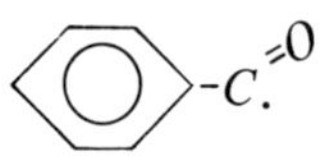

$g = 2.0014.$

Upon heating the Lower Dekoven coal to 873 K, the radical concentration decreases and the g-value increases, suggesting that the sigma-type radicals are participating in recombination reactions. The above conclusions should be considered tentative, since the investigators [157] did not dilute the samples with a nonconducting medium as discussed previously. In a related study [177, 179], the g-values for an Ireland Mine high-volatile bituminous coal exhibited the same trend as the Lower Dekoven coal in Figure 2-38. However, the addition of a nonconducting medium such as potassium bromide resulted in the g-values increasing slightly, leveling off at 523–723 K, and decreasing slightly as the heat-treatment temperature approached 873 K.

D. ESR SPECTRA OF SOLVENT-REFINED COAL

In the solvent-refined coal (SRC-I) process [121] a low-sulfur, low-ash solid fuel is produced from coal. When the g-value and linewidth data for the SRC products from five different coals are viewed collectively, the presence of organic free radicals with only minor interaction between the unpaired electrons and heteroatoms in the samples is observed [174, 177–179].

Although the g-values observed for most SRC-related substances are in a narrow range, 2.0026 to 2.0037, there are significant differences in radical environments [86]. SRC filter-cake solid has a g-value (2.0026) considered typical of most aromatic hydrocarbon radicals of one to six condensed rings [20, 191] and shows no interaction of radicals with heteroatoms. Process solvent and filtrate have g-values (2.0029 and 2.0031, respectively) typical of the high end of the range for hydrocarbons, and probably have some heteroatoms near the radical centers. In contrast, SRC filter-wash solvent has a g-value (2.0037) that is typical of radicals on oxygen heteroatoms, such as ethers [153] or some hydroxy-substituted quinones [7]. All of the above g-values for SRC-related materials [86] are in agreement with most other reported values [177, 179], except for one group of researchers [225] whose results appear much too high. As with coals, the correlation of spectral linewidths with hydrogen content is also good for some of the SRC-derived materials [86].

E. ESR SPECTRA OF COAL-DERIVED ASPHALTENES AND OILS

As discussed in a previous section, asphaltenes are defined as the portion of the coal-liquefaction product that is soluble in benzene and insoluble in pentane (or hexane). Asphaltenes are considered to be one of the key intermediates in the conversion of coal to distillate products [214, 225]. It has been shown that coal-derived asphaltenes are hydrogen-bonded acid-base complexes and that these complexes can be separated into their acidic and basic components [198]. Although charge-transfer interactions have been proposed as an alternative means of bonding between the acidic and basic components, ESR measurements do not support this hypothesis [174, 177].

Coal-derived asphaltenes are *not* the same as petroleum asphaltenes, even though both are defined using the same solvents. The asphaltenes from coal liquefaction are more highly aromatic [172] than asphaltenes from petroleum [122, 234]. The ESR spectra of coal-derived asphaltenes and petroleum asphaltenes are also quite different [174, 177–179].

The *g*-values and spectral linewidths for asphaltenes and their acid and base subfractions, derived from two different bituminous coals, were found to be very similar to those determined for SRC samples [179]. An oil sample from a coal-liquefaction process development unit exhibited a slightly higher *g*-value, a broader spectral line, and a significant lower spin concentration than the source coal [177, 179].

F. CHANGES IN THE QUANTITY AND NATURE OF FREE RADICALS DURING COAL LIQUEFACTION

Although a number of different reaction mechanims have been proposed, all donor-solvent coal-liquefaction processes have in common the formation of preasphaltenes and asphaltenes as intermediate products. ESR measurements on each of the materials in the very general sequence

$$\text{coal} \rightarrow \text{preasphaltene} \rightarrow \text{asphaltene} \rightarrow \text{oil}$$

are shown in Figure 2-39 and lead to some very significant observations [179]. The ESR spectral parameters for the coal and preasphaltenes are very similar, especially the concentration of free spins. This indicates that the large number of free radicals formed during the thermal fragmentation of the coal were effectively "capped," so that the preasphaltene sample showed no significant increase in stable radical concentration over that in the source coal. The spectral linewidths increase in the order coal $<$ preasphaltenes $<$ asphaltenes $<$ oil, which parallels the order of increasing hydrogen content. The spin concentration decreases significantly in the order preasphaltenes $>$ asphaltenes $>$ oil, indicating that radical centers are effectively

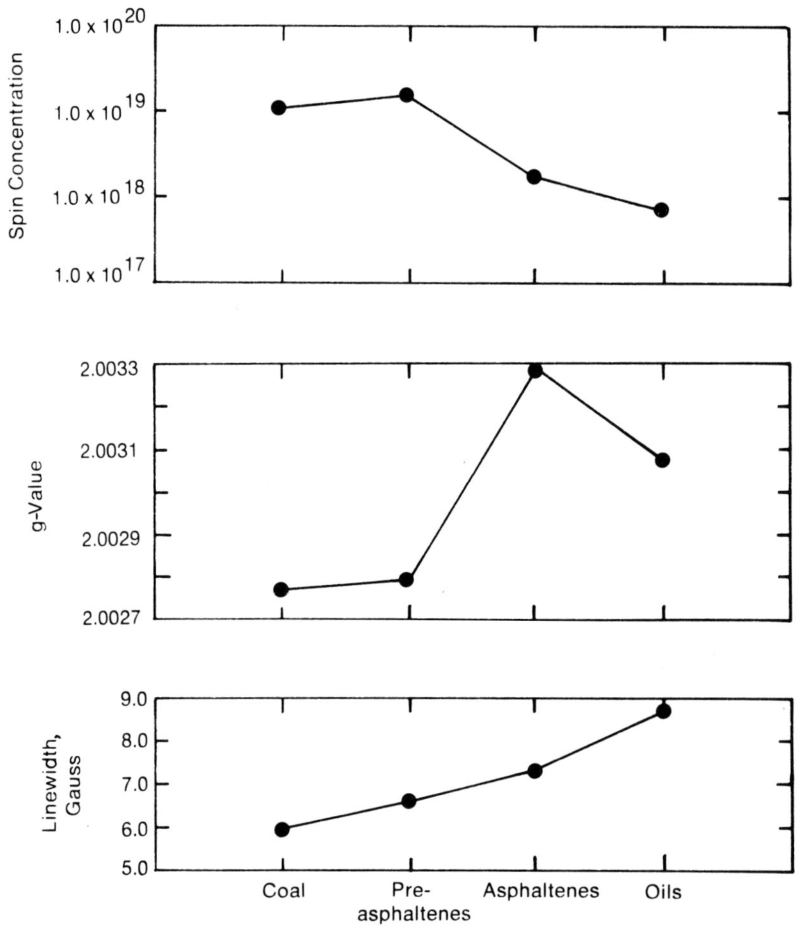

Figure 2-39. 9.5-GHz ESR spectral parameters for West Virginia Ireland Mine hvAb coal and subfractions of its liquefaction products (Retcofsky et al. [179]).

being eliminated at the same time the hydrogen content is increasing. The high asphaltene *g*-value is indicative of the interaction of unpaired electrons with heteroatoms associated with aromatic units, and the intermediate *g*-value for the oil is consistent with significant heteroatom loss, gain in hydrogen, and reduction of free-spin concentration.

The major challenge in ESR spectrometry today is to perform the measurements *in situ,* while the liquefaction reactions are occurring. Several laboratories are known to be developing measurement cells and short-residence-time techniques for this purpose.

2.5 Future Work

From the standpoint of donor-solvent coal-liquefaction process optimization and commercialization, the single most crucial problem is predicting and controlling the behavior of coal feedstocks in the process. A plant designed for one particular coal and set of operating conditions may not be optimum for another coal or may not even be operationally stable. Ash and sulfur levels of coals are known to fluctuate widely, even within the same seam, so that the operational consequences for a plant using coals from various seams would be great. Typical instabilities include lack of solvent balance, viscosity and the buildup of insoluble organic matter in product and recycle streams, and coking tendency in the preheater, dissolver, hot separator, and vacuum-flash column.

A great deal of coal characterization work has been performed by the Pennsylvania State University and Gulf Research and Development Company, as reviewed earlier in this chapter, to correlate the behavior of various coals during liquefaction with physical and chemical properties that can easily be measured. For the most part, the correlations are based on microautoclave data and are not as yet translatable to performance in a continuous-recycle reactor. It is imperative that this type of work be continued and expanded to establish quantitative relationships that can predict a coal's performance in a real process, much the same way petroleum's processability can be fully characterized without processing it in a refinery. The operators of the Wilsonville Solvent Refined Coal pilot plant have developed a microautoclave test procedure with a standard solvent to screen a coal before using it as feedstock. In addition, they monitor the recycle-solvent quality by a similar procedure using a standard coal, and control the process accordingly.

The knowledge and understanding of the structure and chemistry of coal and coal conversion products are greatly advanced, but many questions remain because of the heterogeneous nature of coal and the wide compositional variations caused by widely varied geological history. Work on these questions will greatly influence the second generation of commercial coal-liquefaction processes by giving direction to the development of significant process improvements. One such improvement, which is already being utilized to reduce hydrogen consumption, is short-residence-time coal liquefaction [216], in which the coal is liquefied only enough to permit separation of the residual solids and mineral matter, followed by upgrading in a catalytic process under conditions less severe than exist in the dissolver.

Aside from the coal-feedstock problem, most of the major issues in the first generation coal liquefaction plants (SRC-I, SRC-II, EDS, H-coal) deal with engineering design and materials of construction. The chlorine content of coal can cause havoc in a process because it forms ammonium chloride, which can precipitate out of the gas phase and plug pipelines if the temperature is lowered to near 573 K, or can hydrolyze and cause massive corrosion in the oil-fractionation system in high-temperature zones. Some method of control is needed other than addition of alkali with the coal or dissolver effluent. Vapor-liquid equilibrium data and physical- and thermodynamic-property data on the various process streams, including the water-wash steps, are practically nonexistent compared to the data base for other technologies, such as petroleum refining and chemicals production. Suffice it to say that there are enough design problems to keep the engineering community fully employed for many years.

REFERENCES

1. Abdel-Baset, M. B., Harzab, R. F., and Given, P. H., "Dependence of Coal Liquefaction Behavior on Coal Characteristics. 3. Statistical Correlation of Conversion in Coal-Tetralin Interactions," *Fuel,* **57**, 89 (1978).
2. Abdel-Baset, Z., Given, P. H., and Harzab, R. F., "A Reexamination of the Phenolic Hydroxyl Contents of Coals," *Fuel,* **57**, 95 (1978).
3. Abernathy, R. F., Peterson, M. J., and Gibson, F. H., "Spectrochemical Analyses of Coal Ash for Trace Elements," *U.S. Bureau of Mines Rept. Inv.* 7281, 1969.
4. Abernathy, R. F., Peterson, M. J., and Gibson, F. H., "Major Ash Constituents in U.S. Coals," *U.S. Bureau of Mines Rept. Inv.* 7240, 1969.
5. Aczel, T., Gorbaty, M. L., Maa, P. S., and Schlosberg, R. H., *Fuel,* **54**, 295 (1975).
6. Aczel, R., Gorbaty, M. L., Maa, P. S., and Schlosberg, R. H., "Stability of Adamantane and Its Derivatives to Coal Liquefaction Conditions, and Its Implications Toward the Organic Structure of Coal," *Fuel,* **58**, 228 (1979).
7. Adams, M., Blois, M.S., and Sands, R. H., *J. Chem. Phys.,* **28**, 774 (1958).
8. Anderson, R. P., "The SCR-II Process," *ACS Div. of Fuel Chem. Preprints,* **22** (6), 132 (1977).
9. Anthony, D. B. and Howard, J. B., "Coal Devolatilization and Hydrogasification," *AIChE Journal,* **22**, 625 (1976).
10. Attar, A., "The Fundamental Interactions of Sulfur and Modelling of the Sulfur Distribution in the Products of Coal Pyrolysis," presented at the 83rd National AIChE Meeting, Houston, Texas, March 20–24, 1977.
11. Attar, A. and Corcoran, W. H., *Ind. Eng. Chem., Proc. Res. and Devel.,* **16** (2), 168 (1977).
12. Attar, A., *Fuel,* **57**, 201 (1978).
13. Attar, A. and Dupuis, F., *ACS Div. Fuel Chem. Preprints,* **23**, (2), 44 (1978).
14. Attar, A. and Dupuis, F., "Data on the Distribution of Organic Sulfur Functional Groups in Coal," *ACS Div. of Fuel Chem. Preprints,* **24** (1), 166 (1979).
15. Austen, D. E. G., Ingram, D. J. E., and Tapley, J. G., *Trans. Faraday Soc.,* **54**, 400 (1958).
16. Bartle, K. D., Martin, T. G., and Williams, D. F., *Fuel,* **54**, 226 (1975).
17. Benjamin, B. M., Rasen, V. F., Maupin, P. H., Brown, L. L., and Collins, C. J., *Fuel,* **54**, 269 (1978).
18. Bimer, J., Given, P. H., and Raj, S., "Phenols as Chemical Fossils in Coals," *ACS Div. Fuel Chem. Preprints,* **22** (5), 169 (1977).
19. Blayden, H. E., Gibson, J., and Riley, H. L., "X-Ray Study of the Structure of Coals, Cokes and Chars," in *Proc. Conf. on Ultrafine Structure of Coals and Cokes, BCURA,* London, 1944, p. 176.

20. Blois, M. S., Jr., Brown, H. W., and Maling, J. E., *Free Radicals in Biological Systems,* Academic Press, New York, 1961, p. 130.
21. Bockrath, B. C. and Noceti, R. P., "Coal-Derived Asphaltenes – Relationship Between Chemical Character and Process History," *Fuel Processing Techn.,* **2**, 143 (1979).
22. Brown, J. K., "The Infrared Spectra of Coals," *J. Chem. Soc.,* 744 (1955).
23. Brown, J. K., *BCURA Monthly Bull.,* **23**, 1 (1959).
24. Brown, J. K., Ladner, W. R., and Sheppard, N., "A Study of the Hydrogen Distribution in Coal-Like Materials by High-Resolution Nuclear Magnetic Resonance Spectroscopy I – The Measurement and Interpretation of the Spectra," *Fuel,* **39**, 79 (1960).
25. Brown, J. K. and Ladner, W. R., "A Study of the Hydrogen Distribution in Coal-Like Materials by High-Resolution Nuclear Magnetic Resonance Spectroscopy II – A Comparison with Infrared Measurement and the Conversion to Carbon Structure," *Fuel,* **39**, 87 (1960).
26. Cartz, L. and Hirsch, P. B., "A Contribution to the Structure of Coals from X-Ray Diffraction Studies," *Phil. Trans. Roy. Soc.,* **252A** (68), 557 (1960).
27. Chakrabartty, S. K. and Kretschmer, H. O., *Fuel,* **51**, 160 (1972).
28. Chakrabartty, S. K. and Kretschmer, H. O., *Fuel,* **53**, 132 (1974).
29. Chakrabartty, S. K. and Berkowitz, N., *Fuel,* **53**, 240 (1974).
30. Chakrabartty, S. K. and Berkowitz, N., *Nature,* **261**, 76 (1976).
31. Chakrabartty, S. K. and Berkowtiz, N., *Fuel,* **55**, 362 (1976).
32. Coal Conversion Data Book, IA.50.1, "Composition and Properties of Selected Coal Samples," Prepared for the U.S. DOE by the Institute of Gas Tech. under Contract No. EX-76-C-01-2286, 1978.
33. Collins, C. G., Benjamin, B. M., Raaen, V., Manpin, P. H., and Roark, W. H., "Isotropic Studies of Thermally Induced Reactions of Coal and Coal-Like Structures," *ACS Div. Fuel Chem. Preprints,* **22** (5), 98 (1977).
34. Cronauer, D. C. and Ruberto, R. G., "Investigation of Mechanism of Reactions Involving Oxygen-Containing Compounds in Coal," EPRI Report AF-422, January 1977.
35. Cronauer, D. C., Shah, Y. T., and Ruberto, R. G., "Kinetics of Thermal Liquefaction of Belle Ayr Subbituminous Coal," *I&EC Process Design and Develop.,* **17**, 281 (1978).
36. Cronauer, D. C., Jewell, D. M., Shah, Y. T., Modi, R. J., and Seshadri, K. S., "Isomerization and Adduction of Hydrogen Donor Solvents Under Conditions of Coal Liquefaction," *I&EC Fundamentals,* **18**, 368 (1979).
37. Cudmore, J. F., *Fuel Processing Tech.,* **1**, 227 (1978).
38. Curran, G. P., Struck, R. T., and Gorin, E., "The Mechanism of the Hydrogen Transfer Process of Coal and Coal Extract," *ACS Div. of Petroleum Chem. Preprints,* **C**, 130 (1966).
39. Curran, G. P., Struck, R. T., and Gorin, E., "Mechanism of the Hydrogen Transfer Process of Coal and Coal Extract," *I&EC Process Design and Dev.,* **6**, 166 (1967).
40. Curtis, C. W., Guin, J. A., Jeng, J. F., and Tarrer, A. R., "The Effect of Solvent Quality on Coal Conversion," *ACS Div. of Fuel Chem. Preprints,* **24** (3), 185 (1979).
41. D388, "Standard Specification for Classification of Coals by Rank" 1976 Annual Book of ASTM Standards, Part 26.

42. D2796, "Standard Definitions of Terms Relating to Megascopic Description of Coal and Coal Beds and for Microscopical Description and Analysis of Coal," 1978 Annual Book of ASTM Standards, Part 26.
43. D2797, "Standard Method of Preparing Coal Samples for Microscopical Analysis by Reflected Light," 1978 Annual Book of ASTM Standards, Part 26.
44. D2798, "Standard Method for Microscopical Determination of the Reflectance of the Organic Components in a Polished Specimen of Coal," 1978 Annual Book of ASTM Standards, Part 26.
45. D2799, "Standard Method for Microscopical Determination of Volume Percent of Physical Components of Coal," 1978 Annual Book of ASTM Standards, Part 26.
46. Davis, G. O., Derbyshire, F. J., and Price, R., *J. Inst. Fuel,* **50**, 121 (1977).
47. Deno, N. C., Greigger, B. A., and Strand, S. G., "The Aliphatic Structure in Coal," *ACS Div. of Fuel Chem. Preprints,* **23**, (2), 54 (1978).
48. Deno, N. C., Greigger, B. A., and Strand, S. G., *Fuel,* **57**, 455 (1978).
49. Dreher, G. B., Russell, S. J., and Ruch, R. R., "Chemical and Mineralogical Characterization of Coal Liquefaction Residues," *ACS Div. of Fuel Chem. Preprints,* **24** (1), 309 (1979).
50. Durie, R. A., "The Characteristics of Australian Coals and Their Implications in Coal Liquefaction," *ACS Div. of Fuel Chem. Preprints,* **24** (2), 12 (1979).
51. Elofson, R. M. and Schulz, K. F., *ACS Div. Fuel Chem. Preprints,* **11**, 513 (1967).
52. Ergun, S., McCartney, J. T., and Menster, M., "Physical and Chemical Properties of the Petrographic Components of a High Volatile Bituminous Coal," *Econ. Geol.,* **54**, 1068 (1959).
53. Ergun, S., and Tiensuu, V. H., *Fuel,* **38**, 64 (1959).
54. Ergun, S., and Tiensuu, V. H., *Nature,* **183**, 1668 (1959).
55. Farcasiu, M., Mitchell, T. O., and Whitehurst, D. D., "Asphaltols–Keys to Coal Liquefaction," Chemtech, 680, November 1977.
56. Farcasiu, M., "Fractionation and Structural Characterization of Coal Liquids," *Fuel,* **56**, 9 (1977).
57. Farcasiu, M., "Short Time Reaction Products of Coal Liquefaction and Their Relevance to the Structure of Coal," *ACS Div. Fuel Chem. Preprints,* **24** (1), 121 (1979).
58. Fisher, C. H. and Eisner, A., *Ind. Eng. Chem.,* **29**, 1371 (1937).
59. Fisher, C. H., Sprunk, G. C., Eisner, A., O'Donnell, H. J., Clark, L., and Storch, H. H., "Hydrogenation and Liquefaction of Coal, Part 2 – Effect of Petrographic Composition and Rank of Coal," *Tech. Paper, U.S. Bureau of Mines,* No. 642, 1942.
60. Francis, W., *Coal,* Edward Arnold Publishers, London, 1961.
61. Franz, J. A., "^{13}C, ^{2}H, ^{1}H N.M.R. and G.P.C. Study of Structural Evolution of a Subbituminous Coal During Treatment with Tetralin at 427°C," *Fuel,* **58**, 405 (1979).
62. Friedel, R. A. and Queiser, J. A., "Infrared Analysis of Bituminous Coals and Other Carbonaceous Materials," *Anal. Chem.,* **28**, 22 (1956).
63. Friedel, R. A. and Queiser, J. A., *Fuel,* **38**, 369 (1959).
64. Friedman, S., LaCount, R. B., and Warzinski, R. P., "Oxidative Desulfurization of Coal," *ACS Div. of Fuel Chem. Preprints,* **22** (2), 100 (1977).

65. Furlong, L. E., Effron, E., Vernon, L. W., and Wilson, E. L., paper presented at National AIChE Meeting, Los Angeles, CA, November 18, 1975.
66. Furlong, L. E., Effron, E., Vernon, L. W., and Wilson, E. L., *Chem. Eng. Progr.*, **72**, 71 (1976).
67. Gallo, G., *Chem. Abstr.*, **21**, 1243 (1927).
68. Gan, H., Nandi, S. P., and Walker, P. L., "Nature of the Porosity in American Coals," *Fuel*, **51**, 272 (1972).
69. Garg, D., Tarrer, A. R., Curtis, C. W., Lloyd, J. M., and Guin, J., "The Selectivity of Coal Minerals and SRC Residue Ashes for Hydrodesulfurization in the SRC Process," *ACS Div. of Fuel Chem. Preprints*, **24** (3), 231 (1979).
70. Ghosh, G., Benerjee, A., and Mazumdar, B. K., *Fuel*, **54**, 244 (1975).
71. Gillet, A., *Nature*, **167**, 406 (1951).
72. Gilliam, T. M., Brand, J. I., and Rodger, B. R., "Coal Disolution Studies: Particle Size Distribution in Coal-Derived Liquids as a Function of Degree of Coal Conversion," Oak Ridge National Lab. Report No. TM-6681, U.S. DOE W-7405-ENG-26, May 1979.
73. Given, P. H., "The Distribution of Hydrogen in Coals and Its Relation to Coal Structure," *Fuel*, **39**, 147 (1960).
74. Given, P. H., Peover, M. E., and Wyss, W. F., *Fuel*, **44**, 425 (1965).
75. Given, P. H., "Organic Chemistry of Coals," presented at Short Course on Coal Characteristics and Coal Conversion Processes, Pennsylvania State University, October 29–November 2, 1973.
76. Given, P. H., "How May Coals Be Characterized for Practical Use?" presented at Short Course on Coal Characteristics and Coal Conversion Processes, Pennsylvania State University, October 29–November 2, 1973.
77. Given, P. H., Cronauer, D. C., Spackman, W., Lovell, H. L., Davis, A., and Biswas, B., "Dependence of Coal Liquefaction Behavior on Coal Characteristics. 1. Vitrinite-rich Samples," *Fuel*, **54**, 34 (1975).
78. Given, P. H., Cronauer, D. C., Spackman, W., Lovell, H. L., Davis, A., and Biswas, B., "Dependence of Coal Liquefaction Behavior on Coal Characteristics. 2. Role of Petrographic Composition," *Fuel*, **54**, 40 (1975).
79. Given, P. H., Spackman, W., Davis, A., Walker, P. L., Lovell, H. L., Coleman, M., and Painter, P. C., "The Relation of Coal Characteristics to Liquefaction Behavior," Quarterly Technical Progress Reports for the Period Jan.–Dec. 1977, FE-2494-3/6, U.S. DOE Contract EX-76-C-01-2494, January 1979.
80. Given, P. H., Spackman, W., Davis, A., Walker, P. L., Lovell, H. L., Coleman, M., and Painter, P. C., "The Relation of Coal Characteristics to Liquefaction Behavior," Quarterly Technical Progress Reports for the Period Jan.–June 1978, FE-2494-7/8, U.S. DOE Contract EX-76-C-01-2494, January 1979.
81. Given, P. H., Davis, A., and Jenkins, R. G., "The Geochemistry of Coals: A Characteristic Fundamental in Determining Liquefaction Behavior," paper presented ACS Div. of Fuel Chem. Meeting, Honolulu, HI, April 2–6, 1979.
82. Given, P. H., "Mineral Matter Effects in Coal Liquefaction," presented at the Fall 1979 Information Transfer Session, Cooperative Program in Coal Research, Pennsylvania State University, November 14–15, 1979.
83. Gluskoter, H. J., "Electronic Low Temperature Ashing of Bituminous Coal," *Fuel*, **44**, 285 (1965).
84. Gluskoter, H. J., "Mineral Matter and Trace Elements in Coal," *Advan. Chem. Ser.*, **141**, 1 (1975).

85. Gorin, E., Lebowitz, H. E., Rice, C. H., and Struck, R. T., *Proceedings, Eighth World Petroleum Congress* **4**, 43 (1971).

86. Grandy, D. W. and Petrakis, L., "ESR Investigation of Free Radicals in Solvent-Refined-Coal Materials," *Fuel,* **58**, 239 (1979).

87. Granoff, B. and Thomas, M. G., "Mineral Matter Effects in Coal Liquefaction. 1. Autoclave Screening Study," *ACS Div. of Fuel Chem. Preprints,* **22** (6), 183 (1977).

88. Granoff, B., Baca, P. M., Thomas, M. G., and Noles, G. T., "Chemical Studies on the Synthoil Process; Mineral Matter Effects," SAND-78-1113, Final Report, U.S. DOE Contract AT (29-1)-789, June 1978.

89. Granoff, B., "Mineral Matter Effects in Coal Liquefaction," 1st Quarterly Report, SAND-79-0306, Sandia Laboratories, Albuquerque, NM, October–December 1978.

90. Gray, D., "Inherent Mineral Matter in Coal and Its Effect Upon Hydrogenation," *Fuel,* **57**, 213 (1978).

91. Gray, D., Barrass, G., and Jezko, J., "Relationships Between Coal Liquefaction Behavior and the Composition of South African Coals," *ACS Div. of Fuel Chem. Preprints,* **24** (2), 2 (1979).

92. Guin, J. A., Tarrer, A. R., Taylor, Z. L., and Green, S. C., "A Photomicrographic Study of Coal Dissolution," *ACS Div. of Fuel Chem. Preprints,* **20** (1), 66 (1975).

93. Guin, J. A., Tarrer, A. R., Pitts, W. S., and Prather, J. W., "Kinetics and Solubility of Hydrogen in Coal Liquefaction Reactions," *ACS Div. of Fuel Chem. Preprints,* **21** (5), 170 (1976).

94. Guin, J., Tarrer, A., Taylor, Jr., L., Prather, J., and Green, Jr., S., "Mechanisms of Coal Particle Dissolution," *Ind. Eng. Chem., Process Des. Dev.,* **15** (4), 490 (1976).

95. Guin, J. A., Tarrer, A. R., Lee, J. M., Van Brackle, H. F., and Curtis, C. W., "Further Studies of the Catalytic Activity of Coal Minerals in Coal Liquefaction. 2. Performance of Iron and SRC Mineral Residue, as Catalysts and Sulfur Scavengers," *Ind. Eng. Chem., Process Des. Dev.,* **18**, 631 (1979).

96. Halleux, A. and Tschamler, H., "Extraction Experiments on Coal with Various Pyridine Bases," *Fuel,* **38**, 291 (1959).

97. Hamrin, Jr., C. E., "Catalytic Activity of Coal Mineral Matter," Annual Report, Jan. 1976–March 1977, U.S. DOE Report FE-2233-3, Contract EX-76-C-01-2233.

98. Han, K. W. and Wen, C. Y., "Initial Stage (Short Residence Time) Coal Dissolution," *Fuel,* **58**, 779 (1979).

99. Hayatsu, R. Winans, R. E., Scott, R. G., Moore, L. P., and Studier, M. H., "Oxidative Degradation Studies of Coal and Solvent Refined Coal," *ACS Div. Fuel Chem. Preprints,* **22** (5), 156 (1977).

100. Hayatsu, R., Winans, R. E., McBeth, R. L., Scott, R. G., Moore L. P., and Studier, M. H., "Structural Characteristics of Coal: Lignin-Like Polymers in Coals," *ACS Div. Fuel Chem. Preprints,* **24** (1), 110 (1979).

101. Henley, J. P., Jr., M.S. Thesis, Auburn University, Auburn, AL, 1976.

102. Heredy, L. A. and Neuworth, M. B., "Low Temperature Depolymerization of Bituminous Coal," *Fuel,* **41**, 221 (1962).

103. Heredy, L. A., Kostyo, A. E., and Neuworth, M. B., "Chemical Identification of Methylene Bridges in Bituminous Coal," *Fuel,* **43**, 414 (1964).

104. Heredy, L. A., Kostyo, A. E., and Neuworth, M. B., "Studies on the Structure of Coals of Different Rank," *Fuel,* **44**, 125 (1965).

105. Heredy, L. A. and Fugassi, P., *ACS Adv. in Chem. Ser.,* **55**, Chapter 28, 1966.

106. Heredy, L. A., "The Chemistry of Acid-Catalyzed Coal Depolymerization," *ACS Div. Fuel Chem. Preprints,* **24** (1), 142 (1979).

107. Hill, G. R., Harriri, H., Reed, R. I., and Anderson, L. L., "Kinetics and Mechanism of Solution of High Volatile Coal," in *Coal Science, ACS Adv. in Chem. Ser.,* **55**, 427 (1966).

108. Hooper, R. J. and Evans, D. G., "Deduction of the Structure of Brown Coal by Reaction with Phenol," *ACS Div. Fuel Chem. Preprints,* **24** (1), 131 (1979).

109. Huston, J. L., Scott, R. G. and Studier, M. H., "Reaction of Fluorine Gas with Coal and the Aromaticity of Coal," *Fuel,* **55**, 281 (1976).

110. Ignasiak, B. S. and Gawlak, M., "Polymeric Structure of Coal. 1. Role of Ether Bonds in Constitution of High-Rank Vitrinite," *Fuel,* **56**, 216 (1977).

111. Illig, E., "Effects of Mineral Matter on Coal Liquefaction," presented at the Conference on Mineral Matter Effects and Use of Disposable Catalysts in Coal Liquefaction, sponsored by Sandia Laboratories, Albuquerque, NM, June 14–15, 1978.

112. Ingram, D. J. E., Tapley, J. G., Jackson, R., Bond, R. L., and Murnaghan, A. R., *Nature,* **174**, 797 (1954).

113. Ishii, T., Sanada, Y., and Takeya, G., *J. Chem. Soc. Japan, Ind. Chem. Sect.* **72**, 1269 (1969).

114. Jackson, W. R., Larkins, F. P., Marshall, M., Rash, D., and White, N., "Hydrogenation of Brown Coal. 1. The Effects of Additional Quantities of the Inorganic Constituents," *Fuel,* **58**, 281 (1979).

115. Kang, D., Anderson, L. L., and Wiser, W. H., "Elucidation of Coal Structural Components by Short Residence-Time Extractive Liquefaction," Technical Report to the U.S. DOE, University of Utah, Contract E(49-18)-2006, August 1979.

116. Kawa, W., Hiteshue, R. W., Anderson, R. B., and Greenfield, H., "Reactions of Iron and Iron Compounds with Hydrogen and Hydrogen Sulfide," *Bureau of Mines Report of Investigation* 5690, 1960.

117. Kawa, W. and Hiteshue, R. W., "Iron-Catalyzed Hydrogenation of Phenyl Ether, Benzyl Ether, and Phenyl Sulfide," *Bureau of Mines Report of Investigation* 6179, 1963.

118. Keisch, B., Gibbon, G. A., and Akhtar, S., "Mössbauer Spectroscopy of Iron in Coal and Coal Hydrogenation Products," *Fuel Processing Technology,* **1**, 269 (1977/1978).

119. Kelvin, N. V. P., "Hydroliquefaction of Australian Coals," *ACS Div. of Fuel Chem. Preprints,* **24** (3), 177 (1979).

120. Kevan, L. and Kispert, L. D., *Electron Spin Double Resonance Spectroscopy,* Wiley, 1976, p. 2.

121. Klass, D. L., *Chem. Tech,* **5**, 499 (1975).

122. Koots, J. A. and Speight, J. B., *Fuel,* **54**, 179 (1975).

123. Kreulen, D. J. W., *Fuel,* **31**, 462 (1952).

124. Kwan, J. T. and Yet, T. F., "Aromaticity Determination of Coal, Oil Shale and Their Derivatives by X-Ray Diffraction," *ACS Div. of Fuel Chem. Preprints,* **21** (7), 67 (1976).

125. Ladner, W. R. and Stacey, A. E., *Fuel,* **40**, 295 (1961).

126. Ladner, W. R. and Wheatley, R., *Brit. Coal Util. Res. Assoc. Monthly Bull.,* **29**, 202 (1965).

127. Landelt, R. G., *Fuel,* **54**, 299 (1975).

128. Larsen, J. W. and Kuemmerle, E. W., "Alkylation and Depolymerization Reactions of Coal: A Selective Review with Supplementary Experiments," *Fuel,* **55**, 162 (1976).

129. Larsen, J. W. and Chandhury, P., "Molecular Weight of Depolymerized Bruceton Coal," *J. Org. Chem.,* **44**, 2856 (1979).

130. Larsen, O., Gulf Research & Development, personal communication, 1979.

131. Lee, J. M., Tarrer, A. R., Guin, J. A., and Prather, J. W., "The Selectivity of Coal Minerals as Catalysts in Coal Liquefaction and Hydrodesulfurization," *ACS Div. of Fuel Chem. Preprints,* **22** (6), 120 (1977).

132. Lowry, H. E., Ed., "The Chemistry of Coal Utilization," Vols. I, II, Wiley, New York, 1945.

133. Lowry, H. H., Ed., "Chemistry of Coal Utilization," Supplementary Volume, Wiley, New York, 1963: Dryden, T. G. C., "Chemical Constitution and Reactions of Coal," Chapter 6; Tschamler, H. and DeRuiter, E., "Physical Properties of Coals," Chapter 2.

134. Mayo, F. R., *Fuel,* **54**, 273 (1975).

135. Mayo, F. R., "The Chemistry of Coal Liquefaction," *ACS Fuel Chem. Div. Preprints,* **22** (5), 103 (1977).

136. Mazumdar, B. K. and Lahiri, A., "The Probability Distribution of Carbon-Hydrogen Groupings in Coal," *Fuel,* **41**, 487 (1962).

137. Mazumdar, B. K., Chakrabartty, S. K., and Lahiri, A., "Some Aspects of the Constitution of Coal," *Fuel,* **41**, 129 (1962).

138. Mazumdar, B. K., Ganguly, S., Sanyal, P. K., and Lahiri, A., "Aliphatic Structures in Coal," *Coal Science, ACS Adv. in Chem. Ser.,* **55**, 475 (1966).

139. McCartney, J. T. and Teichmuller, M., "Classification of Coals According to Degree or Coalification by Reflectance of the Vitrinite Component," *Fuel,* **51**, 64 (1972).

140. Medeiros, D. and Petersen, E. E., "Surface Area Variations of Coal During Solvent Extraction," *Fuel,* **58**, 531 (1979).

141. Montano, P. A., "Characterization of Iron Bearing Minerals in Coal," *ACS Div. of Fuel Chem. Preprints,* **24** (1), 218 (1979).

142. Mori, K., Tanuichi, M., Kawashima, A., Okuma, O., and Takahashi, T., "Relationship Between Coal Characteristics and Its Reactivity on Hydroliquefaction," *ACS Div. of Fuel Chem. Preprints,* **24** (2), 28 (1979).

143. Mukherjee, D. K., Sama, J. K., Chowdhury, P. B., and Lahiri, A., "Hydrogenation of Coal with Iron Catalysts: Parts I and II, " in *Proc. of the Symposium on Chemicals and Oil from Coal,* The Central Fuel Research Institute, Dhandad, India, December 6–8, 1969; published June 1972.

144. Mukherjee, D. K and Chowdhury, P. B., "Catalytic Effect of Mineral Matter Constituents in a North Assam Coal on Hydrogenation," *Fuel,* **55**, 4 (1976).

145. Nakata, Y., Veda, S., Maekawa, Y., and Shiboaka, M., "Hydrogenation Reactivity of Petrographic Composition of Coal," *ACS Div. of Fuel Chem. Preprints,* **24** (2), 339 (1979).

146. Neavel, R. C., in *Proc. Symp. on Agglomeration and Conversion of Coal,* Morgantown, WV, 1975.

147. Neavel, R. C., "Liquefaction of Coal in Hydrogen-Donor and Non-Donor Vehicles," *Fuel,* **55**, 237 (1976).
148. Neavel, R. C., "Coal Structure and Coal Science: Overview and Recommendations," *ACS Div. of Fuel Chemistry Preprints,* **24** (1), 73 (1979).
149. Oele, A. P., Waterman, H. I., Goldkoop, M. L., and Van Krevelan, D. W., "Extractive Disintegration of Bituminous Coals," *Fuel,* **30**, 169 (1951).
150. Oelert, H. H., *Fuel,* **49**, 119 (1970).
151. O'Gorman, J. V. and Walker, Jr., P. L., "Mineral Matter and Trace Elements in U. S. Coals," U.S. Dept. of Interior, Office of Coal Research, Rept. No. 61, Interim Rept. 2, 1972.
152. Okutani, T., Yokoyama, S., Yoshida, R., and Ishii, T., "Instrumental Studies of Coal Liquefaction," *Ind. Eng. Chem., Prod. Res. Dev.,* **18** (4), 367 (1979).
153. O'Neill, P., Steenken, S., and Schulte-Frohlinde, D. J., *Phys. Chem.,* **79**, 2773 (1975).
154. Pastor, G. R., Angelovich, J. M., and Silver, H. F., "Study of the Interaction of Solvent and Catalyst in Coal Hydrogenation," *Ind. Eng. Chem., Process Res. Dev.,* **9** (4), 609 (1970).
155. Peel, R. B., Diaz, J. S. V., and Luengo, C. A., "Direct Hydrogenation of High-Ash Brazilian Coals," *Fuel,* **58**, 298 (1979).
156. Peover, M. E., "The Dehydrogenation of Coals with Benzoquinone," *J. Chem. Soc.,* 5020 (1960).
157. Petrakis, L. and Grandy, D. W., *Anal. Chem.,* **50**, 303 (1978).
158. Pines, A., Gibby, M. G., and Waugh, J. S., *J. Chem. Phys.,* **59**, 569 (1973).
159. Pitts, W. S., Tarrer, A. R., Guin, J. A., and Prather, J. W., "Hydrodesulfurization in the Solvent Refined Coal Process," *ACS Div. of Fuel Chem. Preprints,* **22** (2), 214 (1977).
160. Pollack, S. S., "Estimating Mineral Matter in Coal from Its Major Inorganic Elements," *Fuel,* **58**, 76 (1979).
161. Raj, S., "Structural Characteristics of Vitrinites," *ACS Fuel Chem. Div. Preprints,* **24** (3), 251 (1979).
162. Rao, C. P. and Gluskoter, H. J., "Occurrence and Distribution of Minerals in Illinois Coals," Ill. State Geol. Surv. Circ. 476, 1973.
163. Rapoport, I. B., "Chemistry and Technology of Synthetic Liquid Fuels," 2nd ed., Moscow, 1955. Translation published for the National Science Foundation, Washington, D. C. by the Israel Program for Scientific Translations, Jerusalem, 1962, pp. 96–97.
164. Raymond, R., Wender, I., and Reggel, L., "Catalytic Dehydrogenation of Coal," *Science,* **137**, 681 (1962).
165. Reggel, L., Wender, I., and Raymond, R., "Catalytic Dehydrogenation and Reduction of Coal II. Reversibility of the Dehydrogenation and Reduction of Coal," *Fuel,* **43**, 229 (1964).
166. Reggel, L., Wender, I., and Raymond R., "Catalytic Dehydrogenation of Coal III – Hydrogen Evolution as a Function of Rank," *Fuel,* **47**, 373 (1968).
167. Reggel, L., Wender, I., and Raymond, R., "Catalytic Dehydrogenation of Coal Part 4. A Comparison of Exinite, Micrinite and Fusinite with Vitrinite," *Fuel,* **49**, 281 (1970).
168. Reggel, L., Wender, I., and Raymond, R., "Catalytic Dehydrogenation of Coal Part 5. Effect of Air Oxidation on Catalytic Dehydrogenation of Coal," *Fuel,* **49**, 287 (1970).

169. Reggel, L., Wender, I., and Raymond, R., "Catalytic Dehydrogenation of Coal Part 6. Yield of Hydrogen from Whole Coals," *Fuel,* **50**, 152 (1971).
170. Reggel, L., Wender, I., and Raymond, R., "Catalytic Dehydrogenation of Coal Part 7. The Effect of Some Catalyst Variables," *Fuel,* **52**, 162 (1973).
171. Retcofsky, H. L., Stark, J. M., and Friedel, R. A., *Anal. Chem.,* **40**, 1699 (1968).
172. Retcofsky, H. L., Schweighardt, F. K., and Hough, M., *Anal. Chem.,* **49**, 585 (1977).
173. Retcofsky, H. L. and Hough, M. R., "Electron Spin Resonance Studies of Coals and Coal Liquefaction," paper presented at the 29th Pittsburgh Conference on Analytical Chemistry and Applied Spectroscopy, Cleveland, OH, February 27–March 3, 1978.
174. Retcofsky, H. L., Thompson, G. P., Hough, M., and Friedel, R. A., "Electron Spin Resonance Studies of Coals and Coal Derived Asphaltenes," *ACS Div. of Fuel Chem. Preprints,* **22** (5), 90 (1977).
175. Retcofsky, H. L. and Link, T. A., "High Resolution ^{1}H-, ^{2}H-, and ^{13}C-NMR in Coal Research," Chapter 24 in *Analytical Methods for Coal and Coal Products,* Vol. II, Academic Press, New York, 1978.
176. Retcofsky, H. L. and VanderHart, D. L., *Fuel,* **57**, 421 (1978).
177. Retcofsky, H. L., Hough, M. R., and Clarkson, R. B., "Nature of the Free Radicals in Coals, Pyrolyzed Coals and Liquefaction Products," *ACS Div. of Fuel Chem. Preprints,* **24** (1), 83 (1979).
178. Retcofsky, H. L., Thompson, G. P., Raymond, R., and Friedel, R. A., "Studies of ESR Linewidths in Coals and Related Materials," *Fuel,* **54**, 126 (1979).
179. Retcofsky, H. L., Hough, M. R., Maguire, M. M., and Clarkson, R. B., "Nature of the Free Radicals in Coals, Pyrolyzed Coals, Solvent Refined Products and Coal Liquefaction Products" (personal communication).
180. Retcofsky, H. L., Hough, M. R., Link, T. A., Finseth, D. H., Maguire, M. M., and Clarkson, R. B., *Anal. Chem.,* in press.
181. Richardson, J. R., "Thermomagnetic Studies of Iron Compounds in Coal Char," *Fuel,* **51**, 150 (1972).
182. Ross, D., "Homogeneous Catalytic Hydrocracking Processes for Conversion of Coal to Liquid Fuels: Basic and Exploratory Research," Quarterly Reports FE-2202-1, FE-2202-3 prepared by SRI, Menlo Park, CA for ERDA under Contract No. E(49-18)-2202, 1976.
183. Ruberto, R. G., Cronauer, D. C., Jewel, D. M., and Seshadri, K. S., "Structural Aspects of Subbituminous Coal Deduced from Solvation Studies 1. Anthracene-Oil Solvents," *Fuel,* **56**, 17 (1977).
184. Ruberto, R. G., Cronauer, D. C., Jewell, D. M., and Seshadri, K. S., "Structural Aspects of Subbituminous Coal Deduced from Solvation Studies 2. Hydrophenanthrene Solvents," *Fuel,* **56**, 25 (1977).
185. Ruberto, R. and Cronauer, D. C., "Oxygen and Oxygen Functionalities in Coal and Coal Liquids," *ACS Div. Petroleum Chem. Preprints,* **23** (1), 264 (1978).
186. Ruch, R. R., Gluskoter, H. J., and Shimp, N. F., "Occurrence and Distribution of Potentially Volatile Trace Elements in Coal," Ill. State Geol. Surv. Environ. Geol. Note 72, 1974.
187. Russell, S., "Mineral Matter in Residues from Coal Liquefaction," Conference on Mineral Matter Effects and Use of Disposable Catalysts in Coal

Liquefaction, sponsored by Sandia Laboratories, Albuquerque, NM, June 14–15, 1978.

188. Scaroni, A. W. and Essenhigh, R. H., "A Reevaluation of Carbon Ring Structure in Coal," *ACS Div. of Fuel Chemistry Preprints,* **23** (4), 124 (1978).
189. Schlick, S., Narayana, A., and Kevan, L., *J. Am. Chem. Soc.,* **100**, 3322 (1978).
190. Schwager, I. and Yen, T. F., *ACS Div. Fuel Chem. Preprints,* **21**, 199 (1976).
191. Segal, B. G., Kaplan, M., and Fraenkel, G. K., *J. Chem. Phys.,* **43**, 4191 (1965).
192. Seshadri, K. S., Ruberto, R. G., Jewell, D. M., and Malone, H. P., *Fuel,* **57**, 549 (1978).
193. Singer, L. S., *Proc. 5th Conf. on Carbon,* Pergamon, New York, 1963, p. 35.
194. Spackman, W., "What is Coal? The Range of U.S. Coals Available," presented at Short Course on Coal Characteristics and Coal Conversion Processes, Pennsylvania State University, October 29–November 2, 1973.
195. Spackman, W., et al., "Evolution and Development of Special Purpose Coals," Final Report FE-0390-2 under ERDA Contract No. E(49-18)-0390, September 1976.
196. Sternberg, H. W., Delle Donne, C. L., Pantages, P., Moroni, E. C., and Markby, R. E., "Solubilization of a LVB Coal by Reductive Alkylation," *Fuel,* **50**, 432 (1971).
197. Sternberg, H. W. and Delle Donne, C. L., "Solubilization of Coals by Reductive Alkylation," *Fuel,* **53**, 172 (1974).
198. Sternberg, H. W., Raymond, R., and Schweighardt, F. K., *Science,* **188**, 49 (1975).
199. Sternberg, H. W., "A Second Look at the Reductive Alkylation of Coal and at the Nature of Asphaltenes," *ACS Div. of Fuel Chemistry Preprints,* **21** (7), 1 (1976).
200. Storch, H. H., Fisher, C. H., Hawk, C. O., and Eisner, A., *U.S. Bureau of Mines* TP654, 1943.
201. Szladow, A. J. and Given, P. H., "The Role of Oxygen Functional Groups in the Mechanism of Coal Liquefaction," *ACS Div. of Fuel Chem. Preprints,* **23** (4), 161 (1978).
202. Takegami, Y., Kajiyama, S., and Yokokawa, C., *Fuel,* **42**, 291 (1963).
203. Tarrer, A. R., Guin, J. A., Pitts, W. S., Henley, J. P., Prather, J. W., and Styles, G. A., "Effect of Coal Minerals on Reaction Rates During Coal Liquefaction," *ACS Div. of Fuel Chem. Preprints,* **21** (5), 59 (1976).
204. Thurlow, G. G., "Producing Fuels and Chemical Feedstocks from Coal," *Chem. Eng. Prog.,* **76** (3), 81 (1980).
205. Tododa, S. and Honda, H., *Carbon,* **3**, 527 (1966).
206. Uebersfeld, J., Etienne, A. and Combrisson, J., *Nature,* **174**, 614 (1954).
207. VanderHart, D. L. and Retcofsky, H. L., *Fuel,* **55**, 202 (1976).
208. Van Gerven, J., Van Itterbeck, A., and DeWolf, E., *J. Phys. Radium,* **17**, 140 (1956).
209. Van Krevelen, D. W., *Coal,* Elsevier, Amsterdam, 1971, p. 393.
210. Wachowska, H. and Pawlak, W., "Effect of Cleavage of Ether Linkages on Physicochemical Properties of Coals," *Fuel,* **56**, 422 (1977).
211. Weller, S. W., "Catalysis in the Liquid-Phase Hydrogenation of Coal and Tar," Chapter 7 in *Catalysis,* 4, P. H. Emmett, Ed., Reinhold, New York, 1956.

212. Weller, S., Pelipetz, M. G., and Friedman, S., *Ind. Eng. Chem.*, **43**, 1575 (1951).

213. Wender, I., "Catalytic Synthesis of Chemicals from Coal," *ACS Div. Fuel Chem. Preprints*, **20** (4), 16 (1975).

214. Wender, I., "Catalytic Synthesis of Chemicals from Coal," *Catal. Rev. Sci. Eng.*, Marcel Dekker, New York, **14** (1), 9 (1976).

215. Whitehurst, D. D., Farcasiu, M., and Mitchell, T. O., "The Nature and Origin of Asphaltenes in Processed Coals," EPRI Report AF-252, 1st Annual Report Under Project RP-410, February 1976.

216. Whitehurst, D. D. and Mitchell, T. O., "Short Contact Time Coal Liquefaction," *ACS Div. of Fuel Chem. Preprints*, **21** (5), 127 (1976).

217. Whitehurst, D. D., "A Primer on the Chemistry and Constitution of Coal," presented at the ACS Industrial and Engineering Chemistry Division Meeting, Chicago, August 29–September 1, 1977.

218. Whitehurst, D. D., Farcasiu, M., Mitchell, T. O., and Dickert, Jr., J. S., "The Nature and Origin of Asphaltenes in Processed Coals," EPRI Report AF-480, 2nd Annual Report Under Project RP-410, July 1977.

219. Whitehurst, D. D., "The Nature and Origin of Asphaltenes In Processes Coals," EPRI Final Report AF-1298, Project 410, December 1979.

220. Wiser, W. H. and Hill, G. R., "A Kinetic Study of the Thermal Dissolution of a High-Volatile Bituminous Coal," in *Proceedings of the Symposium on the Science and Technology of Coal,* Ottawa, 1967, Mines Branch, Dept. of Energy, Mines & Resources (Canada), pp. 162–167.

221. Wiser, W. H., "A Kinetic Comparison of Coal Pyrolysis and Coal Dissolution," *Fuel,* **47**, 475 (1968).

222. Wiser, W. H., Anderson, L. L., Qader, S. A., and Hill, G. R., "Kinetic Relationship of Coal Hydrogenation, Pyrolysis and Dissolution," *J. Appl. Chem. Biotechnol.*, **21**, 82 (1971).

223. Wiser, W. H., "Some Chemical Aspects of Coal Liquefaction," presented at Short Course on Coal Characteristics and Coal Conversion Processes, Pennsylvania State University, October 29–November 2, 1973.

224. Wiser, W. H. "Some New Evidence Pertaining to the Chemistry and Mechanisms of Coal Liquefaction," presented at the ACS Div. Fuel Chem. Meeting, Miami Beach, FL, September 10–15, 1978.

225. Wooton, D. L., Dorn, H. C., Taylor, L. T., and Coleman, W. H., *Fuel,* **55**, 224 (1976).

226. Wright, C. H. and Severson, D. E., "Evidence for the Catalytic Effect of Mineral Matter," *ACS Div. of Fuel Chem. Preprints,* **16** (2), 68 (1972).

227. Wu, W. and Storch, H. H., "The Hydrogenation of Coal and Tar," *U. S. Bureau of Mines Bulletin* 633, 1968.

228. Yamakawa, T., Imuta, K., Ouchi, K., Tsukasa, K., Morotomi, H., Shimura, M., and Miyazu, T., in *Preprints, 13th Coal Science Conference,* Japan, October 1976, p. 40.

229. Yarzab, R. F., Abdel-Baset, Z. and Given, P. H., "Hydroxyl Contents of Coals: New Data and Statistical Analysis," *Geochim. Cosmochim. Acta.*, **43**, 281 (1979).

230. Yarzab, R. F., Given, P. H., Spackman, W., and Davis, A., "Dependence of Coal Liquefaction Behavior on Coal Characteristics. 4. Cluster Analyses for Characteristics of 104 Coals," *Fuel,* **59**, 81 (1980).

231. Yen, T. F., *Fuel,* **52**, 93 (1973).
232. Yen, T. F. and Young, D. K., *Carbon,* **11**, 33 (1973).
233. Yen, T. F. and Sprang, S. R., *Geochim. Cosmochim. Acta.,* **41**, 1007 (1977).
234. Yen, T. F., Erdman, J. G., and Pollacks, S. S., *Anal. Chem.,* **33**, 1587 (1961).
235. Yergey, A. L., et al., *Ind. Eng. Chem., Process Des. Dev.,* **13**, 233 (1974).

3. Heteroatom Removal During Liquefaction

Y.T. Shah and S. Krishnamurthy

INTRODUCTION

Coal can best be described as a loose polymeric network of di- and triaromatics bridged by carbon-carbon, carbon-oxygen, carbon-sulfur, and carbon-nitrogen bonds. This polymer can be converted to liquid products by hydrogenolysis in the presence of a solvent and under pressure. The basic reactions which occur can be loosely characterized as those of depolymerization and hydrogenation. These reactions are believed to proceed by free-radical mechanisms [11, 18]. The solvent penetrates the pore structure of the coal particle, causing swelling and fracture of the chemical bonds and thereby the disintegration of the coal particle. The fractured bonds can be hydrogenated by hydrogen transfer from solvent molecules or from hydroaromatics formed from the parent coal through a process termed "autogenous" transfer [21]. In this manner, it is possible to convert the polymer-like material to a liquid product which consists predominantly of 2-, 3-, and 4-ring structures. There is some indication that oxygen and sulfur removal occurs in the initial stages of the liquefaction process [29].

Figure 3-1 depicts the changes which occur when coal is converted into liquid products. Straight lines on this plot depict fairly accurately the changes in chemistry which occur during the hydrogenation of coal. For example,

A to *B*	decarboxylation
B to *C*	desulfurization and denitrogenation
A to *D*	deoxygenation
E to *C* to *G*	hydrogenation of aromatic rings

Yatish T. Shah, Reaction Engineering in Direct Coal Liquefaction, ISBN 0-201-07381-1

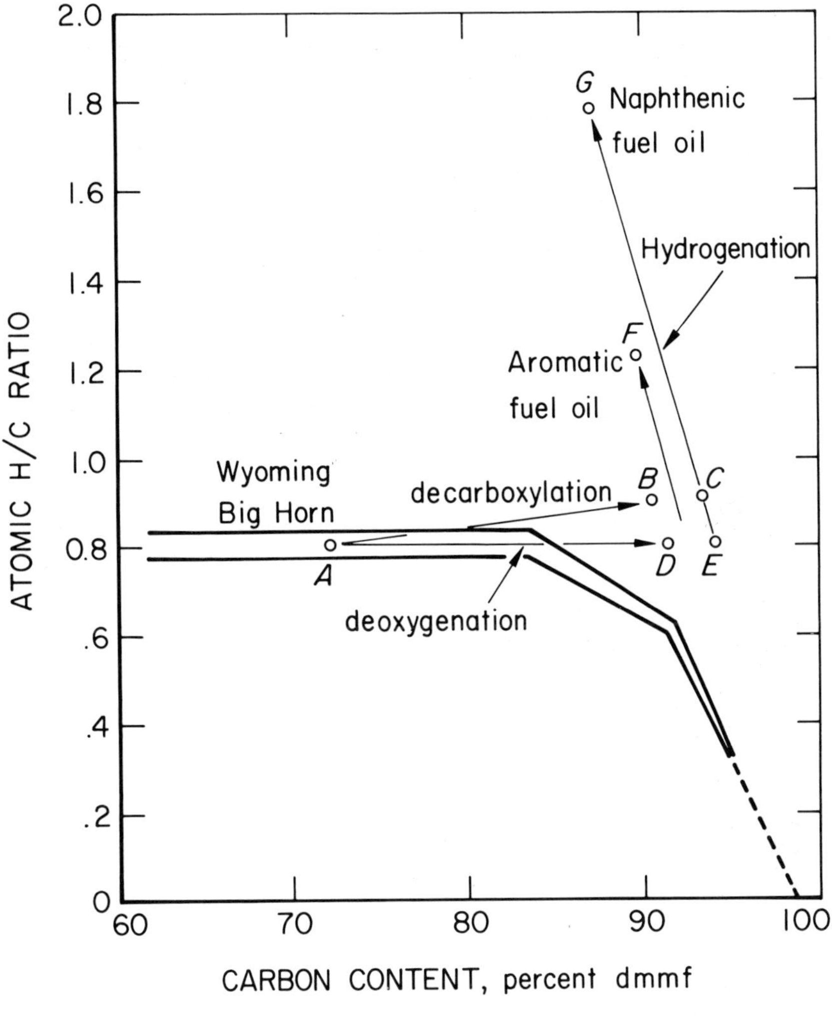

Figure 3-1. Graphical treatment of coal hydrogenation.

Point *F* represents a typical mixture of liquids produced (hydrogen-to-carbon atomic ratio of 1.2) from processing a subbituminous coal under average conditions. Points *C* and *E* represent complete removal of all oxygen, sulfur, and nitrogen. This would result in highly aromatic products. However, the aromatic products within the zone *B-C-E-D* can be hydrogenated to states such as *F* and *G*. The point represented by *G* is the theoretical H/C ratio which would result if all the aromatic rings are completely hydrogenated to naphthenic rings. This assumes a 3-ring condensed system similar to phenanthrene or anthracene.

Coal liquids are chemically different from gas-oil fractions. An important difference between coal liquids and gas-oil fractions can best be explained by considering several specific boiling-range fractions. The structures in Table 3-1 have not actually been found, but enough information exists to infer that these structures are typical of the molecules present. It is probable that the petroleum-derived aromatic distillate at a comparable temperature will contain fewer rings and more alkyl groups surrounding the nucleus. For example, a coal-derived aromatic distil-

Table 3-1

Comparison of Molecular Types between Petroleum and Coal Liquids

		Concentration, %		
	Type	Petroleum gas oils	Coal liquids	Relative concentration*
	Benzenes	9–18	6–9	0.5
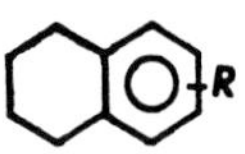	Naphthene-benzenes	6–9	∼27	∼4
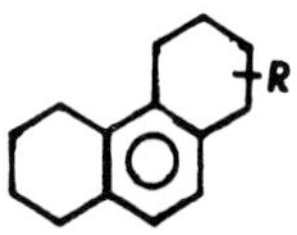	di-Naphthene benzenes	8–13	20	∼2
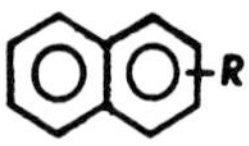	Naphthalenes	3	12	∼4
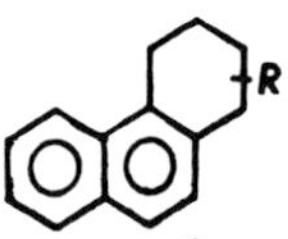	Naphthene naphthalenes	5–12	14–15	∼2
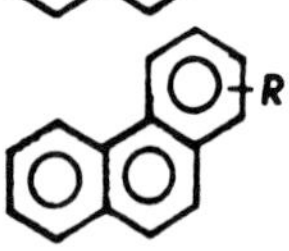	Phenanthrenes	3–15	6	∼0.7
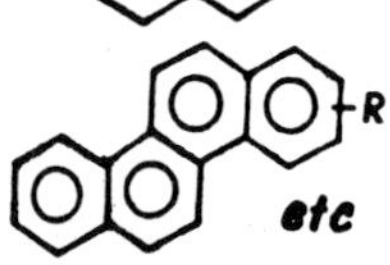	Pyrenes, chrysenes	2–16	∼1	∼0.1
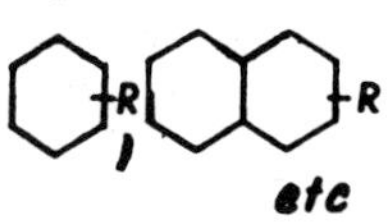	Naphthenes	10–30	∼0	∼0
	Paraffins	10–30	∼0	∼0

*(Conc. in coal liq.)/(conc. in petroleum-gas oils).

late at about 589 K is likely to consist of a naphthalene ring with about 2 alkyl groups. The alkyl groups in coal molecules are probably seldom longer than C_2 (ethyl).

This characteristic of coal liquids appears to prevail throughout the boiling range, as shown in Figure 3-2. At the higher distillation ranges (e.g., above 811 K), it is highly probable that the coal molecule is an asphaltene and is, therefore, relatively insoluble in the solvent. It is the presence of a large number of alkyl groups in petroleum-derived molecules which gives them the special property of being soluble in crude oil at around room temperature.

3.1 Gross Removal of Sulfur, Oxygen, and Nitrogen

In this chapter we briefly evaluate some aspects of the removal of sulfur, oxygen, and nitrogen during liquefaction. The terms percentage desulfurization, deoxygenation, and denitrogenation used in this chapter and in the literature are based upon sulfur, oxygen, and nitrogen contained in the total slurry feed and reactor product. The following equation is used for percentage desulfurization:

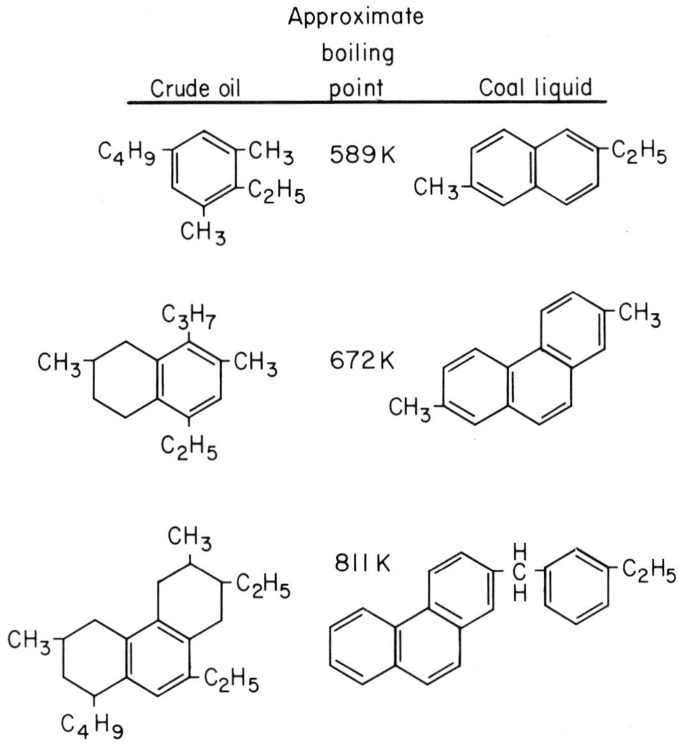

Figure 3-2. Typical molecular structures in petroleum and coal liquids.

$$\% \text{ desulfurization} = \frac{(\text{sulfur in feed slurry})-(\text{sulfur in product slurry})}{\text{sulfur in feed slurry}} \times 100. \quad (3\text{-}1)$$

Similar equations are used for percentage deoxygenation and denitrogenation. It should be noted that this definition includes sulfur removal from the solvent. This is not a serious consideration when using solvent-recycle operation, but when using anthracene oil as solvent, it is not possible to determine the amount of product sulfur in coal-derived molecules or in the solvent molecules, so that the actual percentage desulfurization of the coal cannot be calculated. It should also be noted that the above definition includes inorganic sulfur removal as part of percentage desulfurization. Inorganic desulfurization apparently takes place through reduction of sulfides in the coal.

The terms percentage solvation and percentage hydrocracking are often used in the following discussion as well as in the literature. The former is defined as

$$\% \text{ solvation} = \frac{(\text{maf coal feed})-(\text{maf pyridine insoluble product})}{\text{maf coal feed}} \times 100, \quad (3\text{-}2)$$

where maf stands for "moisture- and ash-free." The percentage hydrocracking is defined as the fraction of the organic matter in the coal which is converted to distillates (fraction of liquid boiling below 797 K) and gas. It should also be noted that the above-described definitions of percentage desulfurization, deoxygenation, and denitrogenation assume that sulfur, oxygen, and nitrogen are removed as H_2S, H_2O, CO, CO_2, and NH_3.

The removal of heteroatoms from coal depends upon a number of operating variables such as the reactor temperature and pressure, the contact or space time (or space velocity), the reactor configuration, and the gas (mainly hydrogen) circulation rate. It also depends upon the nature of catalyst (if present), the rank and nature of the coal, and the nature of the solvent. This subject was recently reviewed by Shah and Cronauer [26]. Here we briefly summarize the important results of this review.

3.1.1 *Thermal Operation (Absence of Externally Added Catalyst)*

Shah and Cronauer [26] and Shah et al. [27] reported the effects of temperature, partial pressure of hydrogen, and gas flow rate on the product distributions during liquefaction of Big Horn coal. All experiments were carried out with 30% maf coal and at a weight hourly space velocity (WHSV) of 2.3. The results indicated that while deoxygenation remained essentially constant in the temperature range of 661 to 722 K, both desulfurization and denitrogenation increased with temperature. Higher pressure (and/or higher gas rate) gave better denitrogenation. While substantial deoxygenation occurred at temperatures as low as 661 K, a substantial degree of desulfurization and denitrogenation (57% and 21% respectively) did not take place until the temperature had reached 700 K. These observations are consistent with the common belief that some carbon-oxygen bonds in coal can be broken rather readily, while breaking of carbon-nitrogen and carbon-sulfur bonds that exist in the form of ring structures require more severe conditions. Furthermore, the carbon-oxygen bonds that are not broken under mild conditions appear to be very stable and cannot be broken under more severe conditions. The results also indicated that while solvation and hydrocracking increased significantly with reactor pressure at both 700 K and 720 K, desulfurization was only mildly affected by the change in pressure. Deoxygenation and denitrogenation were relatively insensitive to the change in reactor pressure.

For Wyodak (Power River Basin) coal the results at high pressure (24.1 MPa) showed that just as in the case of Big Horn coal, higher temperature and space time increased the removal of sulfur and nitrogen; however, deoxygenation remained unaffected by the variations in temperature and space time. Under similar conditions, the percentage desulfurization, deoxygenation, and denitrogenation of Wyodak and Big Horn coals were found to be approximately equal.

Different-rank coals give different results. Shah and Cronauer [26] reported that high-sulfur Kentucky coal (a mixture of Kentucky #9 and Kentucky #14) gave a reasonable level of sulfur removal even under low-pressure conditions. Generally, under similar conditions, bituminous coals give somewhat better solvation and desulfurization than subbituminous coals. Deoxygenation is better achieved for subbituminous coals. Bituminous and subbituminous coals give almost equal values of denitrogenation. Even in the same rank of coal, some coals are more reactive than others. For thermal liquefaction in anthracene oil, the reported results show that under similar conditions Illinois #6 is a more reactive coal (giving somewhat better percentage desulfurization, denitrogenation, deoxygenation, and hydrocracking) than Pittsburgh seam coal.

3.1.2 *Catalytic Operations (Externally Added Catalyst)*

Shah and Cronauer [26] reported that for Big Horn coals, the conventional NiCoMo catalyst improved desulfurization and denitrogenation. Deoxygenation appeared to remain unaffected by the addition of a catalyst. Higher denitrogenation also caused higher hydrogen consumption, which is in general agreement with our understanding of the mechanism of hydrodenitrogenation [15]. In the ranges of operating conditions (a) temperature 644–728 K, (b) pressure 13.79–24.13 MPa, (c) WHSV 1–4.7 (kg slurry)/(hr)(kg catalyst), and (d) coal concentration 30–50 wt%, the following conclusions can be made from the reported experimental results:

1. An increase in temperature generally increases percentage desulfurization and denitrogenation.

2. At constant temperature, pressure, and feed-coal concentration, a decrease in space velocity increases the percentage removal of sulfur, oxygen, and nitrogen. The effect is more pronounced for oxygen and nitrogen than for sulfur.

3. An increase in pressure increases percentage desulfurization, deoxygenation, and denitrogenation. The effect is more pronounced for the nitrogen removal and for the pressure range of 13.79–18.96 MPa.

4. Under otherwise identical conditions, higher coal concentration gives lower percentage desulfurization, deoxygenation, and denitrogenation.

The reported results [26] also showed that for Wyodak (Power River Basin) and Crow coals hydrogenation of solvent and catalyst improved percentage solvation, hydrocracking, and sulfur removal. Higher temperature and space time increased percentage solvation, hydrocracking, desulfurization, and denitrogenation. However, the deoxygenation remained unaffected by temperature and space time. Although Big Horn, Wyodak, and Crow coals have similar physical and chemical properties, the levels of solvation and hydrocracking of these subbituminous coals were different. Under similar reaction conditions, the levels of solvation were in the following order: Crow Indian (96%) > Big Horn (93%) > Wyodak (91%). The levels of hydrocracking are in the order of Big Horn>Crow Indian>Wyodak. Percentage desulfurization and deoxygenation of Wyodak and Big Horn coals were approximately equal under the same system conditions. However, the corresponding values for the Crow Indian coal in general appear to be somewhat higher than those for the other two coals.

For two high-rank Pennsylvania (Clearfield County, PA) bituminous coals, namely Brookville and Upper Freeport coals, the percentage solvations were lower than that for the Pittsburgh seam coal. The levels of hydrocracking of both coals to distillable liquids and gases were also low. Moderate desulfurization and low levels of denitrogenation of these coals were achieved during the liquefaction process. Unlike these coals, high-sulfur Kentucky coal (a mixture of Kentucky #9 and Kentucky #14) gave a high level of solvation and sulfur removal for a wide range of operating conditions.

Some of the unpublished results of Gulf Research and Development Company (GR&DC) for catalytic liquefaction of two lignite samples, a commercially mined (Glenharold mine, North Dakota) and a core sample (Bloomfield area of Montana), indicated that the latter lignite sample can be better desulfurized than the former one. The solvation and hydrocracking of these lignites were more dependent upon operating conditions than those for subbituminous coals. The level of denitrogenation for lignite samples was found to follow that of hydrocracking. The maximum level of denitrogenation in the Bloomfield run was 69%, and that in the Glenharold run was 53%. Desulfurization of the slurry of Bloomfield lignite plus solvent was moderately effective with a maximum up to 80%. Since this lignite contained a low level of sulfur (0.77 wt%), even moderate desulfurization produced a filtrate containing 0.10 wt% sulfur or less. The distillation residue also had a sulfur concentration in this range. The percentage desulfurization of the Glenharold lignite slurry was not as good as that of the Bloomfield slurry, but again the level of sulfur in the coal was low (0.43 wt%). The percentage deoxygenation for Bloomfield and Glenharold lignite coals appeared to be about the same under equivalent system conditions. For both coals, percentage deoxygenation increased significantly with an increase in temperature and space time.

3.1.3 *Catalyst Research*

Most of the experimental data reported thus far on catalytic liquefaction have been with a NiCoMo catalyst. A number of other catalysts have also been examined at GR&DC* to evaluate their initial activity for solvation, hydrocracking, desulfurization, deoxygenation, and denitrogenation. The results are briefly described below. Catalysts tested were nickel-molybdenum-titanium (NiMoTi) on alumina, which had shown an advantage in HDS processing, and on $AlPO_4 \cdot Al_2O_3$; NiCoMo supported on $AlPO_4 \cdot Al_2O_3$ and on Torvex, a glass-bead-supported catalyst; and a NiW chemideposited on Al_2O_3 catalyst.

The NiCoMo on glass beads was significantly less active than the conventional NiCoMo catalyst. Therefore, glass beads do not look promising as a catalyst support for catalytic coal liquefaction. The results of initial-activity test runs comparing $AlPO_4 \cdot Al_2O_3$ against Al_2O_3-supported catalyst with NiCoMo active metals are shown in Figure 3-3. Based on the average results over the operating temperatures and flow rates, the straight Al_2O_3 support appears to be somewhat superior to the $AlPO_4 \cdot Al_2O_3$ support. However, the $AlPO_4 \cdot Al_2O_3$ support needs to be further optimized and investigated. Other results showed that NiMoTi metals were not as effective in coal liquefaction when deposited on $AlPO_4 \cdot Al_2O_3$ as when deposited on Al_2O_3. Thus, a more effective means of depositing Ti on $Al_2PO_4 \cdot Al_2O_3$ support needs to be developed.

As shown in Figure 3-4, NiCoMo metals deposited on Torvex honeycomb showed some activity toward the solvation and hydrocracking of Wyodak coal. The level of coal solvation was within 1–3% of that achieved with a standard NiCoMo catalyst, while the level of hydrocracking was about 10% lower for the Torvex-supported catalyst. Similarly, lower sulfur, nitrogen, and oxygen removal was observed for the Torvex-supported catalyst. However, hydrogen consumption was also lower (slurry hydrogen concentration was 6.7 wt% for NiCoMo on Torvex, and 7.9 wt% for conventional NiCoMo).

Some unpublished data from GR&DC also indicated that the chemideposited NiW catalyst compared favorably with the NiCoMo. The levels of coal solvation and hydrocracking were essentially equal at similar operating conditions. The use of chemideposited NiW catalyst resulted in greater reduction of nitrogen and oxygen in the product streams in addition to promoting a higher level of hydrogenation. After regeneration, the hydrogenation activity of the chemideposited catalyst was approximately equal to that of fresh NiCoMo catalyst. However, its denitrogenation and deoxygenation capacities were somewhat less.

A series of catalysts supported on magnesium aluminate ($Mg_xAl_2O_{3+x}$, $0 < x \leqslant 1$) was prepared for comparison with the NiCoMo on alumina for the liquefaction of Big Horn coal. While the magnesium aluminate catalyst was found to have the same initial catalytic activity as that of NiCoMo on alumina, less coke was deposited on the magnesium aluminate catalyst. This indicates a potential for good

*Personal communication.

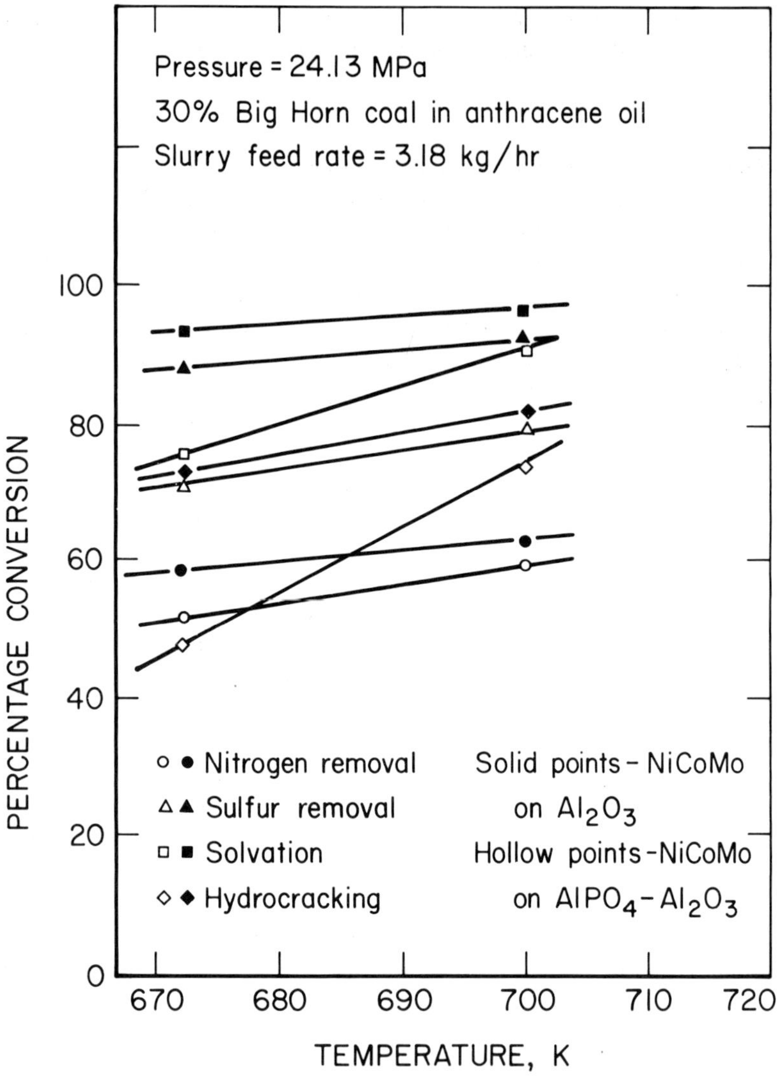

Figure 3-3. Comparison of performance of NiCoMo on Al_2O_3 catalyst with that of NiCoMo on $AlPO_4 \cdot Al_2O_3$ catalyst.

aging characteristics. In addition, the magnesium aluminate catalyst pellets retained greater strength during successive cycles of use and regeneration.

Experiments were also performed at GR&DC to evaluate the effectiveness of shaped (i.e. macaroni) catalyst for liquefaction. The experiments were performed with Big Horn coal and recycle solvent (at weight ratio of 40:60) at operating conditions of 24.1 MPa, 686 K, and WHSV of 0.88 (kg coal)/(hr) (kg catalyst). The

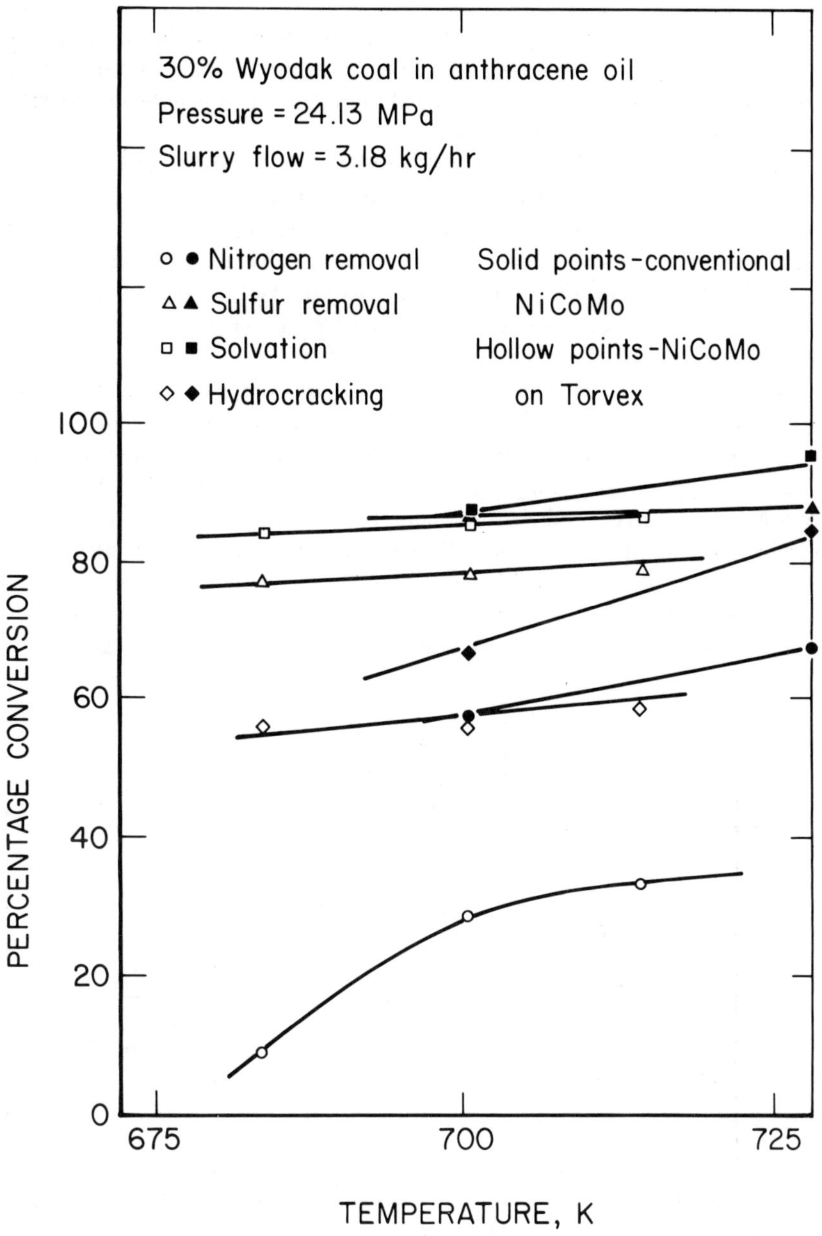

Figure 3-4. Comparison of performance of NiCoMo on Torvex catalyst with that of conventional NiCoMo catalyst.

results indicated that under similar operating conditions, the shaped catalyst gave a 5% increase in coal solvation and equivalent increase in liquid yield, while being less effective with respect to hydrogenation, denitrogenation, and removal of oxygen from the liquid products than conventional extrudates. This may be due to entrapment of solids in a shaped catalyst bed.

Bertolacini et al. [5] evaluated the effects of catalyst properties on the desulfurization of Illinois #6 from the Burning Star mine. The solvent consisted primarily of mono-, di-, and trimethylnaphthalenes. The operating conditions using a 1-liter autoclave and 400-mesh coal were 13.79 MPa, 700 K, with a hydrogen feed rate of 0.17 (standard m^3)/hr, a slurry feed rate of 0.4 kg/hr, a slurry concentration of 20 wt%, a LHSV of 1.33 (g coal)/(hr)(cm^3 catalyst), and a stirrer speed of 1500 rpm.

The effects of varying the cobalt-molybdenum composition on two alumina supports having different surface properties (Cyanamid PA ~300 m^2/g, 60 Å; Kaiser KSA light ~180 m^2/g, 160 Å) on desulfurization were examined. The results for the two supports were indistinguishable. As shown in Figure 3-5, the desulfurization index increased rapidly up to about 10 wt% MoO_3 but then leveled off as molybdenum content was increased further. The optimum cobalt content for desulfurization was also between 2 and 4 wt% CoO. The desulfurization was very sensitive to the variation of CoO concentration when the cobalt content was reduced to zero. Bertolacini et al. [5] also evaluated the effects of surface area and

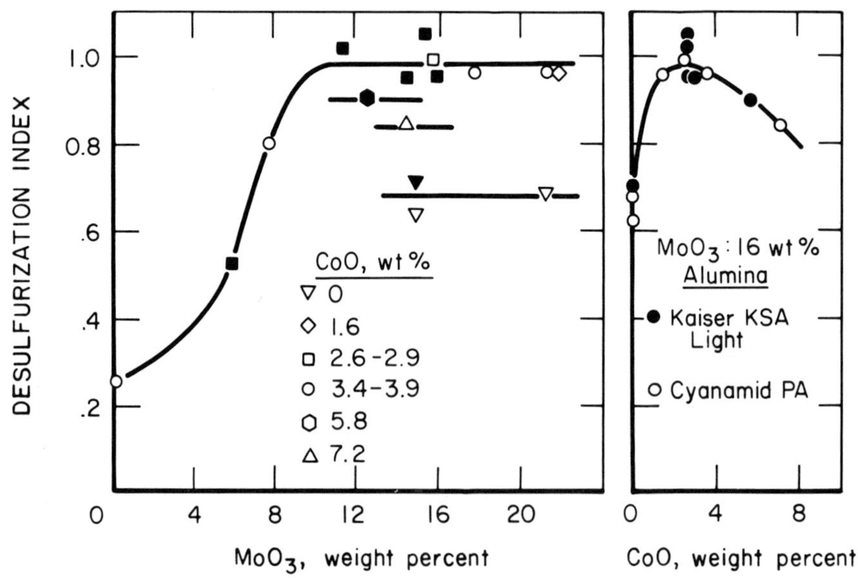

Figure 3-5. Effect of catalyst composition on desulfurization index (after Bertolacini et al. [5]).

average pore diameter of cobalt-molybdenum catalysts on the desulfurization. As shown in Figure 3-6, the desulfurization index increased with increasing surface area. On the other hand, the preferred average pore diameter for desulfurization appeared to be smaller (see Figure 3-7).

Alcorn et al. [2] screened a variety of CoMo, NiMo, and NiW catalysts for their initial desulfurization and denitrogenation activities. Their preliminary conclusions were that NiW catalysts did not perform as well as NiMo or CoMo catalysts. Among the latter two, NiMo had somewhat better desulfurization and denitrogenation activity than CoMo.

Granoff et al. [16] evaluated the effects of both conventional catalysts and some mineral matter on the hydroliquefaction of Illinois #6 (Orient 4 mine) coal using creosote oil at 6.89 MPa pressure and 723 K. Their data showed that while CoMo and pyrrhotite were active for desulfurization, most of the other minerals such as FeS, ZnS, kaolinite, and pyrite were not. The catalytic activity of mineral matter in coal has also been investigated by Tarrer et al. [28] and Hamrin [17].

In a recent study Batchelder and Fu [3] evaluated the use of syngas (H_2 :CO = 2:1) for coal liquefaction. They carried out experiments with West Virginia bituminous coal and SRC process solvent. The reaction was carried out at 20.68 MPa pressure and 723 K temperature. A number of catalysts were examined. The results indicate that the use of syngas resulted in the sulfur concentration in the product oil being very close to the concentration obtained with the use of hydrogen. The study, however, did not illustrate the oxygen and nitrogen removal efficiencies during the use of syngas.

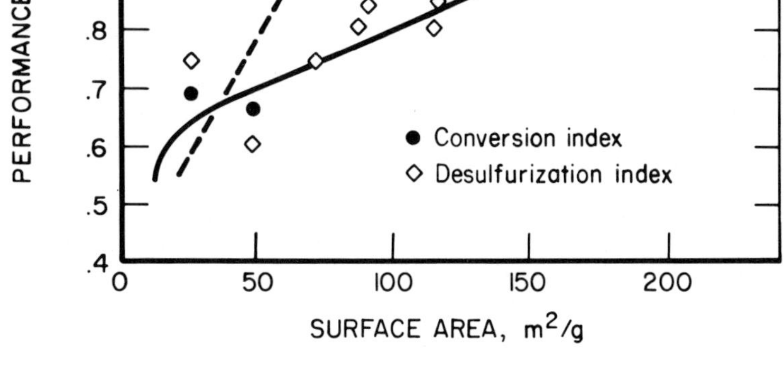

Figure 3-6. Relation of catalyst performance to surface area (W.R. Grace alumina) (after Bertolacini et al. [5]).

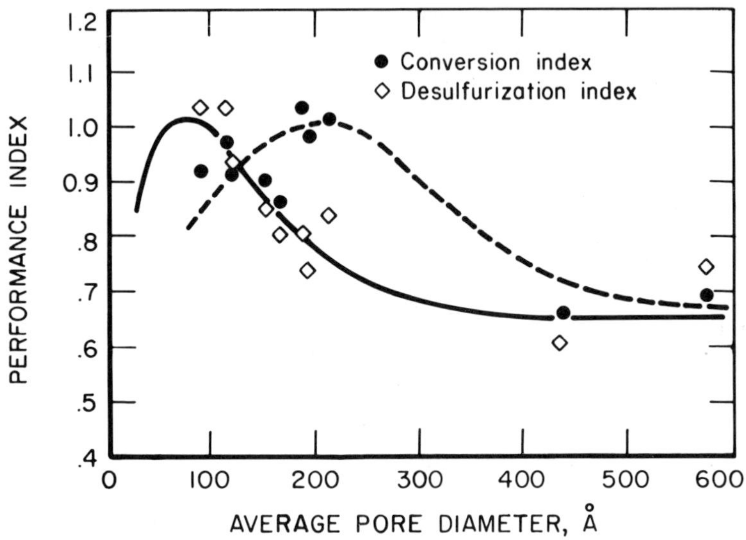

Figure 3-7. Relation of catalyst performance to average pore diameter (W.R. Grace alumina) (after Bertolacini et al. [5]).

3.2 Distributions of Sulfur, Oxygen, and Nitrogen in Various Boiling-Range Fractions of Coal Liquids

It is important to know how the oxygen, nitrogen, and sulfur are distributed in the various boiling-range fractions of coal liquids. For the typical liquefaction of Big Horn coal in recycle solvent under mild conditions, the distribution of sulfur, nitrogen, and oxygen in the liquid products are described in Table 3-2.* As shown in this table, the concentration of nitrogen in the distillate cuts went from a low of 0.08 wt% at the low-boiling end to 0.57 wt% at 699 K. A step change then occurred, with the nitrogen level of the highest-boiling cut being 1.18 wt% while that of the 818-K residue was 1.51 wt%.

The pattern of oxygen concentration in the liquid cuts was different from that of hydrogen and nitrogen. With the exception of the low-boiling cut at 0.75 wt% and the high-boiling cut (1.82 wt%) and residue (2.74 wt%), the oxygen concentration varied randomly between 0.8 and 1.5 wt%, the accuracy of the analysis being ± 0.08 wt%. For the most part, the sulfur content of the residue was 0.23 wt%, and that of the Dry Ice trap oils was 0.18%. Much of the latter value appeared to be due to dissolved H_2S.

Similar results for Illinois coal are described in Table 3-3.* In this case, seventeen distillate cuts plus a residue sample were taken. These cuts were analyzed by combination of separation techniques plus mass spectrometry and gas chromatography. The elemental analysis and the results of the separations shown in Table

*Personal communication with Gulf Research and Development Company.

Table 3-2

Analysis of Distillate Cuts (Low-Severity Run)*

Temperature (K)	Elemental analysis (wt%)					FIA cut (wt%)		
	C	H	N	O	S	Aromatics	Olefins	Saturates
Dry Ice trap (343–463)	85.65	13.34	0.13	0.70	0.18	—	—	—
Light ends								
370–394	85.17	13.93	0.08	0.75	0.07	8.5	8.0	83.5
435–445	86.14	12.37	0.12	1.32	0.05	30.0	6.0	64.0
473–482	87.60	10.68	0.19	1.51	0.03	62.0	3.5	34.5
504–515	88.17	10.67	0.26	0.85	0.05	66.0	4.0	30.0
544–560	89.17	9.15	0.28	1.36	0.04	82.0	2.5	15.5
Podbielniak								
479–502	88.51	10.07	0.33	1.04	0.05	73.5	3.0	23.5
585–599	90.20	8.41	0.34	0.93	0.12	94.5	2.0	3.5
629–636	90.67	7.85	0.54	0.87	0.07	87.2		8.9
Sarnia								
686–699	89.96	8.54	0.57	0.80	0.13			
793–818	89.32	7.61	1.18	1.82	0.07			
818+	89.45	6.07	1.51	2.74	0.23			

*Temp. = 686 K, pressure = 15.51 MPa, (recycle solvent)/(Big Horn coal) = 1.5/1, WHSV = 1.72 (kg coal)/(hr)(kg catalyst); hydrogen flow = 1781 (standard m^3)/m^3 of slurry); catalyst = CoMo.

3-3 indicate that the hydrogen-to-carbon atomic ratio went from 1.70 for the first cut (OP–447 K) to 0.80 for the residue cut. The sulfur content of the distillate cuts varied between 0.04 and 0.23 wt%, with the residue containing 0.43 wt%. The Dry Ice trap oils had a sulfur content of 0.25 wt%, but it appears that most of this was due to dissolved H_2S.

The oxygen concentration of the distillate cuts was a maximum (2.4–3.5 wt%) for the low-boiling cuts and feed, and a minimum (0.6 wt%) for the tenth cut

Table 3-3
Chemical Analysis of Distillate Cuts (Illinois Coal Run at 1.58 WHSV)*

Cut	Temp., K	Light-ends & filtrate yield†	Elemental analysis, Wt%					FIA & SARA analysis, Wt%			
			C	H	N	O	S	Saturates	Olefins	Aromatics	Resins
‡	238–473		82.62	13.74	0.02	1.65	0.25	—	—	—	—
								FIA analysis			
(1)	447	3.43	85.46	12.08	0.16	2.39	0.13	68.0	5.5	26.5	—
(2)	447–476	3.81	86.49	9.87	0.47	3.52	0.04	32.0	2.5	65.5	—
(3)	476–502	3.91	87.66	9.18	0.46	2.29	0.12	20.0	1.5	78.5	—
(4)	502–520	3.91	90.28	9.45	0.48	1.34	0.10	5.5	1.0	93.5	—
(5)	520–536	3.99	89.04	9.09	0.47	1.17	0.04	3.0	1.0	96.0	—
(6)	536–552	4.11	89.19	8.23	0.44	2.16	0.04	0.0	1.0	99.0	—
(7)	552–572	4.19	89.82	7.99	0.29	0.33	0.08				—
(8)	572–584	4.20	90.21	8.06	0.30	1.22	0.15				—
(9)	584–597	4.27	90.65	7.85	0.30	0.93	0.18				—
								SARA analysis			
(10)	597–614	4.54	91.36	7.21	0.34	0.54	0.20	2.5	—	90.9	6.6
(11)	614–623	4.40	90.94	7.28	0.67	0.88	0.19	4.0	—	85.0	10.7
(12)	623–637	4.20	90.44	7.46	0.80	0.97	0.17	5.0	—	83.0	12.0
(13)	637–647	4.51	91.72	6.89	0.52	0.93	0.04	4.9	—	85.1	10.0
(14)	647–664	4.51	91.16	6.88	0.56	0.88	0.18	5.8	—	82.7	11.5
(15)	664–684	4.4	89.67	7.86	0.83	1.46	0.23	9.4	—	75.5	14.2
(16)	684–763	4.60	89.39	7.71	1.01	1.68	0.15	9.2	—	68.8	21.0
(17)	763–814	4.70	89.23	7.59	1.16	2.08	0.19	2.0	—	63.0	34.8
(18)	814+	28.30	88.39	5.93	2.02	2.29	0.43	§	§	§	§

*Temp. = 686 K, pressure = 24.13 MPa, (recycle solvent)/coal = 1.5/1, catalyst = CoMo.
†Yield as wt% of total light ends + filtrate charge.
‡Dry Ice trap.
§Bottoms contained 0.1% saturates + aromatics, 0.3% resins, 60.6% asphaltenes, and 38.9% benzene insolubles.

(597–614 K). The concentration then increased to a high of 2.3 wt% for the residue cut. The phenol and cresol content of the first distillate cut was 1.14%, while those of the second, third, and fourth cuts were 9.11, 14.4, and 10.0%, respectively. The molecular weight of the phenolic compounds increased as the boiling point of the cut increased. Specifically, phenol and *o*-cresol made up most of the yield in the first cut, cresols in the second cut, ethylphenols and xylenols in the third cut, and high-molecular weight compounds in the fourth cut.

The level of nitrogen in the distillate cuts boiling up to 683 K varied non-systematically between 0.16 and 0.83 wt%. This variation was presumably due to isolation of various nitrogen containing compounds by distillation. There was an increase in the nitrogen content up to 1.1% in the distillate cuts boiling above 683 K. The distillation residue contained 2.0 wt% nitrogen, which was marginally above that of the feed coal, considering moisture, ash, and some oxygen removal. Effective extraction of the basic nitrogen compounds could be accomplished for the cuts boiling up to 623 K. The lower-boiling cuts in the naphtha range contained recoverable levels of aniline and toluidines with limited amounts of alkyl pyridines. Quinolines, indoles, and alkyl anilines were observed in the extracted samples of cuts boiling about 475 K.

The nitrogen compounds of distillate cuts boiling above 623 K could not be effectively removed by acid extraction. They can, however, be effectively removed as "resins," using liquid chromatography.* These resins can be classified as polar compounds (presumably containing nitrogen or oxygen) which are retained on ion-exchange resins or iron chloride adsorbent. As noted in Table 3-2, the distillate cuts between 614 and 684 K contained about 12 wt% resins. This level rose to 21 wt% for the 684–763 K cut and further to 35 wt% for the 763–814 K cut.

Results were also obtained for the liquefaction of Big Horn and Pittsburgh Seam coals under medium- and high-severity conditions as illustrated in Table 3-4. A summary of stream properties and elemental analyses is shown in Table 3-5. These results show that the light ends have lower sulfur, oxygen, and nitrogen than the filtrate for both coals. Furthermore, for both coals, high severity gave lower nitrogen and somewhat lower oxygen concentration in the light ends. The sulfur concentration in liquid products for both high- and mid-range severity runs were just about the same for both coals.

Table 3-4

Conditions for Midrange and High Severity for Big Horn and Pittsburgh Seam Coals

Coal:	Big Horn subbituminous		Pitt Seam bituminous	
Run No.:	1	2	3	4
Severity designation:	High	Midrange	High	Midrange
Temperature, K:	713	711	700	700
WHSV, hr^{-1}:	0.91	1.83	1.01	1.73
Pressure, MPa:	24.1	24.1	24.1	24.1
(Recycle solvent)/coal:	1.5	1.5	1.5	1.5

*Separations into saturates, aromatics, and resins.

Table 3-5

Characterization of Liquid Products of the High- and Moderate-Severity Runs

Coal:	Pittsburgh Seam bituminous				Big Horn subbituminous			
Severity:	High		Midrange		High		Midrange	
STREAM DESIGNATION:*	Light ends	Filtrate	Light ends	Filtrate	Light ends	Filtrate	Light ends	Filtrate
Yield: kg/(100 kg coal)	5.5	66.6	5.2	65.5	10.4	34.0	8.1	40.0
10^5 m^3/kg	6.259	54.770	4.695	51.640	12.519	31.297	9.389	32.862
PHYSICAL PROPERTIES:								
Density								
Gravity, API	32.8	1.4	25.2	0.8	31.9	9.7	34.0	4.8
Sp. gr. (289 K)	0.861	1.065	0.903	1.083	0.866	1.002	0.855	1.038
Kin. viscosity, m^2/sec 311 K	1.1×10^{-6}	146×10^{-6}	–	200×10^{-6}	1.2×10^{-6}	7.1×10^{-6}	–	12.5×10^{-6}
373 K	–	8.1×10^{-6}	–	122×10^{-6}	–	1.8×10^{-6}	–	2.4×10^{-6}
Heat of combustion,								
kJ/kg	44.100	40.077	43.008	40.449	43.892	41.752	43.450	40.635
kJ/m^3	37855	42371	38775	43709	37939	41786	37075	42120
Chemical analysis, wt%:								
Hydrogen	11.78	8.21	10.90	7.78	11.69	9.03	11.69	8.32
Nitrogen	0.22	0.82	0.22	0.97	0.19	0.37	0.18	0.66
Oxygen	0.52	1.47	1.41	1.47	0.77	0.88	1.70	1.80
Sulfur	0.04	0.17	0.04	0.11	0.04	0.04	0.04	0.04
Ash	–	0.01	–	0.02	–	0.01	–	0.01

*These yields do not include the figure for Dry Ice trap oil.

A similar comparison of typical yields in liquefaction of Illinois #6 and Big Horn coals indicates that under reasonably mild conditions, the coal liquids derived from Big Horn coal contains lower sulfur and nitrogen than the one obtained from Illinois #6 coal. There appears to be only a marginal effect of space velocity on the levels of desulfurization and denitrogenation of the Illinois coal. The product nitrogen concentration for Big Horn liquefaction increases significantly with WHSV.

3.3 Nature of Oxygen, Sulfur, and Nitrogen Compounds

The compounds containing oxygen are separated in the distillate or in the resin and asphaltene fractions with the exception of ethers. These are included in the neutral oil fraction. The oxygen appearing in the "oils" therefore represents the minimum amount of ether oxygen present in coal liquids, since this type of oxygen function could also be present in other fractions.

The oxygen present in the asphaltenes is multifunctional. The resins are produced predominantly by the hydrogenolysis of coal and are therefore closely related to the oxygen compounds in the coal. If only C-C bonds were broken during solvation, resins would truly be representative of the original coal.

Table 3-6 shows the oxygen balance in the products from liquefaction of a Wyoming subbituminous coal in anthracene oil at 699 K and 24.13 MPa for two hours.* Structures such as

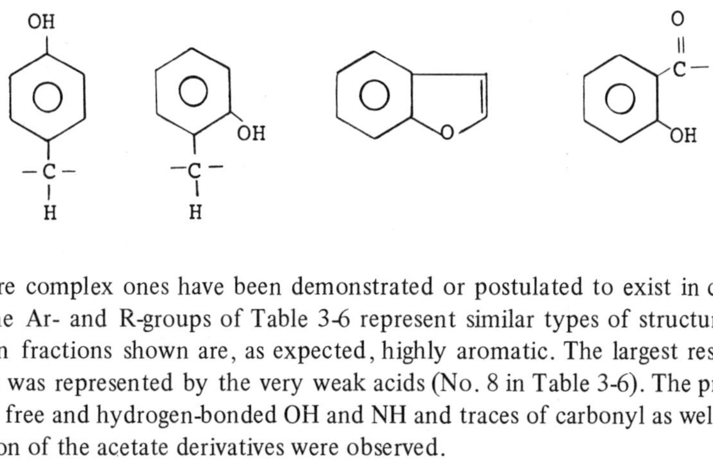

and more complex ones have been demonstrated or postulated to exist in coal [6, 13]. The Ar- and R-groups of Table 3-6 represent similar types of structures. All the resin fractions shown are, as expected, highly aromatic. The largest resin subfraction was represented by the very weak acids (No. 8 in Table 3-6). The presence of both free and hydrogen-bonded OH and NH and traces of carbonyl as well as the formation of the acetate derivatives were observed.

Another interesting fraction was the weak acids (No. 9, Table 3-6), which contained both free and bonded OH and carboxylic acids. The methylester derivatives were obtained when this material was reacted with *N, N*-dimethylformamide dimethyl acetal, $(CH_3)_2$ $NCH(OCH_2)_2$; these derivatives contained only the ester and carbonyl bonds and free as well as bonded OH from secondary functional groups. This indicates that not all the OH in the original fraction was associated with the carboxylic acid functionality.

*Personal communication with Gulf Research and Development Company.

Table 3-6

Oxygen-Compound Types in Coal Liquids (Wyoming Coal in Anthracene Oil)

No.	Fraction	Wt% of fraction	% total oxygen	Possible oxygen function
1	Distillate	13.30	16	R – Ar – OH
2	Oils	69.08	40	R – O – R
3	Asphaltenes	1.93	3	Multifunctional
4	Resins, total	15.68	41	Multifunctional
5	Very weak bases	2.94	6	HO – Ar – NH
6	Weak bases	1.36	4	HO – Ar – NH
7	Strong bases	3.37	5	HO – Ar – N
8	Very weak acids	5.04	13	Ar – OH
9	Weak acids	1.31	6	Ar – COOH + Ar – OH
10	Other acids + bases }	1.62	7	
11	Neutral N-compounds }			

From this discussion and from the data in Table 3-6, it is evident that at most only half of the oxygen present in Wyoming coal liquids was free phenolic oxygen. In fact, almost half of the oxygen was present as ethers, and all the carboxylic acids were probably esters in the original coal. Hydrogenolysis produced hydroxyl but not ether functions, implying that an even smaller quantity of phenolic groups were in the original coal. The largest unknown and indeterminate parameter of solid coals was oxygen incorporated in water of hydration, which is erroneously assumed to be "organic" in nature.

The oxygen compounds present in lower-boiling fractions (b.p. < 643 K) were hydroxyaromatics such as phenols, indoles, and naphthols. The smallest ether structure detected was dibenzofuran. The only dioxy compounds detected were in trace amounts in the higher-boiling fractions. Yields of phenols and cresols for the Illinois bituminous coal were lower than for either the Big Horn subbituminous or Pittsburgh Seam bituminous coals. The yield of cresylic acids was comparable to the Big Horn subbituminous coal, but consisted largely of complex hydroxyl aromatics. These were unusually complicated mixtures and would probably not be suitable for applications utilizing cresylic acids from carbonization processes.

Dryden [14] indicated that most of the oxygen is present in coal in the form of hydroxylic (–OH), ketonic (C=O), and to a lesser extent heterocyclic or etheric (C–O–C) groups. However, products with significantly reduced molecular weight result only when etheric oxygens are eliminated. Elimination of an OH group requires two hydrogen atoms, but elimination of other groups requires four hydrogen atoms. The elimination of inorganically bound oxygen is not believed to occur under most hydrogenation conditions.

The most detailed analysis of chemical structures of oxygen, nitrogen, and sulfur compounds present in coal liquids is given by Aczel et al. [1]. The liquefaction products of three coals, i.e., Kentucky Homestead Mine, (Clearfield, PA), Manor No. 44, and Ireland Mine (WV) feed coals by the SYNTHOIL process were analyzed. The elemental analyses obtained on the clay-gel fractions and on the asphaltenes indicated that:

1. Heteroatoms are concentrated in the asphaltenes and polar fractions, with exception of sulfur in the feed-coal fractions. Most of this sulfur is elemental sulfur; the rest is in aromatic thiophenes.
2. Oxygen is the most abundant heteroatom.
3. Considerably less sulfur is present in the product fractions than in the feed fractions. The product fractions contained more nitrogen than the feed fractions. Possibly the nitrogen compounds in the feed coals are more polar and therefore less soluble in the benzene and cyclohexane used in solvent extractions.
4. Basic nitrogen comprises about one-third to one-half of the nitrogen in the product fractions, and less in the feed fractions. The reason might be analogous to that proposed above for total nitrogen.
5. The heteroatom content of the polar and asphaltene fractions is very similar. An analysis of hydrogen content indicates that the depolymerization of asphaltenes might precede hydrogenation of heteroatoms for the molecules found in the polar fractions. From 80 to 90% of the oxygen and nitrogen compounds were segregated in the polar fractions. Oxygen compounds in the aromatic fractions were mainly aromatic furans.

Minor amounts of components with more than one heteroatom per molecule are present in coal liquids. The most abundant of these components are the structures derived from attaching additional thiophenic and/or furanic groups to the most abundant aromatic monothiophenes and monofurans. These components thus include thiodibenzothiophenes, phenanthrenodithiophenes, phenanthreno- and pyrenodefurans, etc. Typical sulfur- and oxygen-containing compounds present in aromatic fractions of SYNTHOIL products are given by Aczel et al. [1].

The main sulfur compound types in the feed extracts are the $C_nH_{2n-10}S$, $C_nH_{2n-16}S$, and $C_nH_{2n-22}S$ series, corresponding to thiophenic rings attached to benzenes, naphthalenes, and phenanthrenes. Within the products, those with higher naphthenic and hydroaromatic contents contain relatively less thiophenes. The more exotic sulfur compounds, containing two sulfur atoms per molecule or an SO group, are found mostly in the feed extracts. There was little change in the relative amounts of oxygenated compounds among feeds and products, or among the products themselves. The major types are those corresponding to the addition of a furanic group to fully atomatic rings, such as dibenzofurans and phenanthrenes ($C_nH_{2n-16}O$ and $C_nH_{2n-22}O$ series). Aromatic fractions also contained small or trace amounts of N compounds, mainly carbazoles.

3.3.1 *Composition of Polar Fractions*

Polar aromatic components comprise a significant portion of both feed and coal extracts. One of the salient characteristics of the polar fractions is the high oxygen and nitrogen contents with respect to the aromatic and saturate fractions. Most of the components identified in the polar fractions by mass spectrometry are indeed aromatic heterocompounds, such as hydroxyaromatics and aromatic pyrroles and pyridines. The presence of OH, NH, and other functional groups was also indicated from the infrared spectra of these fractions. The high heteroatom content increases considerably the complexity of the polar fractions. Most compounds detected have at least one oxygen, sulfur, or nitrogen atom, and many contain elemental groupings such as O_2, O_3, NO, NO_2, SO, SO_2, NS, and NSO.

The main compound types in polar fractions are one- to six-ring condensed hydroxyaromatics, dihydroxyaromatics, pyridines, and pyrroles. Sulfur-containing compounds are present in minor amounts, mostly in feed extracts. The number of homologs per series ranges up to 20 or 25; the most abundant components are those with three or four carbon atoms in side chains. Structure assignment is considerably more difficult for the polar components than for the aromatics. In assigning these structures, it is assumed that most oxygenated compounds in the polar fractions from the products are hydroxy-substituted ones. Phenolic structures are dominant in the liquid products. The presence of other oxygen types, such as hydroxyfurans, ethers, and minor amount of furans, and of ketonic structures such as fluroenone and hydroxyfluorenone, is also possible.

The nitrogen types are assumed to be mostly one- to eight-ring pyrroles and pyridines. Most aromatic pyrroles and pyridines have isomeric structures. Simple structures such as pyridines, indoles, quinolines, carbazoles, acridines, and benzocarbazoles are believed to be present. The mass spectroscopy of feed extracts indicate a predominance of $C_nH_{2n-15}N$ and $C_nH_{2n-21}N$ types, assumed to be mostly carbazoles and benzocarbazoles. Chemical analysis indicates higher relative amounts of nonbasic nitrogen for two of the feed extracts. Components with NO groups could be either hydroxypyrroles or hydroxypyridines, or amides. Aromatic mercaptans (thiophenols, thionapthols, etc.) are preferred for some of the sulfur types.

3.3.2 *Heteroatomic Elements and Functional Groups in Asphaltenes*

Oxygen in asphaltenes ranges from just under 1 to more than 3 per molecule, with the coal extracts having the higher numbers. Nitrogen ranges from about ½ to about 1, with little difference between feeds and corresponding products. In general, sulfur is quite low in all product samples by comparison with oxygen and nitrogen, ranging from 0.03 to 0.16 per molecule.

Infrared analysis can account for less than half of the total oxygen as phenols, carboxylic acids, and amides. This contrasts with over two-thirds of the total accounted for in the polars. Phenols remain the predominant oxygenated type. Phenols were found to be at about the same concentration in feeds and products. Less phenols were found in asphaltenes than in the polar fractions of the products, but the reverse was true for coal extracts. The amount of nitrogen as pyrroles ($\sim$ 20%) in product asphaltenes was comparable to that found in many of the polar fractions of the products. Pyrroles and amides in the coal-extract asphaltenes were less than those found in the corresponding polar fraction of the feed coal. As with the polars, there were carboxylic acids, amides, and other carbonyl-containing types in the asphaltenes from coal extracts, but they seemed to be absent, or present only in trace concentrations, in the product asphaltenes. Chemical analysis showed that basic nitrogen types are about equal to or slightly higher than nonbasic types in the product asphaltenes, and just the reverse in the coal extracts. The asphaltenes may contain a greater proportion of furans than the polars and therefore a lower proportion of phenols.

3.4 Related Model-Compound Studies

Benjamin and coworkers [4] studied a large number of model compounds in an effort to determine the kinds of bonds that rupture during coal liquefaction. Rapid thermal breakdown of tri- and tetra-substituted complex alkyl-aryl structures was found in this study. They also found that the bond cleavage is enhanced in the presence of vitrinite.

3.4.1 *Carbon-Carbon Linkages*

The ease of cracking of the –C–C– bond in dibenzyl makes it an important reaction with relevance to coal liquefaction. Horrex and Miles [19] found the pyrolysis of dibenzyl to proceed with an activation energy of 201 kJ/mol in the temperature range 903–1063 K and a reaction pressure in the range of 2.67–40 MPa. Cronauer et al. [8] carried out a detailed kinetic study of dibenzyl in the presence of various donor solvents. They found the scission of the –C–C– bond leading to the formation of benzyl radicals to be thermal and rate controlling. The above reaction was found to be independent of the nature of the donor solvent. The product distribution was, however, found to depend on the solvent type. Assuming first-order irreversible kinetics, they calculated an activation energy of 201 kJ/mol (Figure 3-8). Sato et al. [25] also found the rate of cracking of dibenzyl to be independent of the nature of the solvent (Table 3-7). They reported an activation energy of 251 kJ/mol for this reaction in the presence of tetralin. Whitehurst et al. [30] found the benzyl radicals formed during the decomposition of dibenzyl to extract hydrogen from the parent compound (dibenzyl) even in the

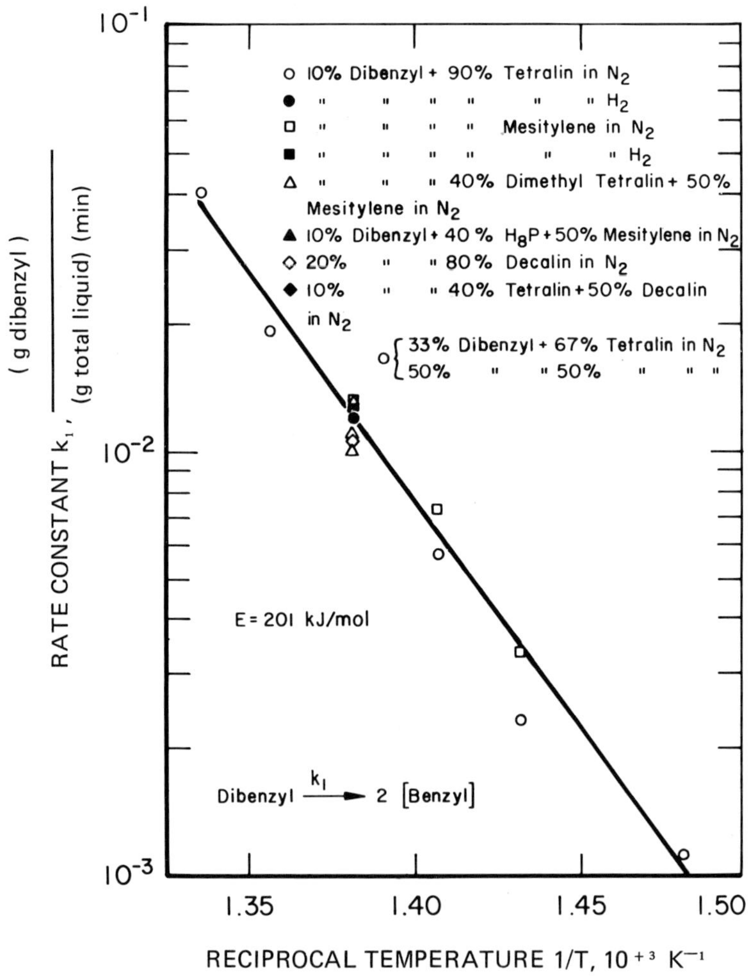

Figure 3-8. Arrhenius plot for the conversion of dibenzyl (Cronauer et al. [8]).

Table 3-7

Effect of Solvent on the Gas-Phase Decomposition of Dibenzyl at 697 K for 90 min*

Solvent	Solvent/reactant mole ratio	Conversion	Yield of by-products, † mol %
Tetralin	3	32.7	2.9
Methylcyclohexane	3	34.4	13.7
	30	36.5	3.3
Indan	3	32.2	12.6
	10	30.9	9.9
Cyclohexane	3	37.3	14.4
	30	28.2	4.3
Cis-1,2-Dimethyl-cyclohexane	3	39.2	18.9
	30	35.7	2.9

*Sato et al. [25].
†9, 10-Dihydrophenanthrene, trans-stilbene, and phenanthrene.

presence of a good donor solvent. Accordingly they proposed the following mechanism for this reaction:

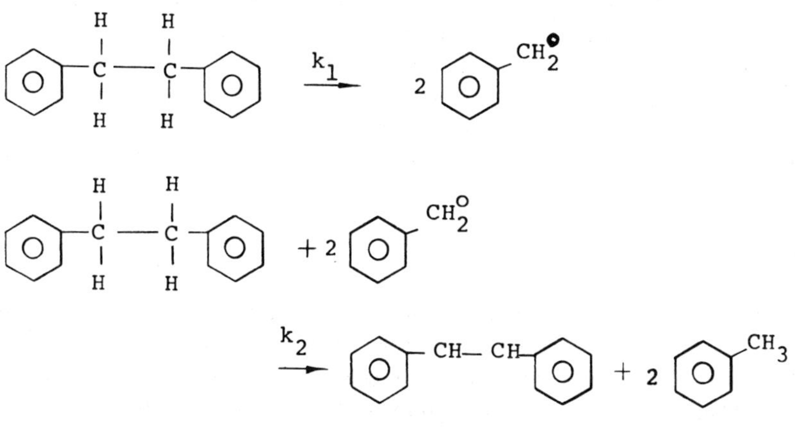

Cronauer et al. [10] also found the reduction of stilbene to occur only in the presence of tetralin. Linkages longer than two carbons such as 1-phenyl hexane,

1,4-diphenyl butane, and diphenyl methane were found to be less reactive than dibenzyl. This was attributed to the presence of only one activating phenyl group. Ethyl benzene, toluene, *n*-butane, and methane were formed from the cracking of 1-phenyl hexane. To explain these results the authors proposed the following mechanism:

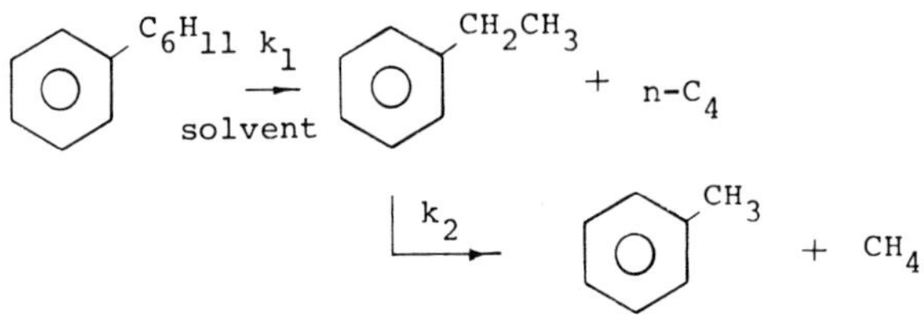

Sato et al. [25] found the thermal decomposition of 1,2-di-1-naphthyl ethane to follow first-order kinetics in the presence of tetralin. An activation energy of 236 kJ/mol was reported. Whitehurst et al. [30] found condensed systems, such as anthracene, phenanthrene, and chrysene, as well as diphenyl, acenaphthene, and acenaphthalene, to be stable at 700 K. However, substitution was found to enhance bond cleavage.

Ruberto and Cronauer [24] reported the reduction of carboxylic, carbonylic, and etheric functional groups in coal during hydrogenative liquefaction. From an examination of the liquids from a subbituminous Wyoming coal, Shah and Cronauer [26] reported the oxygen to be present predominantly as phenols, ethers, and carboxylic acids.

Benjamin et al. [4] examined the reactions of several oxygenated compounds in tetralin for 18 hours at 673 K. They observed that with increasing complexity of the aromatic structure the aromatic nature of the products increases. In several cases the formation of intractable residues was observed even in the presence of tetralin. Cronauer et al. [10] examined the reaction kinetics of a variety of oxygen compounds in donor solvents and found the relative reactivities of the various groups to be in the following order: furans $<$ phenols $<$ aldehydes $<$ chain ethers $<$ thioethers.

3.4.2 *Ethers*

The study of etheric linkages is necessitated by the fact that such bonds are readily cleaved under liquefaction conditions. Brucker and Kolling [7] and Reifarth [22] found toluene and benzyl alcohol to be stable end products in the decomposition of dibenzyl ether in tetrahydroquinoline. Cronauer et al. [10] found the major products for the same reaction to be benzaldehyde and toluene with traces of benzyl alcohol. This was also found to be true in the presence of tetralin. On this basis the above authors proposed the following mechanism for dibenzyl ether:

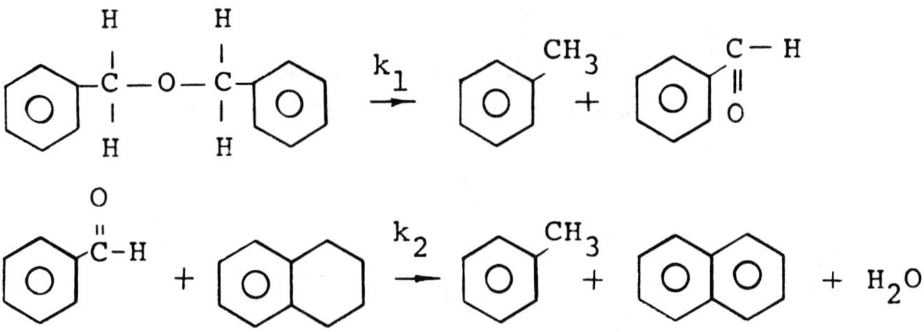

Based on first-order irreversible kinetics with respect to dibenzyl ether for the first reaction and benzaldehyde and tetralin for the second reaction, the activation energies for the above two reactions were found to be 152 and 134 kJ/mol respectively (see Figure 3-9). Kamiya et al. [20] examined the thermal decomposition of several ethers using different solvents. Results from their study on dibenzyl ether at 673 K using tetralin showed the formation of benzaldehyde and toluene with benzaldehyde exhibiting the behavior of a reaction intermediate. From Table 3-8 it is seen that linkages between aliphatic carbon and oxygen, such as those found in dibenzyl ether, are cleaved at much lower temperatures than linkages between aromatic carbon and oxygen such as those present in diphenyl ether. It is also seen that the decomposition of polynuclear aromatic ethers increases with increasing number of rings. The effect of phenols on the thermal decomposition of ethers is shown in Tables 3-9 and 3-10. There is clearly an enhancement of the conversion of dinapthyl and benzyl phenyl ether with increasing complexity of the phenol structure. This effect, which was not observed in the case of dibenzyl, is attributed to the solvation of the transition radicals and subsequent hydrogen transfer to these alkoxy radicals. The rates of decomposition of certain ethers were also found to be affected by the presence of mineral matter from coal as shown in Table 3-11. The decomposition of dibenzyl ether was found to be enhanced in the presence of coal ash, with benzyl tetralin as a dominant product.

Sato et al. [25] also found the decomposition of benzyl phenyl ether to be enhanced by the presence of phenol. Results from their study are shown in Table 3-12. Assuming first-order kinetics, they reported an activation energy of 218 kJ/mol for the above reaction in tetralin. Ross and Blessing [23] found diphenyl ether to be decomposed at 673 K to a mixture of polymethyl phenol in the presence of a mixture of MeOH/KOH. Whitehurst et al. [30] proposed the following mechanism for the decomposition of benzyl phenyl ether:

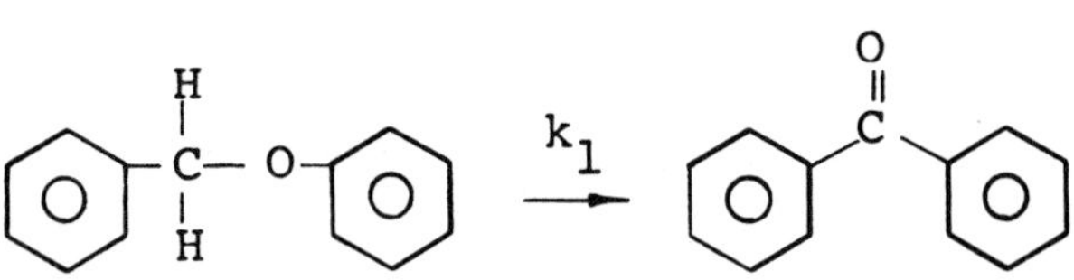

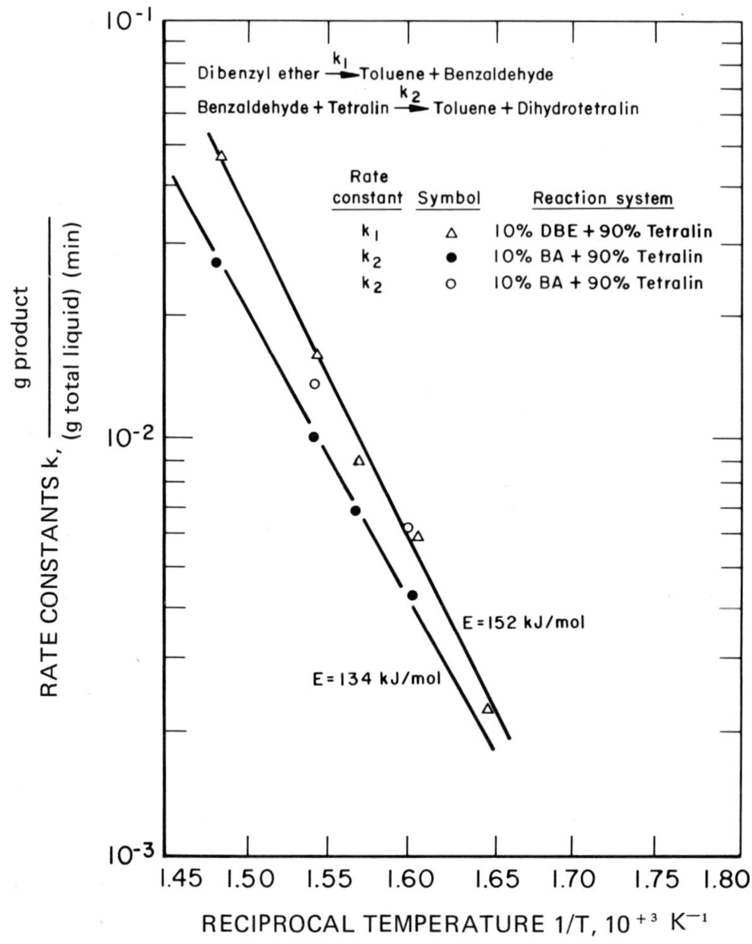

Figure 3-9. Arrhenius plots for the conversion of dibenzyl ether and benzaldehyde (Cronauer et al. [10]).

The formation of benzophenone was observed only under mild conditions (672 K). Compounds such as anisole, benzodioxole, and benzofuran were unreactive at 700 K in the above studies. 2,3-dihydrobenzofuran was, however, found to be very active and was postulated to decompose by the following mechanism:

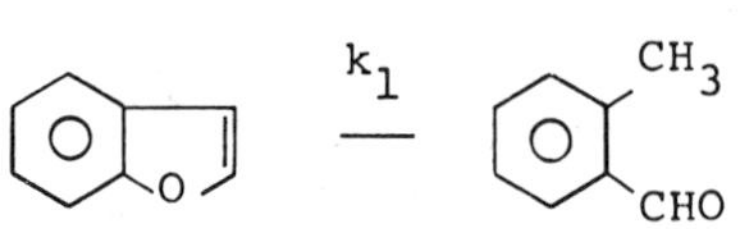

Table 3-8

Thermal Decomposition of Diaryl Ethers in Tetralin*

Diaryl ether	Temp. K	Time, min	Conversion, %	Products, mol% of reacted ether
Diphenyl ether	723	30	0	–
	723	120	2	–
Dibenzofuran	723	30	3.3	–
Phenyl 1-naphthyl ether	723	120	25	PhOH 66
2,2′-Dinaphthyl ether	723	60	12.5	2-Naphthol 84
	723	120	23.3	2-Naphthol 63
Phenyl 9-phenanthryl ether	723	120	45.5	PhOH 50, Phenanthrene 15
Phenyl benzyl ether	593	30	31.4	$PhCH_3$ 55, PhOH 50, Benzylphenol 40
	673	30	100	$PhCH_3$ 61, PhOH 66, Benzylphenol 27
Dibenzyl ether	673	30	65	$PhCH_3$ 106, PhCH 73, PhH 12

*Kamiya et al. [20].

Table 3-9

Effect of Phenols on the Thermal Decomposition of 2,2′-Dinaphthyl Ether at 723 K for 60 min*

Additive	Conversion, % Dinaphthyl ether	Phenol
1-Methylnaphthalene	12.6	–
Phenol	17.1	1.5
p-Cresol	21.0	0
2, 4, 6-Trimethylphenol	26.0	3.5
1-Naphthol	33.7	22.1
p-Phenylphenol	34.1	0
p-Methoxyphenol	49.5	100
Hydroquinone	63.5	57.7

*Kamiya et al. [20].

Table 3-10

Effect of Phenols on the Thermal Decomposition of Phenyl Benzyl Ether at 593 K for 30 min*

Additive	Conversion, %	
	Benzyl phenyl ether	Phenols
1-Methylnaphthalene	31.4	–
p-Cresol	40.2	0
p-Methoxyphenol	44.1	1.1
1-Naphthol	43.0	0

*Kamiya et al. [20].

The aldehyde thus formed can condense with a phenol, which is consistent with the observations of the above authors during the liquefaction of coal.

3.4.3 *Ketones*

Ketones can be very reactive, can consume hydrogen, and can also react with donor-solvent molecules. Therefore, it is essential to delineate its chemistry with particular reference to liquefaction conditions. Cronauer et al. [10] showed the presence of an alcohol intermediate in the decomposition of acetophenone. The following mechanism was proposed:

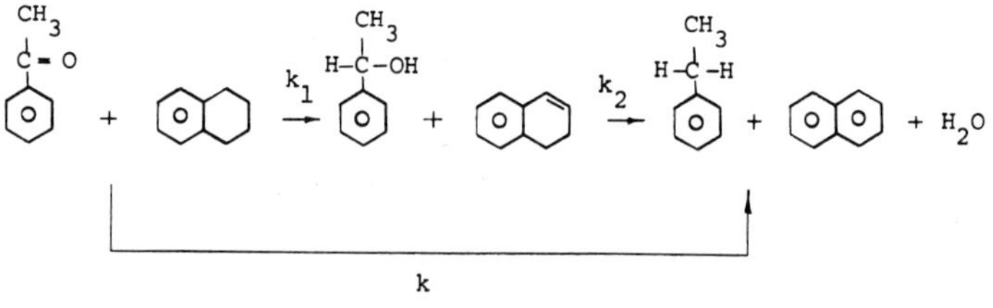

The formation of the alcohol was rapid and that of dehydration slow below 673 K. The conversion of acetophenone in tetralin was found to be first order and irreversible with an activation energy of 140 kJ/mol. Whitehurst et al. [30] found benzophenone to be unreactive in benzene. In the presence of *p*-cresol, however, reaction did proceed with the formation of diphenyl methane. It is speculated that the hydrogen for the reaction may have come through alkylation reactions with *p*-cresol resulting in the formation of heavier products. Sato et al. [25] examined the thermal reactions of benzyl phenyl ketone in the presence of tetralin and tetralin-phenol mixtures. The decomposition of ketones were also found to be en-

Table 3-11

Effect of Coal Ash on the Thermal Decomposition of Diaryl Ethers*

Ethers	Added ash, g	Temp., K	Time, min	Conversion, %	Products, mol% of reacted ether
Dibenzyl ether	—	673	30	65	$PhCH_3$ 110, PhCHO 77, PhH 13 Benzyltetralin 2.2
	0.015	673	30	100	$PhCH_3$ 6, PhH trace, Benzyltetralin 152
Phenyl benzyl ether	—	673	30	100	$PhCH_3$ 61, PhOH 66, $PhCH_2Ph$ 6, Benzylphenol 27, Benzyltetralin 2
	0.015	673	30	100	$PhCH_3$ 38, PhOH 58, $PhCH_2Ph$ 3, Benzylphenol 39, Benzyltetralin 13
Phenyl 9-phenanthryl ether	—	723	120	46	PhOH 50, Phenanthrene 17
	0.05	723	120	55	PhOH 45, Phenanthrene 16

*Kamiya et al. [20].

hanced by the presence of phenols. Assuming first-order irreversible kinetics, they reported an activation energy of 245.72 kJ/mol for the above reaction.

Cronauer et al. [10] also examined the reactivity of α-tetralone in tetralin, 2,6-dimethyl tetralin, and mesitylene. The reduction of the ketone to naphthalene was found to proceed even in the absence of a good donor solvent. Based on first-order irreversible kinetics, an activation energy of 104 kJ/mol was reported for the decomposition of α-tetralone.

Table 3-12

Effect of Solvent on the Liquid-Phase Decomposition Rate of Benzyl Phenyl Ether at 573 K for 90 min*

Solvent	Solvent/reactant mole ratio	Rate constant, sec^{-1}
Tetralin	25.1	8.058×10^{-6}
Trans-Decalin	15.8	6.164
Tetralin Cyclohexanol	13.2 17.2	8.063
Tetralin 9, 10-Dihydroanthracene	22.1 2.1	8.188
Tetralin Phenol	18.9 11.3	9.374
Tetralin Phenol	15.4 18.1	13.52

*Sato et al. [25].

3.4.4 *Aldehydes*

Aldehydes can be formed by the cleavage of ketones, ether rearrangement, or elimination of hydrogen from alcohols. It is therefore instructive to examine the thermal decomposition of some representative aldehydes under coal-liquefaction conditions. Benzaldehyde is a highly reactive aldehyde whose kinetics was discussed earlier in connection with the decomposition of dibenzyl ether. The conversion of benzaldehyde was found to be enhanced by the presence of *p*-cresol [30]. In the presence of poor solvents or at high aldehyde concentrations, benzaldehyde was found to react with itself with the formation of benzoic acid [10].

3.4.5 *Alcohols and Phenols*

Benzyl alcohol is known to be a highly reactive alcohol. It has also been reported to be formed during the thermal decomposition of dibenzyl ether in tetralin and tetrahydroquinoline. The products from the cracking of benzyl alcohol range from benzaldehyde, toluene, dibenzyl ether, dibenzyl, and stilbene to several solvents adducts [10]. Benzaldehyde was always formed as an intermediate in the reaction, with subsequent reaction to form toluene. Deuterium tagging studies indicated hydrogen exchange between the alcohol and the donor solvent. The overall reaction was therefore postulated as

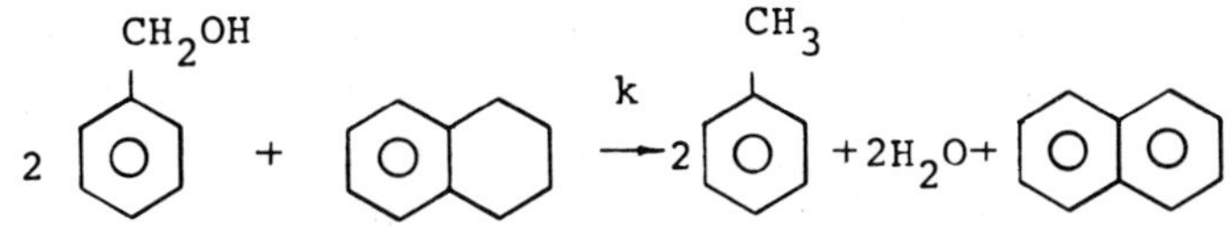

The activation energy for the above reaction, based on first-order irreversible kinetics, was found to be 110 kJ/mol. Side reactions include those that lead to the formation of benzaldehyde, dibenzyl, and dibenzyl ether.

In the same study α-naphthol was found to react with solvents, such as tetralin, 2,6-dimethyl tetralin, and tetrahydroquinoline, to form naphthalene. The disappearance of α-naphthol in tetralin under a nitrogen atmosphere was found to follow first-order kinetics with an activation energy of 142 kJ/mol. The compounds cyclohexanol, sorbitol, and dimethyl phenols were found to be essentially unreactive under coal-liquefaction conditions.

3.4.6 *Hydroaromatics*

Hydroaromatics, otherwise referred to as hydrogen-donor solvents, serve a useful purpose during liquefaction, in that they surrender hydrogen and revert to their fully aromatic state upon mild heating. However, they are also known to rearrange under the same conditions. This rearrangement, which appears to be initiated thermally as well as by free radicals, is detrimental to the role of such solvents as hydrogen donors. Another consequence of rearrangement is the formation of adducts wherein a solvent combines with fragments of the acceptor molecules.

Tetralin has been commonly employed as a hydrogen-donor solvent. Curran et al. [12] observed that tetralin decomposed to "C_4 benzenes and indan." Whitehurst et al. [30] pointed out that rearrangement of tetralin occurred with the formation of 1-methyl indan. This rearrangement, which was considered to be a free-radical process, was found to be solely dependent on temperature. Cronauer et al. [8] proposed the following mechanism for the decomposition of tetralin above 673 K:

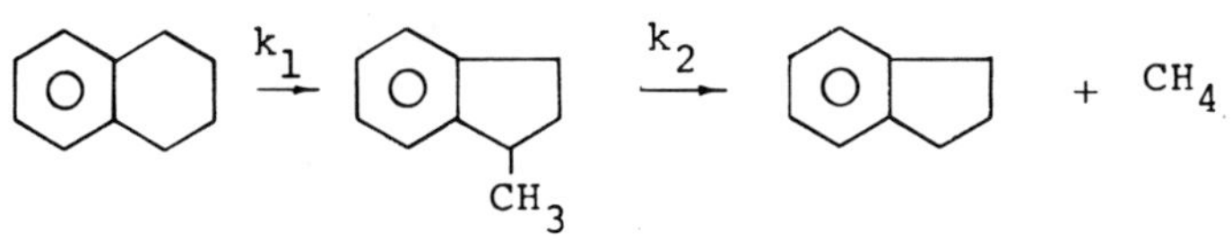

along with the expected reaction

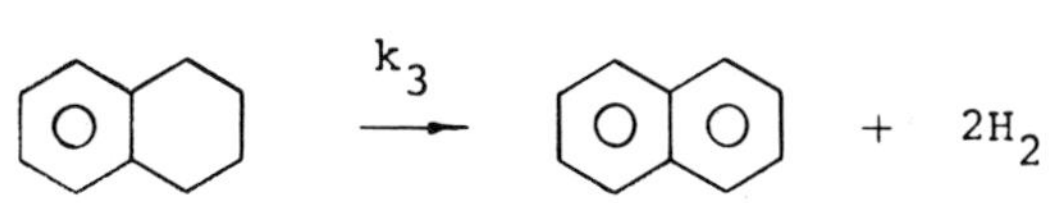

Figure 3-10 is an Arrhenius plot for the first of the above reactions based on first-order kinetics. It is seen from this figure that whereas the presence of phenanthrene alters the value of the rate constant to a mild degree, a significant change is caused by the addition of dibenzyl. Also the activation energy in the presence of dibenzyl is 42 kJ/mol lower, indicating a change in the reaction mechanism. Rearrangement reactions were also found to occur in the presence of alkylated tetralins, resulting in the formation of di- and trialkyl indans.

The dealkylation of methyl indan was found in the above study to follow zero-order kinetics at low levels of indan formation. The rate of this reaction was enhanced by the presence of dibenzyl. The activation energy, however, was not

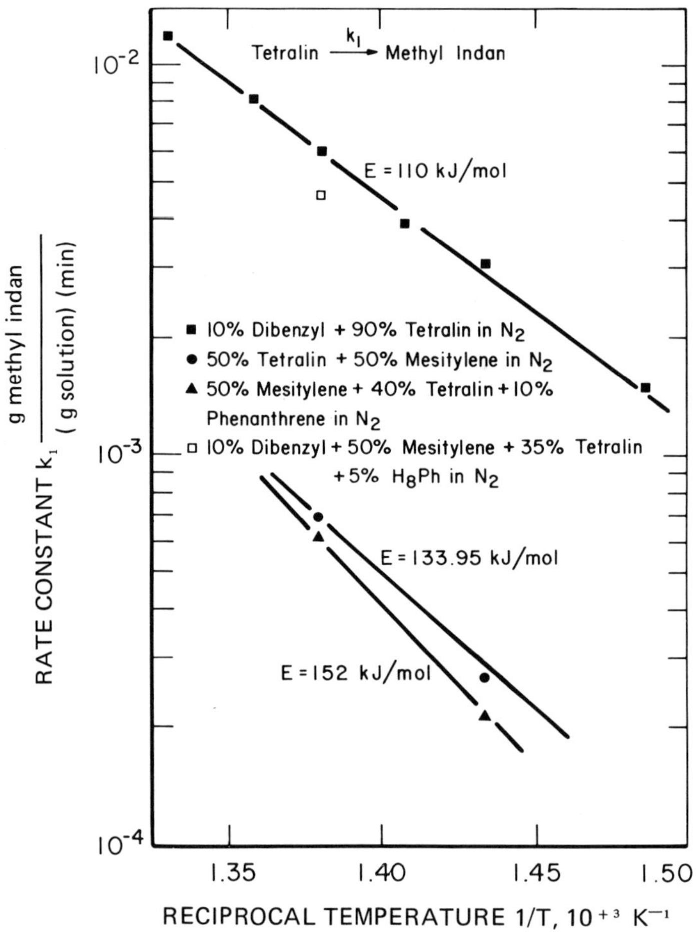

Figure 3-10. Arrhenius plot for the reaction of tetralin $\overset{k_1}{\rightarrow}$ methyl indan (Cronauer et al. [8]).

changed significantly. At high concentrations of dibenzyl (>25%), as well as at 723 K with 10% dibenzyl and 90% tetralin, in the presence of hydrogen, the reaction was found to deviate from first-order kinetics. In the latter case an enhancement in the rate of reaction was observed. This is attributed to the presence of molecular hydrogen. The dehydrogenation of tetralin to form naphthalene has been found to follow zero-order kinetics, and the reaction was only mildly enhanced by the presence of dibenzyl. The presence of phenanthrene appeared to hinder the reaction.

Cronauer et al. [9] discussed the rearrangement of *sym*-octahydrophenanthrene (H_8Ph) in great detail. At 723 K over 60% of the parent compound appeared to have isomerized at the end of 3 hours. The overall isomerization and dehydrogenation reactions were modeled by the following scheme

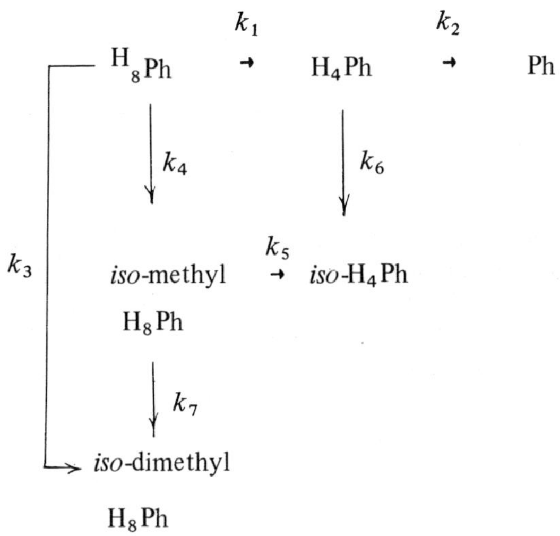

where

H_8Ph = octahydrophenanthrene,

H_4Ph = tetrahydrophenanthrene,

Ph = phenanthrene.

The rate constants for the various steps at 723 K based on first-order irreversible kinetics are given in Table 3-13. These results indicate that whereas the initial dehydrogenation to H_4Ph is rapid, subsequent dehydrogenation and isomerization reactions are competitive. Thus, it is seen that despite the fact that H_8Ph is an excellent donor solvent and possesses many desirable properties, rearrangement reactions, such as those described above, cause it to lose its potency. Such rearrangement reactions above may very well prove to be significant in polynuclear hydroaromatics.

Table 3-13

First-Order Rate Constants for the Various Steps in the network for H_8Ph*

	min^{-1}
k_1	0.0059
k_2	0.003
k_3	0.0017
k_4	0.0029
k_5	0.0001
k_6	0.002
k_7	0.0035

*Cronauer et al. [9].

Another type of secondary reaction that occurs during hydrogen transfer reactions and that is undesirable is solvent adduction. This reaction, which has been examined by Cronauer et al. [9], is enhanced by the presence of oxygen and sulfur functional groups on the free radicals and is likely to occur even at low temperatures, especially at long reaction times. At high temperatures (>673 K) such reactions are likely to involve only poor donor solvents. A case in point is the mesitylene-dibenzyl system. In this case the following mechanism was postulated:

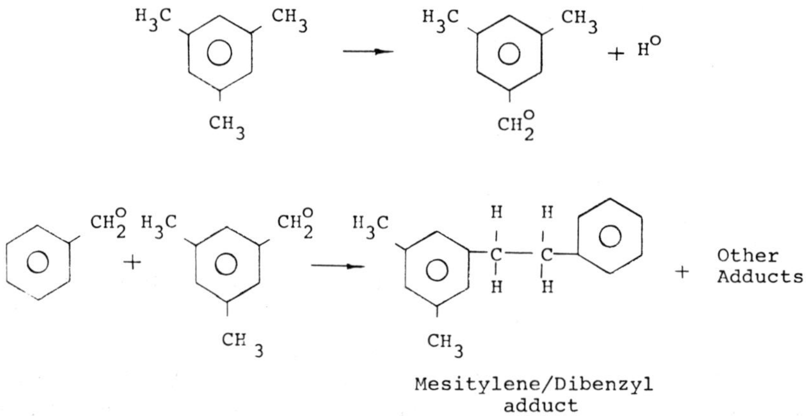

Reactions of the type shown above result in the formation of high-molecular-weight products. In the above case the presence of hydrogen was found to promote the cracking of mesitylene to form xylene radicals, thereby complicating the mixture of products. Adducts were not observed in the case of the dibenzyl-tetralin system. Thus adduction can be minimized by using sufficiently high concentrations of hydroaromatics with good hydrogen-donating capability. Good donor solvents can, however, become adducted to acceptor radicals which are generated from coal.

3.4.7 *Nitrogen and Sulfur Functionalities*

Cronauer et al. [10] examined the thermal decomposition of dibenzyl sulfide in mesitylene and tetralin at 573 K. In addition to the primary yield of toluene and H_2S, significant amounts of both stilbene and dibenzyl were observed. The yields of the latter compounds were found to be greater in the presence of a poor solvent. At low conversions the thermal decomposition of dibenzyl sulfide was found to follow first-order kinetics with an activation energy of 79 kJ/mol.

Whitehurst et al. [30] examined the reactivity of a few sulfur- and nitrogen-containing compounds in a synthetic solvent. The authors found benzylphenyl sulfide to be 99% converted both in a synthetic solvent and in pyridine. Compounds such as 2-phenylindole and 5,6-benzoquinoline were found to be fairly stable. Table 3-14 lists some of the substituted aromatics and oxygen-, sulfur-, and nitrogen-containing aromatic compounds that Benjamin et al. [4] found to be stable in tetralin at 673 K for 18 hours. A few nitrogen and sulfur compounds that were found to be reactive by the above authors are shown below along with probable reaction schemes:

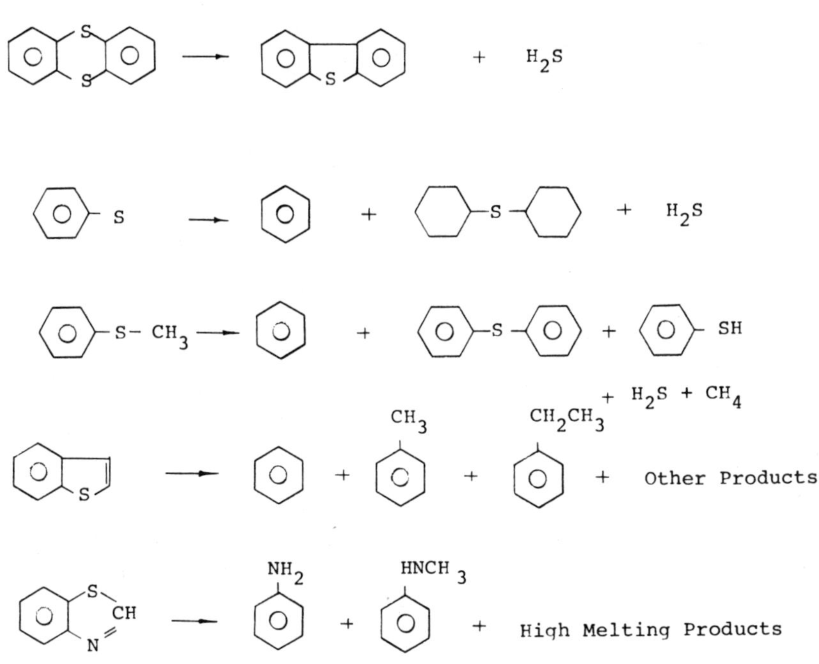

Table 3-14

Compounds Stable in Tetralin at 673 K for 18 hr*

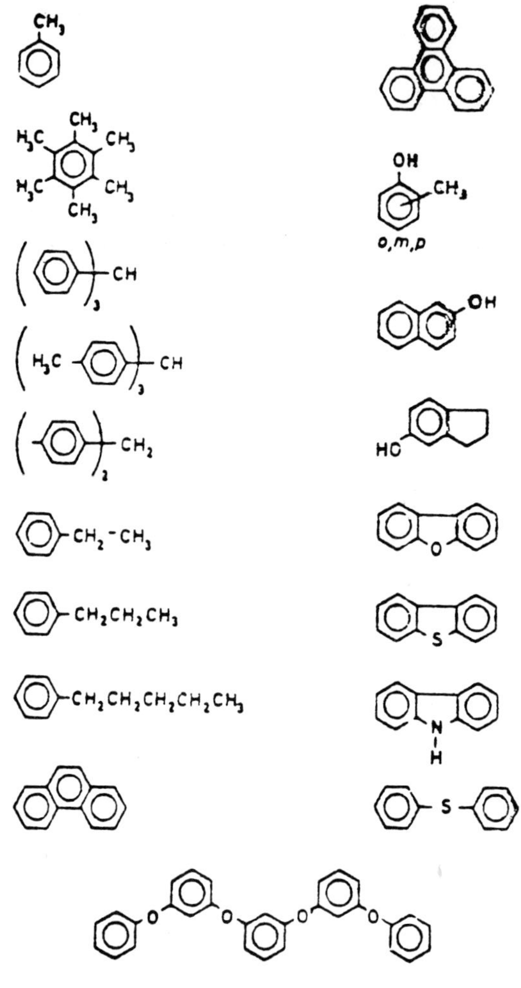

*Benjamin et al. [4].

3.4.8 *Implications for Coal Liquefaction*

1. The rate-controlling step in the decomposition of dibenzyl-type carbon linkages is the formation of the benzyl free radical. This reaction is thermal in nature and is independent of the hydrogen-donor solvent. The product distribution, however, depends upon the nature of the solvent. The benzyl free radical has been found to extract hydrogen even from the unreacted dibenzyl.

2. Linkages longer than two carbons cleave predominantly beta to the aromatic ring to generate benzyl radicals, which then react as above.

3. The presence of sulfur or oxygen heteroatoms in a linkage increases the rate of cleavage over carbon-carbon linkages.

4. Hydrogen-transfer cracking of linkages between aliphatic carbon and oxygen, such as those found in dibenzyl ether, can occur at a much lower temperature than aromatic –C–O– linkages such as those found in diphenyl ether. The cleavage of etheric linkages are enhanced by the presence of phenols.

5. The decomposition of aldehydes and ketones is also enhanced by the presence of phenols.

6. In the presence of poor donor solvents or at high concentrations, aldehydes such as benzaldehyde react with themselves, forming benzoic acid.

7. Phenols are fairly stable up to 723 K.

8. The order of group reactivity was found to be furans $<$ phenols $<$ ketones $<$ aldehydes $<$ aliphatic ethers $<$ thioethers.

9. Hydroaromatics rearrange to form methyl-indan compounds, which result in a solvent of reduced hydrogen-donor capacity. The major rearrangement appears to be that of isomerization of six- to five-membered rings rather than simple ring opening.

10. The presence of free radicals enhances solvent degradation by greatly promoting both the above rearrangement reaction and subsequent dealkylation.

11. Good donor solvents can form adducts with coal free radicals at moderate temperatures ($<$ 723 K). These become increasingly stable due to ring rearrangement at high temperatures ($>$723 K).

12. Poor donor solvents can form stable adducts with coal free radicals at all temperatures.

13. Condensed aromatics such as dibenzofuran, dibenzothiophene, carbazole, and phenanthrene are stable in hydrogen-transfer solvents at temperatures up to 748 K.

REFERENCES

1. Aczel, T., Williams, R. B., Pancirov, R. J., and Karchmer, J. H., "Chemical Properties of Synthoil Products and Feeds," Analytical Project Report BRD. 1BA. 76, Exxon Research and Engineering Co., Baytown, TX 77520, prepared for ERDA Contract No. E 46-1-8007, September 1976.
2. Alcorn, W. R., Elliott, G. E., and Cullo, L. A., "Batch Screening of Coal Liquefaction Catalysts with a Falling Basket Reactor," *ACS Div. of Fuel Chemistry,* **23** (1), 11, preprints of papers presented at Anaheim CA (March 12–17, 1978).
3. Batchelder, R. F. and Fu, Y. C., "Evaluation of Use of Syn Gas for Coal Liquefaction," *ACS Div. of Fuel Chemistry,* **23** (1), 30, preprints of papers presented at Anaheim, CA (March 12–17, 1978).

4. Benjamin, B. M., Raaen, V. F., Maupin, P. H., Brown, L. L., and Collins, C. J., "Thermal Cleavage of Chemical Bonds in Selected Coal-Related Structures," *Fuel,* **57**, 269 (1978).
5. Bertolacini, R. J., Gutberlet, L. C., Kim, D. K., and Robinson, K. K., "Relation of Coal Liquefaction Catalyst Properties to Performance," *ACS Div. of Fuel Chem.,* **23** (1), 1, Anaheim, CA (March 12–17, 1978).
6. Brigges, G. C. and Lawson, G. J., *Fuel,* **49**, 39, London (1970).
7. Brucker, R. and Kolling, G., *Bren. Chemie,* **46**, 41 (1965).
8. Cronauer, D. C., Jewell, D. M., Shah, Y. T., and Kueser, K., *I&EC Fundamentals,* **17** (4), 291 (1978).
9. Cronauer, D. C., Jewell, D. M., Shah, Y. T., Modi, R. J., and Seshadri, K. S., "Isomerization and Adduction of Hydrogen Donor Solvents under Conditions of Coal Liquefaction." *I&EC Fundamentals,* **18**, 368 (1979).
10. Cronauer, D. C., Jewell, D. M., Shah, Y. T., and Modi, R. J., "Mechanism and Kinetics of Selected Hydrogen Transfer Reactions Typical of Coal Liquefaction," *I&EC Fundamentals,* **18**, 153 (1979).
11. Curran, G. P., Struck, R. T., and Gorin, E., "The Mechanisms of the Hydrogen Transfer Process to Coal and Coal Extract," preprint, *Div. Fuel Chem.,* **10** (2), C-130 (1966).
12. Curran, G. P., Struck, R. T., and Gorin, E., *I&EC Proc. Res. & Dev.,* **6**, 166 (1967).
13. Davies, L. and Lawson, G. J., *Fuel,* **46**, 95, London (1967).
14. Dryden, I. G. C., in *Chemical Constitution of Reactions of Coal,* Suppl. Vol., H. H. Lowry, Ed., Wiley, New York, 1963.
15. Flinn, R. A., Larson, O. A., and Beuther, H., "The Kinetics of Hydrodenitrogenation," *ACS Preprints Div. of Petroleum Chemistry* **7** (1), 163–172 (1962).
16. Granoff, B., Thomas, M. G., Baca, P. M., and Noltes, G. T., "Effects of Mineral Matter on the Hydroliquefaction of Coal," *ACS Div. of Fuel Chemistry* **23** (1), 23, preprints of papers presented at Anaheim, CA (March 12–17, 1978).
17. Hamrin, C. E., Jr., "Catalytic Activity of Coal Mineral Matter," Interim Report for the period of April through September 1976, FE-2233-2, February 1977.
18. Hill, G. R., "Experimental Energies and Entropies of Activation – Their Significance in Reaction Mechanism and Rate Prediction for Bituminous Coal Dissolution," *Fuel,* **45**, 329 (1966).
19. Horrex, C. and Miles, S. F., *Discussions Faraday Soc.,* **10**, 187 (1951).
20. Kamiya, Y., Tao, T., and Oikawa, S., *ACS Div. of Fuel Chemistry Preprints,* 116 (1979).
21. Neavel, R. C., *Fuel,* **55**, 237 (1976).
22. Reifarth, K., "Reaction of Tetralin and Dibenzyl Ether," Ph.D. Thesis, University of Berlin 1977.
23. Ross, D. S., and Blessing, J. E., *ACS Div. of Fuel Chemistry Preprints,* **24** (2), 125 (1979).
24. Ruberto, R. G. and Cronauer, D. C., *Organic Chemistry of Coal,* ACS Symposium Series 71, American Chemical Society, Washington, D. C., 1978.
25. Sato, Y., Yamakawa, T., and Kameyama, H., paper presented at the joint ACS-CSJ Chemical Congress, Honolulu, HI, April 2–6, 1979.
26. Shah, Y. T. and Cronauer, D. C., "Oxygen, Nitrogen and Sulfur Removal Reactions in Donor Solvent Coal Liquefaction," *Catalysis Reviews Science and Engineering,* **20** (2), 209–307 (1979).

27. Shah, Y. T., Cronauer, D. C., McIlvried, H. G., and Paraskos, J. A., "Kinetics of Catalytic Liquefaction of Big Horn Coal in a Segmented Bed Reactor," *I&EC Process Design Dev.*, **17** (3), 288 (1978).

28. Tarrer, A. R., Guin, J. A., Pitts, W. S., Henley, J. P., Prather, J. W., and Styles, G. A., *Preprints of Div. of Fuel Chemistry, ACS,* **21** (5), 59 (1976).

29. Whitehurst, D. D., Farcasiu, M., and Mitchell, T. O., Annual Reports submitted by Mobil Research & Development Corp., Princeton, NJ, to EPRI, under Research Project 410-1 (EPRI AF-248), July 1977.

30. Whitehurst, D. D., Farcasiu, M., Mitchell, T. O., and Dickert, J. J., Jr., "The Nature and Origin of Asphaltenes in Processes Coal," EPRI Report AF-480, Second Annual Report under Project RP-410, July 1977.

4. Kinetic Models for Donor-Solvent Coal Liquefaction

Y.T. Shah, S. Krishnamurthy, and R.G. Ruberto

INTRODUCTION

It is now generally accepted that the donor-solvent liquefaction of coal proceeds through a sequence of steps that commence with the scission of weaker bonds within the coal structure, resulting in the formation of very reactive fragments. The initial stabilization of these reactive fragments occurs through hydrogen transfer from the hydroaromatic portions of the coal as well as donor solvents. The intermediates thus formed react further through additional bond rupture and hydrogen transfer to form molecules with a wide spectrum of molecular weights.

In modeling the kinetics of coal liquefaction it must be borne in mind that the process of liquefaction depends not only on operating parameters such as pressure, temperature, and nature of solvent, but also upon the nature and rank of the feed coal. The dissolution of bituminous coking coal in anthracene oil has been observed to occur at temperatures of 613–623 K, where the organic structure disintegrated almost entirely [34]. On the other hand a low-volatile bituminous coal could not be dissolved in the same solvent at similar temperatures. The sensitivity of the extraction yield to the rank of the coal is shown in Table 4-1, where the yields from subbituminous coal and lignite were substantially lower than from bituminous coal, while graphite and anthracite were essentially insoluble [35]. Thus, models developed for a given type of coal do not possess global applicability and must therefore be used with caution.

In attempting to model the kinetics of coal liquefaction on a limited scale one is confronted with the problems of establishing the reaction networks and subsequent estimation of a set of optimum rate constants associated with the networks. A brief review of the relevant models that have appeared in the literature has been

Yatish T. Shah, Reaction Engineering in Direct Coal Liquefaction, ISBN 0-201-07381-1

Table 4-1 [35]

Extraction of Various-Rank Coals with Phenanthrene at 613 K

Substrate	Extract yield, % maf coal	Benzene-soluble yield, % maf coal
Graphite	Trace	0
Anthracite	1	0
Bituminous coal	95	19
Subbituminous coal	27	15
Lignite	23	17

presented by Lee [26]. This chapter attempts to provide a critical review of the various reported kinetic models for both short- and long-residence-time reactors.

4.1 Laboratory Reactors for Rate Measurements

The rate measurements at large residence times can generally be carried out in a continuous or a batch-stirred tank reactor. However, in much of the early extraction work [13, 28, 29, 34, 45] it was apparent that the initial stage of extraction proceeded at a very high rate at temperatures above 573 K. Because the thermal inertia of the heavy-walled autoclaves used in some of this work required long heat-up and cooldown times, it was impossible to study the significant dissolution reactions occurring during heatup or at very short residence time.

In more recent works [10, 11, 14, 20, 22, 33, 50, 52, 54, 56] various investigators have devised reactors that overcome the thermal-inertia problem and allow measurements to be performed at residence times as short as 0.5 min. Table 4-2 summarizes the characteristics of these types of reactors. The microautoclave or tubing-bomb type is made from stainless-steel tubing and is simply plunged into a preheated sand bath with vertical agitation applied in one dimension. In the rapid-injection autoclave, a portion or all of the solvent is preheated in a standard autoclave to the reaction temperature. A separate cylinder above the autoclave containing the coal and remaining solvent is then discharged into the agitated solvent in the autoclave. The thermal recovery time of the autoclave to the desired temperature constitutes the sample heatup time and is usually of the order of a minute or two. Variations on the rapid-injection concept have been used where dry coal is injected using some of the hydrogen charge. The rapid-injection autoclave is more cumbersome and time consuming to use than the tubing-bomb reactor but has two distinct advantages: the capability of maintaining a constant pressure during reaction by adding makeup hydrogen, and the ability to measure gaseous-hydrogen consumption and gaseous products.

Another novel reactor has been developed, based on a differential conversion and recycle principle. In one study [22], approximately 25 g of coal and 225 to

Table 4-2
Short-Residence-Time Reactors for Studying Coal Dissolution*

Reactor type	Reactor dimensions, Description	Volume, cm^3	Heatup/cooldown times	Charge
Microautoclave [10, 11] and fluidized sand bath	1.59 cm i.d. × 15.24 cm Vertical agitation 2300 cpm	30	2.5 min/30 sec	2 g coal, 4–8 cm^3 solvent
Tubing bomb [33] and fluidized sand bath	7.6 mm i.d. × 17.8 cm Vertical Agitation 40 cpm	8	28 sec/10 sec	1.5 g coal, 3 g solvent
Rapid-injection [50, 52] Autoclave	Standard stirred autoclave Agitation 1200 rpm	300	0.5–1 min/10 sec	Injector: 20 g coal, 40 g solvent Vessel: 60 g solvent
Same [50]	1.11 cm i.d. × 15.25 cm No agitation	16	Similar to above	Injector: 1 g coal or 1 g coal, 1 g solvent Vessel: 10^3 cm solvent
Same [56]	Standard stirred autoclave Agitation 1500 rpm	1000	∼ 2 min/ Sampled during tests	Injector: 50 g coal, 50 cm^3 solvent Vessel: 450 cm^3 solvent
Differential-recycle reactor [22]	Preheater: 1.5 mm i.d. × 6.0 m Reactor: 9.5 mm i.d. × 43.8 cm Quench: 1.5 mm i.d. × 6.1 m Reservoir	11.4 31.2 14.6 200	2.4 sec/3.4 sec	25 g coal, 225–350 g solvent

*Personal communication with J.A. Grey.

350 g of solvent were charged to a 200-cm^3 high-pressure autoclave which was used as a slurry reservoir at a temperature of 673 K and system pressure. Slurry was pumped through a horizontal helical-coil preheater where the heatup time to reaction temperature was approximately 2.4 sec. This was followed by a downflow tubular reactor with a nominal hold time of 5.7 sec. Finally, a concentric tube quench heat exchanger was used to lower the temperature to 323 K in approximately 3.4 sec. Table 4-2 lists the dimensions of the key vessels used in this study. Slurry was recycled through the system and back to the reservoir at a nominal rate of 14.2 liters/hr. Only a differential amount of reaction occurred per pass, and the reservoir was considered perfectly mixed so that the residence time was computed as the run time multiplied by the mass ratio of reactor holdup to system holdup less reactor holdup. About 79 sec was required for a complete volume turnover. One of the major advantages of this system is that the products can be sampled at various times during the reaction, thus allowing several measurements to be made during a single experiment.

4.2 Rate Models

A summary of the relevant kinetic studies reported in the literature is shown in Table 4-3. Oele et al. [34] proposed one of the earliest models for the liquefaction of bituminous Dutch coals using anthracene oil and β-naphthol as solvents. The rate data obtained in the temperature range of 523–623 K were correlated assuming the liquefaction to be a reversible process. The forward reaction was assumed to be zero order, while the reverse reaction was assumed to be first order with respect to the fraction of extracted material.

The rate equation on the above basis is given by

$$\frac{d\alpha}{dt} = k - k'\alpha. \tag{4-1}$$

where α is the fraction of maf coal extracted, k and k' the forward and reverse reaction rate constants, and t the time in minutes. An integration of Equation (4-1) yields

$$\ln\left(1 - \frac{k'\alpha}{k}\right) = -k't. \tag{4-2}$$

In the limit, $t \rightarrow \infty$. Assuming that $\alpha \rightarrow \alpha_{eq}$ results in

$$\frac{k'}{k} = \frac{1}{\alpha_{eq}}. \tag{4-3}$$

Activation energies in the range of 20–40 kcal/mole were obtained for the forward reaction using different coals.

Table 4-3

Kinetic Studies

Coal type	Solvent	Catalyst	Pressure MPa	Temp., K	Time, hr	Reactor type	Reference
Pittsburgh Seam bituminous from Bruceton mine	–	SnS	41.37	673–713	3	1-liter batch autoclave	Weller et al. [49]
UT Spring Canyon high volatile bituminous	Tetralin	–	34.47	623–723	25–30	1-liter batch autoclave	Hill et al. [20]
Pittsburgh Seam bituminous and shallow extract from Ireland Mine, northern WV	Tetralin & other donor solvents	–	NA	597–713	2	Micro-autoclave; 300-ml & 1-liter autoclave	Curran et al. [10]
UT high-volatile bituminous	Tetralin	–	NA	623–723	20	NA	Wiser [54]
Bituminous	Tetralin	–	16–20	653–713	5	Batch autoclave	Liebenberg & Potgieter [27]
KY No. 9 bituminous	Creosote oil, tetralin, decalin, paraffin oil	–	6.89	603–673	2	300-cm^3 batch autoclave	Guin et al. [14]
UT high-volatile B bituminous; KY high-volatile B bituminous	Tetralin	–	10.35	673	0.5	Differential-recycle reactor	Kang et al. [22]

Table 4-3 (Continued)
Kinetic Studies

Coal type	Solvent	Catalyst	Pressure MPa	Temp., K	Time, hr	Reactor type	Reference
High-volatile bituminous, low-volatile bituminous, subbituminous, & lignite	Tetralin, dodecane, naphthalene, hydrogenated creosote oil	–	13.79	673	13	Tubular reactor	Neavel [33]
West KY 9/14, IL #6–Burning Star, IL #6–Monterey bituminous; Wyodak–Belle Ayr subbituminous	2% 4-picoline, 17% *p*-cresol, 38% tetralin, 43% 2-methyl naphthalene	–	0–13.79	673–733	5–500	Batch autoclave	Whitehurst et al. [50, 51]
Subbituminous from Belle Ayr mine, Amax Coal Co., WY	Hydrogenated anthracene oil; hydrogenated phenanthrene & anthracene oil	–	13.79	673–743	$\sim$ 1	Laboratory-scale CSTR	Cronauer et al. [8]
IL #6		–	10–17	644–756	NA	Tubular-flow reactor	Brunson [5]
Western KY 9 & 14 seam, Colonial Mine, Pittsburgh & Midway Coal Co.	SRC recycle solvent & tetralin	–	2.07–12.41	686–727	5–60	2-liter batch autoclave	Langanbach et al. [25]

Table 4-3 (Continued)

Kinetic Studies

Coal type	Solvent	Catalyst	Pressure MPa	Temp., K	Time, hr	Reactor type	Reference
Japanese, Australian, & Indonesian	Unhydrogenated & hydrogenated solvents	Mixture of Mo oxides & Fe hydroxides	0.69	623–723	2	0.3- & 0.5-liter batch autoclave	Morita et al. [31, 32]
Roland seam Wyodak	Methanol & other solvents	$ZnCl_2$	5.51	498–548	1	0.6-liter batch autoclave	Shinn [41, 42]
High volatile bituminous from Fies Mine, KY	Tetralin	–	13.79	623–673	3	0.3-liter batch autoclave	Shalabi et al [40]
From Big Horn Mine, WY	Recycle solvent	Patented catalyst	24.13	686	2 (space time)	Segmented bed	Shah et al. [39]

Hill et al. [19, 20] proposed the following kinetic expression for the liquefaction of a high-volatile bituminous coal in tetralin:

$$\frac{dx}{dt} = k\,(1 - x), \tag{4-4}$$

where x is the fraction of coal liquefied and k the first-order rate constant. An interesting variation of their model from the model proposed by Oele et al. [34] is the dependence of the rate constant k on the fraction liquefied. This relation is given by

$$k = k_o\,(1 - ax), \tag{4-5}$$

where k_o is a pseudo second-order rate constant and a is the reciprocal of the maximum liquefiable fraction at a given temperature.

The fraction of bituminous coal liquified in tetralin determined by Soxhlet extraction using benzene, as a function of reaction time at various temperatures, is shown in Figure 4-1. With increasing temperature, up to a maximum of 723 K, the fraction liquified increases rapidly to an "equilibrium" value beyond which it re-

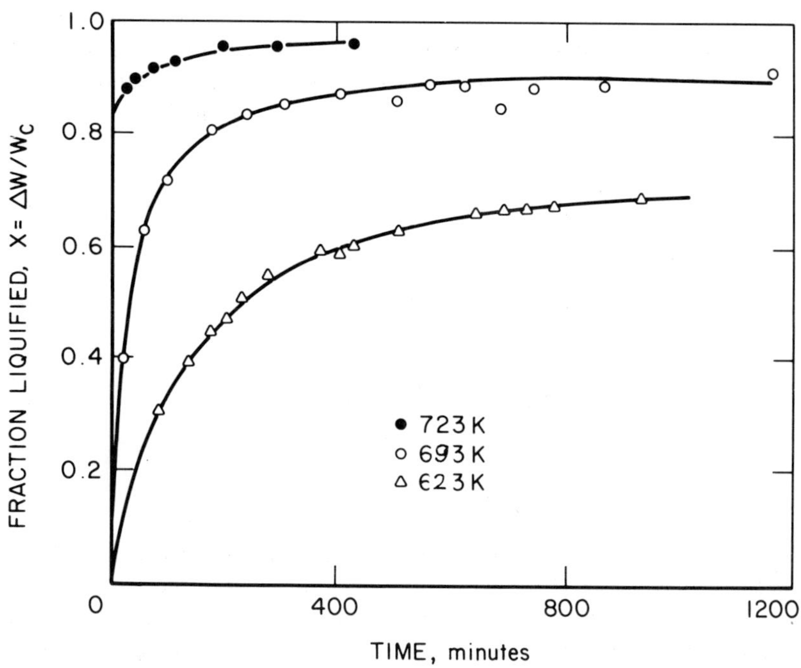

Figure 4-1. Fraction of bituminous coal liquefied in Tetralin at 623, 693, and 723 K (Hill et al. [20]).

mains constant. The time required to reach this "equilibrium" value is seen to decrease with increasing temperature, while the "equilibrium" value attained increases with increasing temperature. The finite "equilibrium" value does not mean that products of coal liquefaction react back to yield the feed coal. It does, however, represent the maximum amount of coal that is soluble and therefore characterizes the reactive portion of coal. Besides being a function of temperature (and possibly pressure), as described above, it also depends on the nature of the donor solvent. Cronauer et al. [8] have observed an increase in "equilibrium" value for a given coal with increasing hydrogen-donor capacity of the solvent used to liquefy the coal. The "equilibrium" value concept has found application in recent models for the donor-solvent liquefaction of coal (Cronauer et al. [8], Shalabi et al. [40]). In the temperature range of 623–723 K Hill et al. [19, 20] reported the activation energy for the reaction involving the breakdown and dissociation of the coal structure to be 133.9 kJ/mol.

Mechanistically speaking, Hill et al. [19, 20, 21] assumed high-volatile coals to consist of large, alkylated, polynuclear, oxygenated, aromatic and heterocyclic nuclei held together by oxygen, sulfur, and carbon-carbon bridges to form the basic porous structure of the coal matrix. At the beginning of the reaction the materials that enter the liquid phase are those that are trapped in the pores and may be weakly bonded to the main coal structure. These have the lowest activation energy for dissociation from the main coal matrix. The residue formed subsequently dissolves through the breaking of chemical bonds which have a higher activation energy for bond cleavage. This sequence can be continued, the activation energy for each step being higher than the preceding step, and can be formulated by the following series of first-order reactions:

$$\text{coal} \rightarrow X_o{}^* \xrightarrow{k_o} R_1 + L_1 + G_1, \tag{4-6a}$$

$$R_1 \rightarrow X_1{}^* \xrightarrow{k_1} R_1 + L_2 + G_2, \tag{4-6b}$$

$$R_2 \rightarrow X_2{}^* \xrightarrow{k_2} R_3 + L_3 + G_3, \tag{4-6c}$$

$$R_n \rightarrow X_n{}^* \xrightarrow{k_n} R_{n+1} + L_{n+1} + G_{n+1}, \tag{4-6d}$$

where X_i* is the *i*th activated complex, k_i is the *i*th rate constant, and R_i is the residue, L_i the liquid product, and G_i the gas product from the *i*th reaction. The authors emphasize that the R_i's are new products and not the unreacted coal, thereby indicating that the above reactions do not occur consecutively. Although the fractions R_i and L_i are not defined, it is surmised that they are comparable to asphaltene and oil fractions in coal liquids.

Another model along the lines of Hill et al. [20] has been proposed by Wiser and coworkers [54, 55]. Using solubility of reaction products in benzene as a basis for determining the conversion of a bituminous coal, they proposed the following kinetic expression:

$$\frac{dx}{dt} = k_2 \ (a - x) \ (b - x), \tag{4-7}$$

where a is the maximum weight fraction of coal that can be converted at the temperature of reaction, b is the initial weight fraction of tetralin ($b = 1.0$), and k_2 the second-order rate constant. The above equation was found to represent the early stages of donor-solvent liquefaction. At long times a first-order expression of the following form was found to be more appropriate:

$$\frac{dx}{dt} = k_1 \ (a - x). \tag{4-8}$$

This model, which represents the primary reaction step was found to correlate the data reasonably well up to a temperature of 723 K.

The kinetic model proposed by Curran et al. [10] considers the coal to be composed of two constituents of different reactivities as shown below:

$$C_T - C = \gamma C_T e^{-k_1 t} + (1-\gamma) \ C_T e^{-k_2 t}, \tag{4-9}$$

where C is the weight percentage of coal decomposed, C_T is the maximum weight percentage of the coal which can be decomposed, γ is the fraction of the coal which decomposes at the faster rate, and k_1 and k_2 are the first-order rate constants ($k_1 > k_2$), displaying the classical Arrhenius temperature dependence. The conversion of coal in different mixtures of hydroaromatics was calculated on the basis of the solubility of reaction products in xylenol at 267 K. A comparison of model predictions with experimental results is shown in Figure 4-2. The activation energies for the fast and slow reactions were reported to be 125.6 and 160.1 kJ/mol respectively. The conversion of coal (to pyridine-soluble materials) appears to increase with temperature up to 985 K. Also, the rapid approach to equilibrium coal conversion as well as a higher equilibrium conversion with increasing temperature is evident.

Curran et al. [10] have also noted, as shown in Figure 4-3, that the amount of hydrogen transferred and not the nature of the solvent appears to be a critical factor. However, examination of the composition of the solvents indicates that tetralin is the "hydrogen donor" in all cases. Therefore, more evidence is needed with different hydrogen-donor solvents before a definite conclusion about the effect of the nature of the solvent can be reached. From Figure 4-3 it is also seen that the first 50% conversion consumes less than 0.2% hydrogen by weight. This is not surprising in the light of Neavel's [33] findings and can be attributed to transfer of hydrogen from within the hydroaromatic portions of the coal.

In the presence of excess donor solvent the following mechanism of hydrogen transfer has been proposed by the above authors:

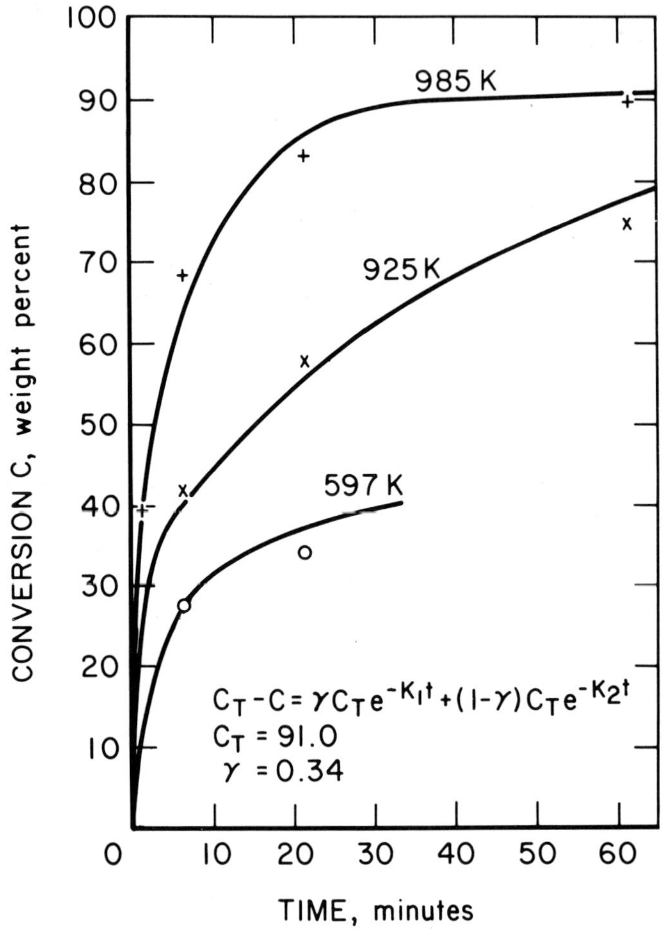

Figure 4-2. Conversion of bituminous coal as a function of reaction time at 596.8, 925, and 985 K (Curran et al. [10]).

$$M \xrightarrow{k_o} 2R\cdot \,, \tag{4-10a}$$

$$R + D \xrightarrow{k_1} RH + D\cdot \,, \tag{4-10b}$$

$$R + D\cdot \xrightarrow{k_2} RH + (D - H_2) \,, \tag{4-10c}$$

$$D\cdot \xrightarrow{k_3} (D - H_2) + H\cdot \,, \tag{4-10d}$$

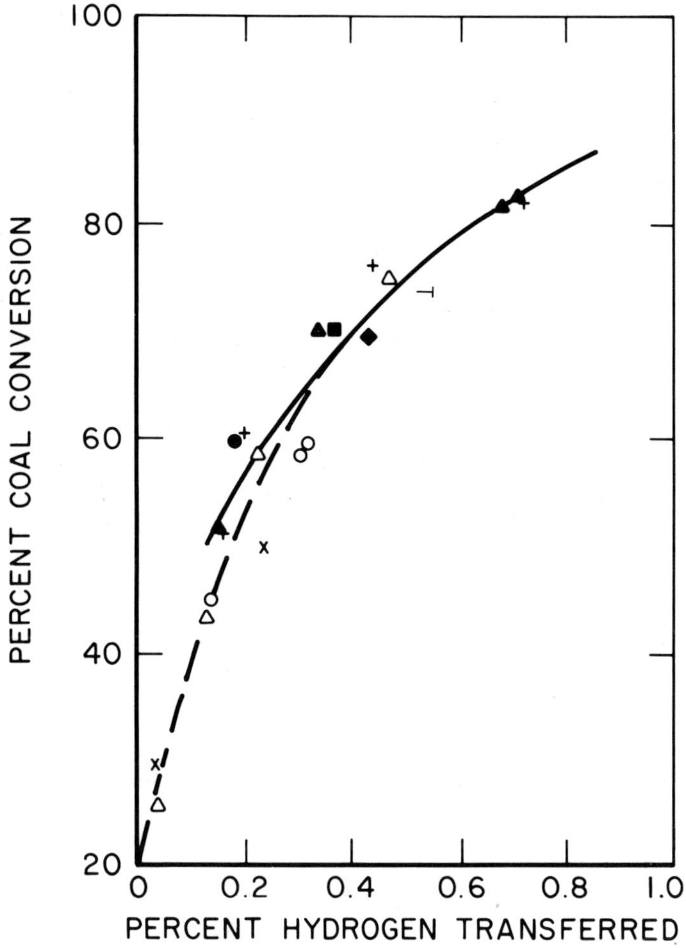

Figure 4-3. Conversion of bituminous coal in various solvents at 925 and 985 K as a function of % H_2 transferred (Curran et al. [10]). Key:

Temp., K	925 (– –)	985 (——)
Tetralin	△	▲
1:1 tetralin:decane	○	●
1:1 tetralin:decalin		⊣
1:3 tetralin:decalin	×	♦
3:1 tetralin:xylenol	+	■

$$H\cdot + D \xrightarrow{k_4} D\cdot + H_2 \,, \tag{4-10e}$$

where M is the feed coal which yields free radicals R· upon thermal dissociation, D· corresponds to a radical produced by abstraction of a hydrogen atom from tetralin, and ($D - H_2$) represents dihydronaphthalene. Although the kinetic model proposed by the above authors has the same drawbacks as that of Weller et al. [49], some of their findings given below are noteworthy. They postulated the initial reaction in coal liquefaction to be one of thermal decomposition. This reaction is assumed to produce a host of free radicals that are stabilized by capture of hydrogen atoms from donor molecules or from within the coal itself. Thus, if hydrogen donors of sufficient activity are present, stable molecules of lower molecular weight are formed.

The differential-recycle reactor described earlier was used to study the dissolution reactions of two different coals, Hiawatha hvBb Utah coal and Kentucky #14 hvBb coal, in tetralin at 673 K and a hydrogen pressure of 10.35 MPa [22]. Most of the conversion of Hiawatha coal to pyridine-soluble products occurred in less than 5 minutes, while the less reactive Kentucky coal required up to about 10 minutes for completion of the rapid-dissolution stage. Initially preasphaltenes (asphaltols) were rapidly formed and exhibited a maximum concentration within the period of rapid dissolution for each coal. Rate data were also obtained at 623 and 648 K for the Hiawatha coal; however, conversion was measured and correlated in terms of benzene solubility. As before, these results gave a conservative estimate of the actual rate of dissolution. The rate expression used to fit the data was

$$\frac{dX_1}{dt} = k_1 \, (X_m - X_1) + k_2 \, (X_m - X_1)^2 \, , \tag{4-11}$$

where the X's are defined in terms of benzene-soluble products. Integration of Equation (4-11) with the conditions $X_1 = 0$ at $t = 0$ and $X_1 = X_m$ at $t \to \infty$ gives

$$1 - \frac{X_1}{X_m} = \frac{k_1 e^{-k_1 t}}{k_1 + k_2 \, (1 - e^{-k_1 t})} \, . \tag{4-12}$$

The activation energies and frequency factors for the first- and second-order rate constants that best fitted the data were

$$k_1 = 4.575 \times 10^4 \ \text{sec}^{-1}, \qquad E_1 = 34.1 \ \text{kcal/mole},$$

$$k_2 = 2.064 \times 10^6 \ \text{sec}^{-1}, \qquad E_2 = 20.2 \ \text{kcal/mole}.$$

The models of Hill et al. [20], Curran et al. [10], Kang et al. [22], and Wiser [54] are similar in that they are all representations of only the primary reaction of coal dissolution in the presence of excess donor solvent. The information that is currently desired from lumped kinetic models, and that is noticeably lacking in the above efforts, is the effect of operating variables on the yields of various molecular-weight fractions from the primary as well as the secondary reactions.

Neavel [33] found that when a highly volatile C bituminous coal is rapidly heated to liquefaction conditions, 90% of the coal is converted to pyridine-solubles in the presence of tetralin. He also found that in the presence of solvents such as naphthalene and dodecane the initial conversion of the feed coal to benzene-solubles is high. However, at longer times, retrogressive reactions become predominant, transforming some of the benzene-solubles formed at short times into benzene-insolubles. On the basis of his results Neavel postulated the following mechanism for donor-solvent liquefaction. At very short times the coal structure is fragmented due to formation of lower-molecular-weight products from the scission of weak bonds. The fragmented particles are subsequently dispersed in the solvent vehicle. At this stage the quality of the donor solvent is not critical, and aromatics are as effective in transferring hydrogen as hydroaromatics. This has also been borne out by the results of Heredy and Fugassi [18], who examined the liquefaction of a bituminous coal using phenanthrene.

The above-described insensitivity of coal liquefaction to the quality of the solvent at short contact times is apparently due to transfer of hydrogen from the hydroaromatic portions of the coal to the locations of bond rupture, the solvent vehicle merely facilitating such transfer. With increasing residence time, the molecules formed initially by the thermal rupture of chemical bonds seek stabilization through reaction with the surrounding molecules. It is at this stage that the nature of the solvent vehicle becomes critical. A good donor solvent provides hydrogen which acts as a free-radical terminator. In the case of a poor hydrogen-donor solvent the free radicals tend to recombine among themselves, resulting in products that are insoluble in extracting solvents such as benzene or pyridine. Neavel [33] also examined the effect of coal rank upon conversion to benzene solubles and gases at 673 K, using a hydrogenated creosote oil as a solvent. From his results, shown in Figure 4-4, it is seen that the lowest-ranked coal has a higher conversion than that of the high-volatile C bituminous coal at all residence times. An interesting result that is noticeable in this figure is the almost identical conversion to benzene-solubles of the high-volatile C bituminous coal in tetralin (H/C atomic ratio 1.2) and hydrogenated creosote oil (H/C atomic ratio 1.0). Thus the H/C atomic ratio alone is not a true indicator of the hydrogen-donating capacity of the solvent, as only the hydrogen present in the naphthenic portions of the hydroaromatics is available to acceptors. Additionally, hydroaromatics can undergo rearrangement reactions, thereby rendering the hydrogen unavailable. The effect of donor solvent upon pyridine solubles is, however, not reported.

Although Neavel [33] did not attempt to provide a kinetic model, his study at short residence times has provided critical experimental evidence of the mechanisms that are operative during donor-solvent liquefaction of coal.

A significant contribution to the understanding of donor-solvent liquefaction, especially at short contact times, has been made by Whitehurst and coworkers [12, 50, 51, 52]. These investigators reacted several coals in a variety of solvent mixtures at a wide range of contact times both in the presence and absence of gaseous hydrogen. The products of reaction were separated by means of sequential elution by solvents chromatography (SESC) using silica gel.

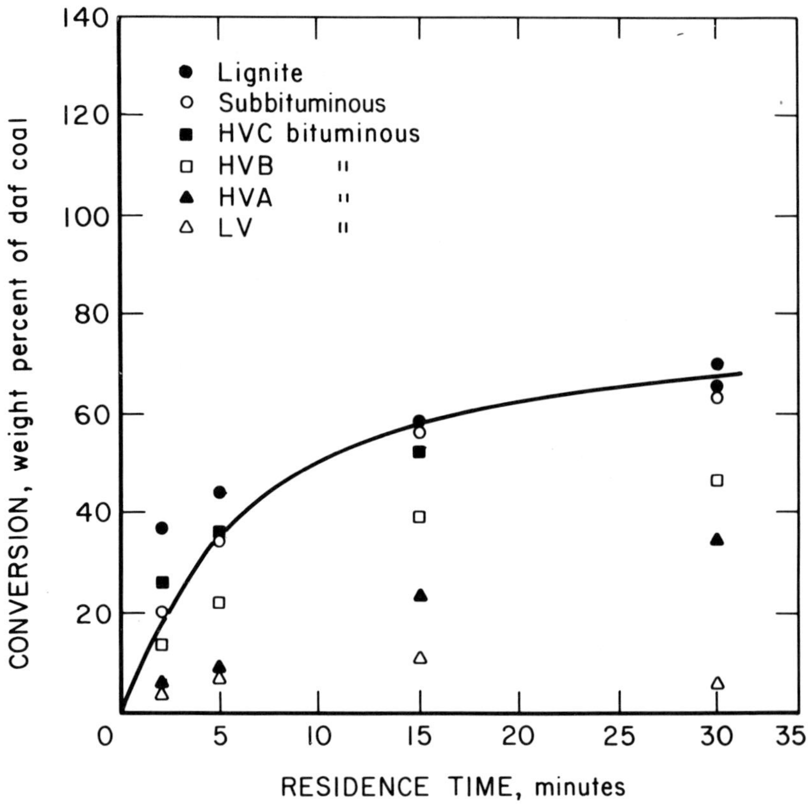

Figure 4-4. Conversion of coal to benzene-solubles and gases using hydrogenated creosote oil at 673 K (Neavel [33]).

The solvents used and the fractions separated are given in Table 4-4. The fraction that was left behind in the column after elution with pyridine was classified as unknown material which can be considered as unconverted coal. The above authors found the initial reaction products to be independent of the nature of the solvent or the presence of hydrogen. This is in agreement with Neavel's results [33]. Using ^{13}C-NMR spectroscopy, they found the initial reaction products to be more aromatic than the feed coal, together with a 40% removal of oxygen and sulfur. The reactions apparently are those of hydrodesulfurization and hydrodeoxygenation–the hydrogen for the reactions coming from the parent coal. At longer times they found the aromatic content of the solvent-refined coal (a product of coal liquefaction which is soluble in pyridine but boils over at approximately 755 K) to increase, as did also the number of polyaromatic rings. The percentage removal of sulfur and oxygen was found to level off. Thus, the hydrogen present in the gas phase during the reaction was apparently used to rejuvenate spent

Table 4-4

Solvents Used in SESC by Whitehurst et al. [50, 51, 12, 52]

	Solvent	Fraction eluted
(1)	Hexanes (mixed isomers)	Saturates
(2)	Hexanes with 15% benzene	Aromatics
(3)	Chloroform	Polar aromatics: nonbasic nitrogen, oxygen, and sulfur heterocyclics
(4)	Chloroform with 10% diethyl ether	Simple phenols
(5)	Diethylether with 3% ethanol	Basic nitrogen heterocyclics
(6)	Methanol (absolute)	Highly functional molecules, with > 10% heteroatoms
(7)	Chloroform with 3% ethanol	Polyphenols
(8)	Tetrahydrofuran with 3% ethanol	Increasing oxygen content and increasing basicity of nitrogen
(9)	Pyridine with 10% methanol	More of the above

solvent and enhance gas consumption. The above conclusion implies that any consumption of hydrogen for the removal of heteroatoms in this fraction far exceeds the supply of hydrogen through the donor solvent.

Some of the results reported by Whitehurst et al. [51] are shown in Figures 4-5 and 4-6. The initial conversions of the Kentucky, Illinois #6 Burning Star, and Illinois #6 Monterey bituminous coals at 700 K are higher than that of the subbituminous Wyodak coal. The "equilibrium" conversion in the latter case is also found to be lower than that of the bituminous coals. Han and Wen [17] have analyzed the data of Whitehurst et al. [50] as well as data from the preheater section of the Wilsonville pilot plant using Illinois #6 Monterey [6] and Indiana V coals [7]. Assuming that the initial thermal reaction to be represented as coal $\xrightarrow{k_1}$ asphaltenes they formulated the following first-order kinetic expression:

$$\ln\left(\frac{X_m - X_1}{X_m}\right) = -k_1 t\ , \tag{4-13}$$

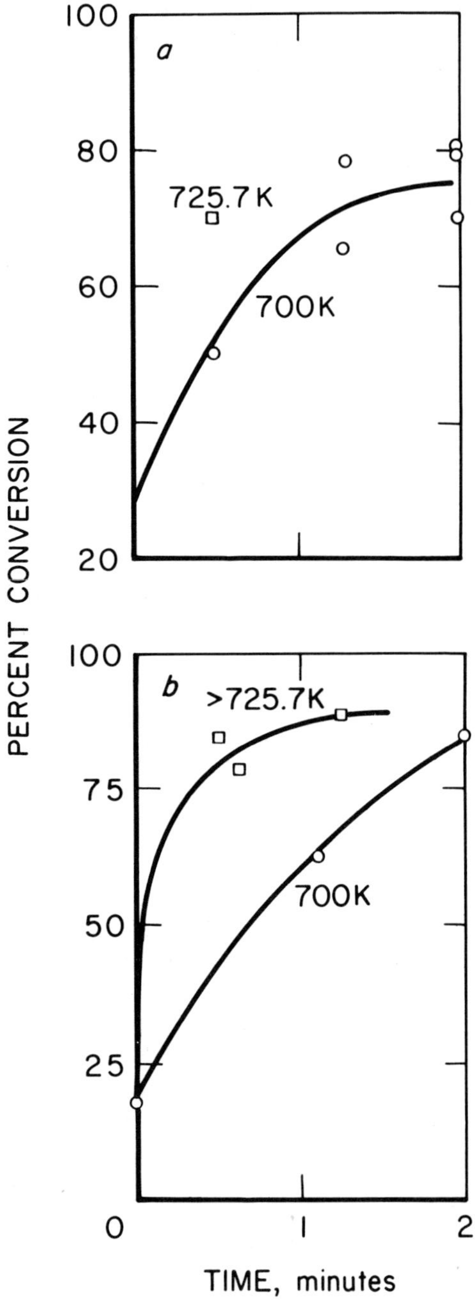

Figure 4-5. Conversion of Kentucky and Illinois #6 Burning Star coal as a function of time at 700 and 725.7 K (Whitehurst et al. [51]).

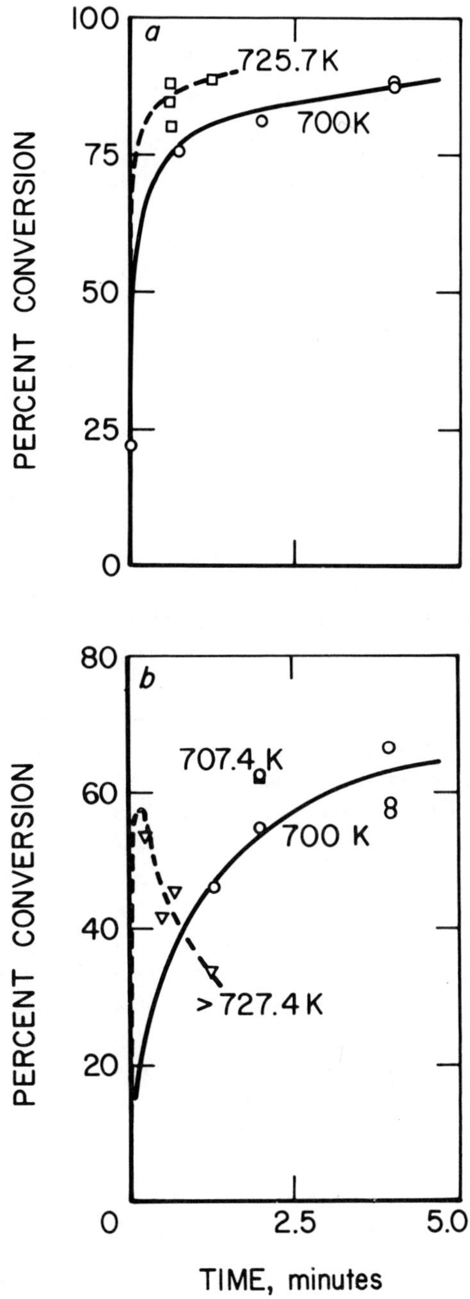

Figure 4-6. Conversion of Illinois #6 MT coal and Wyodak coal as a function of time at 700, 707.4, and >727.4 K (Whitehurst et al. [51]).

where X_1 = moisture- and ash-free conversion based on pyridine solubles,

X_m = maximum possible conversion,

k_1 = initial-stage first-order rate coefficient, hr^{-1},

t = reaction time, hr, for batch-reactor data; average residence time, hr, for pilot-plant data.

Average activation energies of 200 ± 20 kJ/mol and 230 ± 60 kJ/mol were reported for the batch-reactor and pilot-plant data respectively for the initial coal-dissolution step.

Whitehurst et al. [50] found the conversion of bituminous coals to increase with increasing temperatures up to 727 K and that of subbituminous coals to exhibit retrogressive behavior. After attaining a conversion of 58% in the first few seconds, the conversion of Wyodak coal decreased considerably with time. This behavior, which is shown in Figure 4-6, is attributed to secondary reactions occurring in the liquid phase that transform the initially converted coal into insoluble material.

There have been several other reports on the occurrence of retrogressive reactions under certain conditions of liquefaction. Guin et al. [14] noted that the quality of the donor solvent is an important factor in the liquefaction process. Kang et al. [23] have reported that with increasing reactor temperature, other conditions remaining the same, a deterioration in solvent quality is observed. They indicated that this could be due to an increase in the rate of thermal cracking over the rate of donor-solvent-aided hydrogenation. Langanbach et al. [25] have also reported polymerization of the initial reaction products above 713 K, during the liquefaction of a bituminous coal using the recycle solvent obtained from the solvent-refined coal process. Such reactions were found to be inhibited in the presence of tetralin and gaseous hydrogen. Mitchell et al. [30], Anderson and Kang [2], and Walker et al. [47] have observed the formation of a pseudocrystalline mesophase in liquefaction residues. They described this formation as anisotropic carbon which had grown on and bonded to undissolved solids, including mineral and fusinitic matter, resulting in intractable deposits. There is some evidence which indicates that large aromatic molecules are precursors of the mesophase.

In summary, the relevant features of the above studies which have given new directions in lumped modeling of donor solvent coal liquefaction kinetics are as follows (see Figure 4-7):

1. At very short reaction times, a wide range of reaction intermediates are formed. The formation of these intermediates is independent of both the nature of the solvent and the presence or absence of gaseous hydrogen, implying that these reactions are purely thermal. The only requirement on a solvent during this step may be that it should possess the ability to "shuttle" hydrogen from the hydrogen-rich to the hydrogen-depleted portion of the coal under circumstances in which direct interaction between hydrogen-donating and hydrogen-accepting coal fragments is not practicable.

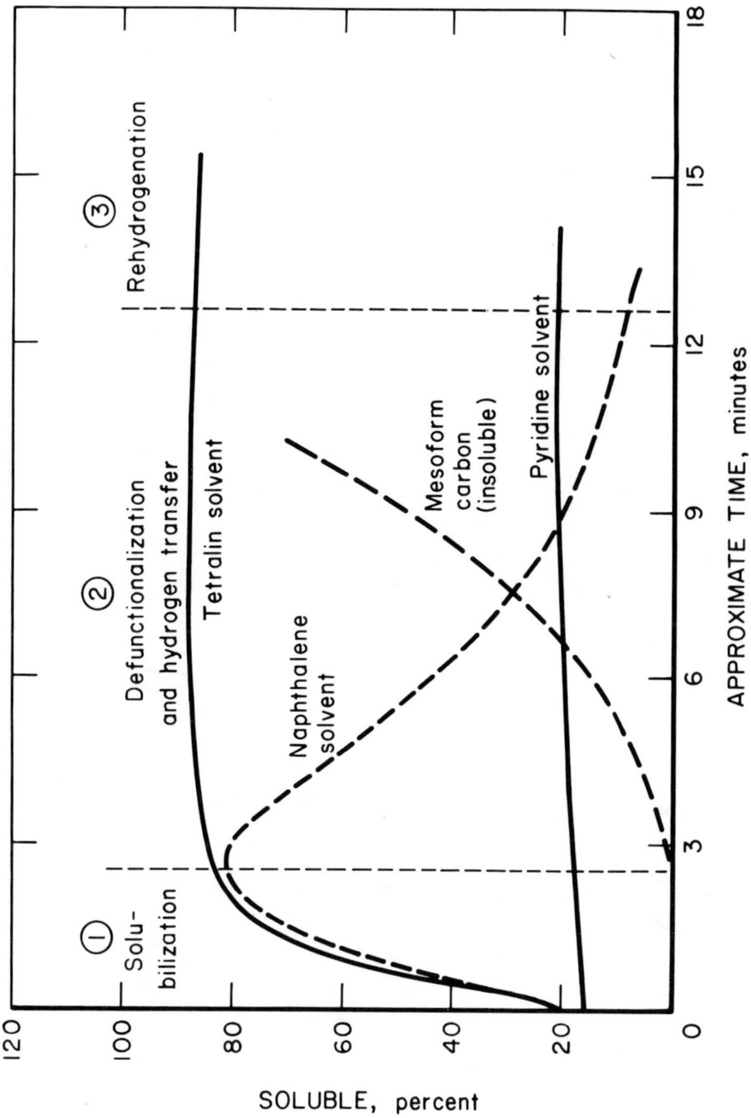

Figure 4-7. Conversion of coal as a function of time during donor-solvent liquefaction (Whitehurst et al. [52]).

2. At intermediate and long reaction times, the intermediates initially formed seek stabilization through termination of their free-radical state by combining with other radicals and/or atoms. It is at this juncture that the nature of the donor solvent plays an important role. The relative rates of progressive and retrogressive reactions depend upon the quality of the donor solvent used. Rates of progressive

reactions are enhanced in the presence of good donor solvents such as tetralin, whereas retrogressive reactions are dominant in the presence of poor donor solvents such as naphthalene.

3. At any temperature, the conversion of coal to pyridine-solubles attains a maximum value which has some of the characteristics of an equilibrium value, even though the reaction is not reversible in the normal sense. This value increases with increasing temperature up to a certain temperature, beyond which retrogressive reactions predominate. Also, the time required to attain this equilibrium value decreases with increasing temperature.

Squires [43] proposed the model shown in Figure 4-8, based in part on the findings of Neavel [33], Mitchell et al. [30], Whitehurst et al. [50, 51], Farcasiu et al. [12], and Whitehurst and Mitchell [51]. The various fractions used in the model of Squires can be identified with the following SESC fractions defined earlier (Whitehurst et al. [52]).

$$\text{oils} = \text{fractions 1 and 2,}$$
$$\text{asphaltenes} = \text{fractions 3, 4, and 5,}$$
$$\text{preasphaltenes} \equiv \text{asphaltols} = \text{fractions 6, 7, 8, and 9.}$$

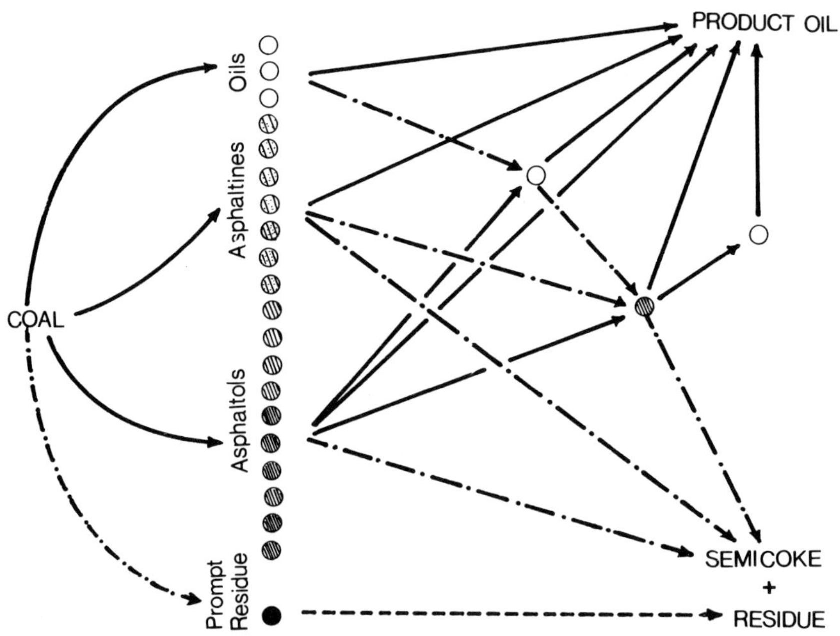

Figure 4-8. Proposed conceptual mechanism for the donor-solvent liquefaction of bituminous coal. Curved lines at left indicate rapid reactions. The continuous and discontinuous straight lines indicate slower progressive and retrogressive reactions respectively (Squires [43]).

The model of Squires considers the feed coal to break down initially into asphaltols, asphaltenes, and oils. The asphaltols and asphaltenes react further to produce oil, as well as undesirable material such as coke. The relative rates of the progressive and retrogressive reactions depend upon the temperature, the coal-to-solvent ratio, and the hydrogen-donating ability of the solvent. The formation of gases, which probably occurs primarily due to breaking of "scissile" bonds, appears to be overlooked in this model.

A model such as the above assumes that many of the elementary reactions are similar and result in similar products which can be lumped together into a small number of dependent variables that can be easily measured. Such an approach reduces the complexities involved in product analyses for kinetic purposes and thereby facilitates the representation of the actual reactions in terms of simple working models. It may be argued that the very step of simplification could result in the loss of valuable information that might otherwise be obtained. However, in view of the fact that there is a current need for practical models that provide adequate information for comparison purposes, the use of such an approach is justified. What is then needed is an examination of the validity of lumping individual fractions that possess similar chemical characteristics.

It is evident from the models described up to now that the lumped fractions can be defined in many ways. The most commonly used lumps in the literature are the following:

1. oils,
2. asphaltenes,
3. preasphaltenes or asphaltols,
4. unreacted coal.

The concentrations of these fractions in the liquefaction products can be used to establish reaction networks and estimate rate constants.

Having formulated an approach to modeling the kinetics of coal liquefaction, one is then confronted with the lack of unity that currently exists in the definition of the various lumped groups. For example, the unconverted coal is defined as benzene insolubles by Weller et al. [49] and Hill et al. [20], as xylenol insolubles by Curran et al. [10], and as tetrahydrofuran insolubles by Shalabi et al. [40]. Whitehurst et al. [50, 51] defined the reaction products as specific chromatographic fractions, while Shah et al. [39] classified the products into various boiling fractions.

In addition to these differences (which result in fractions that can easily have different chemical compositions) there is the added problem that even data obtained by apparently identical methods may not be comparable. Ruberto and Cronauer [37] have shown that the sequence in which the various solvents are used affects the final weight fraction of the various products.

In view of the above differences it is unlikely that agreement will be found in the kinetic data reported by various authors in the literature. The question then arises whether the chemical composition of the liquefaction products obtained under different conditions will be similar. The answer lies in the work of Aczel et al. [1], who have analyzed various SYNTHOIL products listed in Table 4-5 using the same analytical techniques. Two of the four samples analyzed by them were from different batches of the same liquefaction run, FB-39. The other two samples were also from the same run, FB-40, but batch 11 was obtained from the preheater and batch 62 from the reactor. In addition, runs FB-39 and FB-40 were made using H_2 at 6.893 and 13.786 MPa respectively.

Each sample was separated into discrete fractions by solvent extraction and liquid chromatographic techniques, and the fractions were subsequently examined by various spectroscopic techniques. Figure 4-9 is a schematic of the analytical approach used by Aczel et al. [1]. The solvent fractionation results of the four samples are shown in Table 4-6. The oils range from 52 to 90%, the asphaltenes from about 3 to 25%, and the benzene insolubles from 1 to 25%. These data are consistent with the processing conditions. In fact, more oils would be expected from FB-39 (hydrogen partial pressure 13.786 MPa) than from FB-40 (hydrogen partial pressure 6.893 MPa), and more preasphaltenes and less asphaltenes and oils would be expected from the preheater than from the reactor in run FB-40. However, if the chemical composition of these oils, asphaltenes, and preasphaltenes is not the same, or at least similar, then the above comparisons, and any kinetic

Table 4-5

Conditions for the Liquefaction of SYNTHOIL Products*

Run Conditions:	Feed coal		Kentucky Homestead Mine	
	Reactor temperature		723 K	
	Preheater temperature		723 K	
	Paste composition		35% coal, 65% oil	
	Reactor size		2.54 cm i.d. X 4.27 m	
	Catalyst		$CoMo/SiO_2$-Al_2O_3	
Run No.	FB 39	FB 39	FB 40	FB 40
Batch No.	34	94	11	62
Hydrogen pressure, MPa	13.79	13.79	6.89	6.89
Remarks	CFG product oil	CFG product oil	SPT sample from preheater. "This sample was not put through the reactor, but represents any chemical change due to the preheater."	CFG product oil

*Aczel et al. [1].

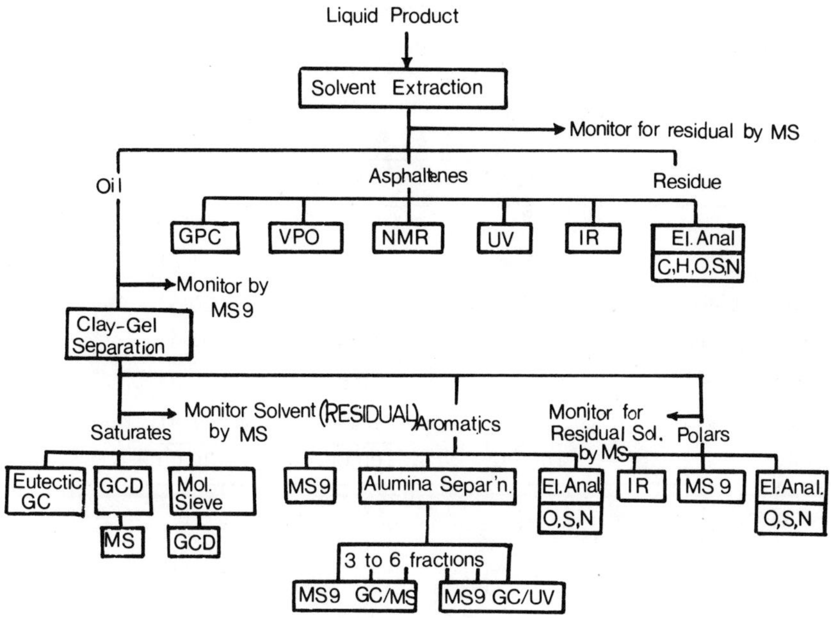

Figure 4-9. Analytical Scheme of Aczel et al. [1]. Key:

GCD	Gas-chromatographic distillation
MS	Low-resolution mass spectrometry
Eutectic GC	Eutectic gas chromatography
Msg	High-resolution, low-voltage mass spectrometry
GC/MS	Directly coupled GC/MS
GC/UV	UV on trapped GC cuts
GPC	Gel permeation chromatography
VPO	Vapor-pressure osmometry
UV	Ultraviolet-visible spectrometry
IR	Infrared spectrometry
NMR	Nuclear magnetic resonance

model developed therefrom, are meaningless. It is therefore necessary to examine the composition of each of these fractions. This is briefly discussed below.

4.3 Characteristics of the Oil

The results of the chromatographic separation of the oils are shown in Table 4-7. Some large differences are apparent from this table, but there are no definitive trends that can be correlated with processing conditions. This is especially true of batches 11 and 67, indicating that the quality of the oil produced in the preheater and the reactor is the same, even though the amounts are different.

Table 4-6

Solvent Fractionation of Original Samples*

	Weight percent			
Run No.:	FB-39		FB-40	
Batch No.:	34	94	11	67
Fraction:				
Oil	89.65	68.33	51.81	65.15
Asphaltenes	3.26	15.18	20.33	24.80
Preasphaltenes	0.87	5.49	24.90	8.53
Water	< 0.05	< 0.05	0.05	0.30
Loss	6.22	11.00	2.91	1.22
Total	100.00	100.00	100.00	100.00

*Aczel et al. [1].

Table 4-7

Saturates, Aromatics, and Polars in Oil Fractions*

	Weight percent			
Run No.:	FB-39		FB-40	
Batch No.:	34	94	11	67
In oil:				
Saturates	15.08	8.92	6.46	7.72
Aromatics	59.85	48.72	49.91	43.54
Polars	23.44	41.89	40.34	45.90
Loss	1.63	0.47	3.29	2.84
Total	100.00	100.00	100.00	100.00
In sample:				
Saturates	13.52	6.10	3.35	5.03
Aromatics	53.66	33.29	25.86	28.37
Polars	21.01	28.62	20.90	29.90
Loss	1.46	0.32	1.70	1.85
Total	89.65	68.33	51.81	65.15

*Aczel et al. [1].

The compound type distribution of the saturate fractions is shown in Table 4-8. The major components of the four saturate samples are the paraffins and 1- to 4-ring naphthenes. The paraffins consist of C_{15} to C_{35} normal and isoalkanes. The naphthenes contain decalins, tricyclodecanes, perhydroacenaphthenes, perhydrofluorenes, perhydrophenanthrenes, perhydropyrenes, and similar compounds.

The compound type distribution of the aromatic fractions is shown in Table 4-9. Each of the four aromatic samples contain mostly 1- to 3-ring aromatic hydrocarbons. The 2-ring aromatic furans are the most abundant oxygen-containing compounds, and only very small amounts of thiophenes are present. Table 4-10 shows the structure of the major ring systems expected in coal liquefaction products. The thiophenes and furans have structures similar to those shown in Table 4-10 except that one of the ring carbons (usually that of a 5-member ring) is replaced by a sulfur or an oxygen atom. Hydroaromatic compounds with similar structures are also present.

Finally, each of the aromatic and hydroaromatic structures and the furans and thiophenes represent the base compound of the homologous series that is obtained by the addition of alkyl substituents. Most of these substituents contain only 1 to 4 carbons. However, compounds containing 20 or more carbon substituents, either as one large substituent or as many small substituents, have been identified (Aczel et al. [1], Ruberto et al. [38]).

The elemental analyses and the molecular weights of the four aromatic samples are shown in Table 4-10. The data in Tables 4-9 and 4-11 show that the

Table 4-8

Compound Type Distribution of Saturate Fractions*

	Weight percent			
Run No.	FB-39		FB-40	
Batch No.	34	94	11	67
In saturates:				
Paraffins	8.13	16.40	20.40	34.20
1-ring naphthenes	23.11	15.10	19.18	18.38
2-ring naphthenes	27.21	17.05	19.68	16.30
3-ring naphthenes	24.23	24.59	20.77	14.64
4-ring naphthenes	11.86	16.20	12.83	10.35
5-ring naphthenes	2.55	5.79	3.74	2.97
6-ring naphthenes	0.00	2.92	0.77	1.09
Monoaromatics†	2.91	1.95	2.63	2.07
Total	100.00	100.00	100.00	100.00

*Aczel et al. [1].

†Value probably erroneously high. The MS matrix used, C_{24} *n*-paraffin, ASTM-D2786, is known to give high monoaromatic values. Inspection of MS spectra excludes significant amounts of aromatics.

Table 4-9

Ring Distribution in Aromatic Fractions*

	Weight percent in aromatics			
Run No.:	FB-39		FB-40	
Batch No.:	34	94	11	67
Aromatic hydrocarbons:	97.39	91.59	93.43	90.45
1 ring	37.17	22.21	24.51	21.36
2 ring	31.77	35.27	34.46	37.57
3 ring	13.57	16.13	16.71	15.94
4–8 ring	14.88	17.98	17.75	15.58
Aromatic furans:	2.37	7.47	5.83	8.15
1 ring	0.34	1.21	0.82	1.51
2 ring	2.01	4.68	3.65	4.85
3 ring	0.00	0.94	0.89	1.11
4–8 ring	0.02	0.64	0.47	0.68
Aromatic thiophenes:	0.24	0.94	0.74	1.40
1 ring	0.09	0.33	0.23	0.54
2 ring	0.09	0.43	0.31	0.61
3 ring	0.02	0.12	0.16	0.14
4–8 ring	0.04	0.06	0.04	0.11
Totals:	100.00	100.00	100.00	100.00
1 ring	37.60	23.75	25.56	23.41
2 ring	33.87	40.38	38.42	43.03
3 ring	13.59	17.19	17.76	17.19
4–8 ring	14.94	18.68	18.26	16.37

*Calculated from Table 27, Aczel et al. [1].

four aromatic fractions are fairly similar, even though their amount in the total coal liquids is different. Again, this is evidence that the quality of the lump fractions obtained from the same coal under different conditions is essentially the same, and only their quantity is affected by the process.

The last component of the "oils" is the "polars." The polars are made up of molecules which contain N, O, and S atoms in addition to C and H. They are, therefore, more polar than the saturates and the aromatics and consequently are held up in the clay-gel column, thus making the separation possible. The effectiveness of the clay-gel column separation can actually be determined by the amount of nitrogen and oxygen content of the saturate and aromatic fractions. Table 4-11 shows the elemental analysis of the aromatics and polars; the aromatics contain very little nitrogen and oxygen, while the amount of these elements in the polars is considerable, indicating that the separation was quite effective.

Table 4-10

Major Ring Systems Expected in Coal-Liquefaction Products*

Name	Formula	Structure	Mol. wt.
Benzene	C_6H_6		78
Indene	C_9H_8	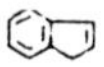	116
Naphthalene	$C_{10}H_8$	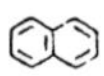	128
Acenaphthylene	$C_{12}H_8$	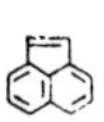	152
Acenaphthene	$C_{12}H_{10}$	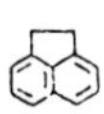	154
Fluorene	$C_{13}H_{10}$	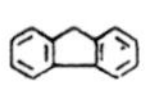	166
Phenanthrene	$C_{14}H_{10}$	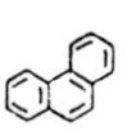	178
Anthracene	$C_{14}H_{10}$	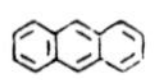	178
Cyclopentanophenanthrene	$C_{15}H_{10}$	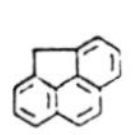	190
Pyrene	$C_{16}H_{10}$	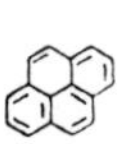	202
Fluoranthene	$C_{16}H_{10}$		202
Chrysene	$C_{18}H_{12}$	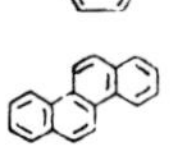	228
Benzanthracene	$C_{18}H_{12}$	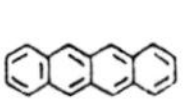	228
Benzo [*ghi*] fluoranthene	$C_{18}H_{10}$	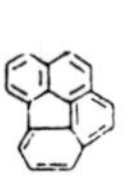	226
Cholanthrene	$C_{19}H_{12}$	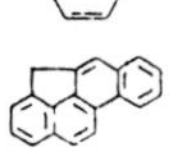	240
Benzopyrene	$C_{20}H_{12}$	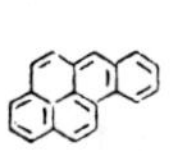	252
Benzo [*c*] chrysene	$C_{22}H_{14}$	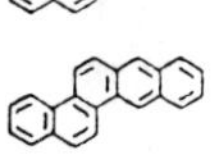	278

Table 4-10 (Continued)

Major Ring Systems Expected in Coal-Liquefaction Products*

Name	Formula	Structure	Mol. wt.
Anthanthrene	$C_{22}H_{12}$	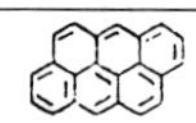	276
Benz [*ghi*] perylene	$C_{22}H_{12}$		276
Dibenzopyrene	$C_{24}H_{14}$	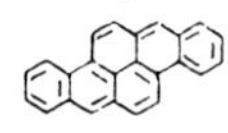	302
Coronene	$C_{24}H_{12}$		300
Dibenzo [-*cd, lm*] perylene	$C_{26}H_{14}$	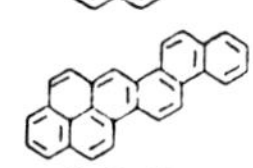	326
Dibenzo [a, o] perylene	$C_{28}H_{16}$	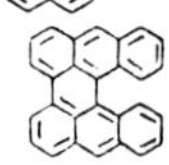	352
Benzocoronene	$C_{28}H_{14}$	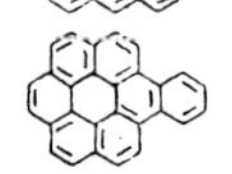	350
Tribenzo [*de, kl, rst*] perylene	$C_{30}H_{16}$		376

*Aczel et al. [1].

The data in Table 4-11 also provides evidence that the four polar fractions are very similar in composition. This conclusion is further supported by the data of Table 4-12. Here the polars are shown to contain mainly mono- and dihydroxy-aromatics, pyrroles, and pyridines. These compounds are mostly made up of 1–3 aromatic rings.

The fact that only compounds containing hydroxylic and etheric oxygens are found in these samples is in agreement with the results of Cronauer and Ruberto [9] that the other oxygen functional groups found in coal (carboxylic, carbonylic, and acyclic ethers) are readily removed even under noncatalytic donor-solvent liquefaction conditions.

4.3.1 *Asphaltenes*

The elemental analyses (Table 4-13) and the molecular weight distributions (Table 4-14) of the four asphaltene fractions show that their composition is very similar. This is also shown by the NMR analysis (Table 4-15). Various parameters (hydrogen on carbon α and β to aromatic rings, number of asphaltene rings and alkyl links, etc.) have a very uniform distribution in the four asphaltene samples.

Table 4-11

Molecular Weights and Elemental Analyses of Aromatics and Polars*

Run No.:	FB-39		FB-40	
Batch No.:	34	94	11	67
Aromatics				
Molecular weight	224	241	216	217
Wt% of aromatics:				
Carbon	90.81	90.40	91.00	90.64
Hydrogen	8.92	8.64	8.24	8.22
Nitrogen	0.06	0.26	0.16	0.24
Oxygen	0.17	0.53	0.41	0.66
Sulfur	0.04	0.17	0.19	0.24
Polars				
Wt% of polars:				
Oxygen	3.95	4.12	4.26	4.98
Nitrogen, total	1.92	1.79	1.87	1.51
basic	0.65	0.63	0.86	0.73
Sulfur	0.13	0.31	0.25	0.19

*From Tables 10 and 28, Aczel et al. [1].

4.3.2 *Preasphaltenes and Unconverted Coal*

Very few data on these two fractions are available in the literature, and there are no available data that could be used for this discussion. On the basis of the characteristics of the other two "lumped" fractions discussed above, it should be safe to assume that the quality of these fractions is also not affected by process conditions.

In summary, the results of Aczel et al. [1] indicate that when a coal is liquefied by a particular process and the products are analyzed by the same technique, the quality of corresponding lumped fractions is very similar. Therefore, lumped fractions can be used to model liquefaction reactions and estimate rate parameters.

4.4 Reaction Networks Based on Lumped Fractions

Weller et al. [49], on the basis of their studies in the absence of a solvent or liquid vehicle, proposed the following mechanism for coal liquefaction:

$$\text{coal} \xrightarrow{k_1} \text{asphaltenes} \xrightarrow{k_2} \text{oils}. \tag{4-14}$$

Table 4-12

Compound Type Distribution in the Polar Fractions*

	Wt% of polar			
Run No.:	FB-39		FB-40	
Batch No.:	34	94	11	67
Mono-oxygen	52.41	41.76	49.29	58.96
Dioxygen	3.71	7.02	6.89	7.66
Miscellaneous oxygen	0.52	1.40	0.81	2.17
N-compounds	23.85	21.17	21.10	18.76
N-O compounds	0.95	1.64	1.00	0.57
Miscellaneous N-compounds	0.14	0.70	0.19	1.94
S-compounds	0.10	0.07	0.10	0.
Hydrocarbons	1.71	1.54	0.0	1.04
Residue	16.61	24.70	20.62	8.90
Total	100.00	100.00	100.00	100.00

*From Table 41, Aczel et al. [1].

Table 4-13

Elemental Analyses of Asphaltene*

	Weight percent			
Run No.:	FB-39	FB-39	FB-40	FB-40
Batch No.:	34	94	11	67
Asphaltenes:				
C	–	86.82	87.71	87.94
H	–	6.29	6.03	5.84
O	2.89	4.39	3.77	3.20
S	0.18	0.36	0.62	0.35
N	1.35	2.11	2.04	2.13
(Basic N)	(0.97)	(1.17)	(1.06)	(1.19)
Total	–	99.97	100.17	99.46

*Aczel et al. [1].

Table 4-14

Molecular-Weight Distribution of Asphaltenes by Gel-Permeation Chromatography (THF Solvent)*

	Weight percent or value			
Run No.:	FB-39	FB-39	FB-40	FB-40
Batch No.:	34	94	11	67
Molecular weight:				
Number average	691	784	650	596
Weight average	1180	1433	1394	1061
Range (0.1–99.9%)	105–3450	146–8548	144–12105	166–4568
Weight percent material in MW ranges:				
< 500	20.6	18.4	27.4	30.6
500–1000	28.4	26.9	24.6	29.8
1000–2000	41.0	37.2	30.4	30.1
2000–3000	9.4	11.2	10.3	7.6
3000–4000	0.6	3.6	3.9	1.5
>4000	0.0	2.7	3.4	0.4
Total	100.0	100.0	100.0	100.0

*Aczel et al. [1].

Conversions of the reacting species were determined by measuring the product solubilities in benzene and hexane. The authors found both reactions to be first order with respect to coal and asphaltenes respectively, and the ratio k_1/k_2 = 10–25 in the temperature range 673–723 K. This model is an oversimplification in that it classifies the wide range of molecules into three distinct categories, thereby precluding an insight into various mechanisms that are operative during coal liquefaction. Although the hypothesis that oil is only a secondary product of coal liquefaction (in the presence of donor solvent or otherwise) is not true [33], the above model has provided direction in the lumped modeling of the kinetics of donor solvent liquefaction.

Cronauer and Ruberto [9] have reported the liquefaction of a subbituminous coal from the Belle Ayr mine in Wyoming as well as a bituminous coal from the Burning Star mine in Illinois (#6 seam) using hydrogenated anthracene oil (HAO), hydrogenated phenanthrene (HPh), and anthracene oil (AO) as solvents. The hydrogen contents of the above three solvents were 6.86, 8.83, and 5.84 wt% respectively. The various fractions formed during liquefaction are defined as follows:

Table 4-15

Calculations on Asphaltenes Using NMR, Elemental Analyses, and Molecular Weight (No Corrections Applied)*

Run No.:	FB-39		FB-40	
Batch No.:	34	94	11	67
% H_a on AR†	2.60	2.52	2.35	2.57
% H_α on C α to AR	1.59	1.95	1.87	1.81
% $H_{\alpha 2}$ on C α to 2 ARs	0.26	0.36	0.21	0.23
% $H_{\rho+}$ on C β & β^+ to AR	1.71	1.82	1.81	1.46
% C_s saturate (alkyl)	19.7	22.47	21.93	19.49
% C_α α to AR	9.48	11.62	11.15	10.79
% $C_{\alpha 2}$ α to 2 ARs	1.55	2.15	1.25	1.36
% C_a total AR	70.0	64.35	65.78	68.45
% C_1 peripheral AR	40.47	41.62	39.16	41.42
Total rings	0.218	0.198	0.215	0.215
Naphthene rings & alkyl links	0.053	0.067	0.064	0.061

*Aczel et al. [1].
†Aromatic ring.

$$
\begin{aligned}
\text{oils} &= \text{(total sample)} - \text{pentane-insolubles,}\\
\text{asphaltenes} &= \text{pentane-insolubles} - \text{benzene-insolubles,}\\
\text{preasphaltenes} &= \text{benzene-insolubles} - \text{pyridine-insolubles.}
\end{aligned}
$$

For the liquefaction of Belle Ayr coal they found that the percentage solvation of coal as well as the yield of oils increased with increasing hydrogen content of the donor solvent. This is shown in Figures 4-10 and 4-11. With the unhydrogenated anthracene oil as a solvent at 723 K it is seen that the yield of oils at low space times was actually negative. This is probably due to the occurrence of retrogressive reactions which result in the solvents being taken up by the coal and/or asphaltenes. The yield of asphaltenes and preasphaltenes was also high at all space times, indicating that very little reaction to form oils was occurring. Retrogressive reactions leading to the formation of coke appear to predominate at 743 K and at long space times for the Belle Ayr coal even in the presence of a hydrogenated anthracene oil (HAO) (Figure 4-12). The percentage solvation and yield of oils for the Burning Star coal are shown in Figures 4-13 and 4-14. At 733 K the percentage solvation is lower than at 723 K in the presence of HAO solvent. The yield of oils at the former

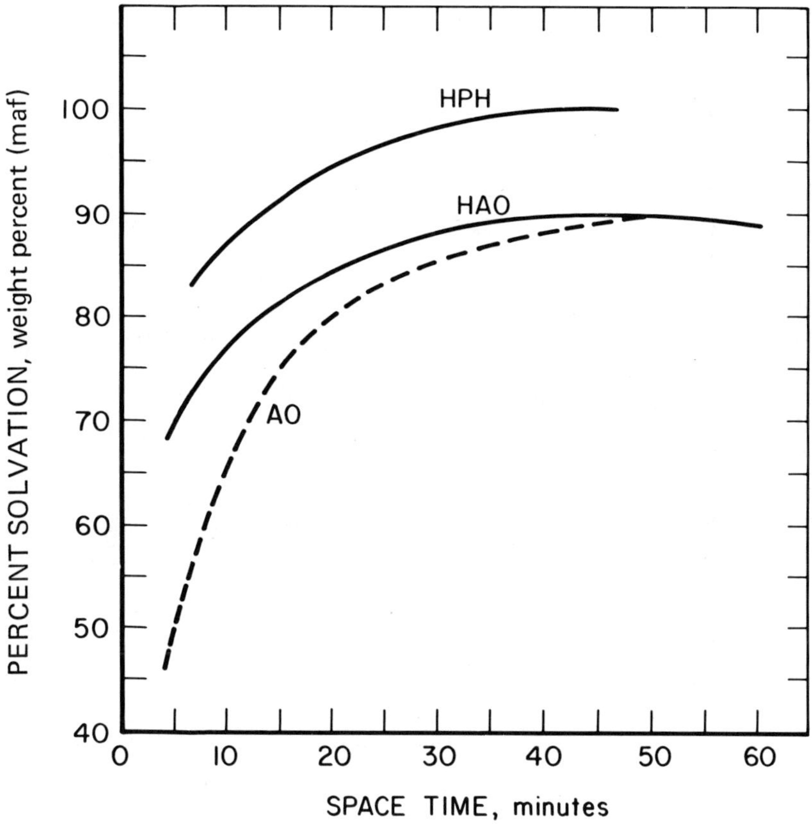

Figure 4-10. Percentage solvation of Belle Ayr coal in various solvents at 723 K as a function of space time (Cronauer and Ruberto [9]).

temperature is again negative at short space times using the HAO solvent. As before, this is attributed to solvent entrapment or reaction with the coal or preasphaltenes, resulting in the formation of polymeric material. The authors have proposed the following kinetic model for the liquefaction of Belle Ayr coal in the presence of hydrogenated anthracene oil and hydrogenated phenanthrene:

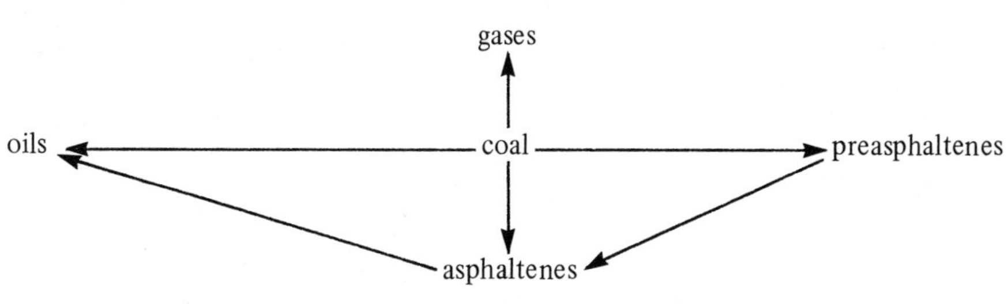

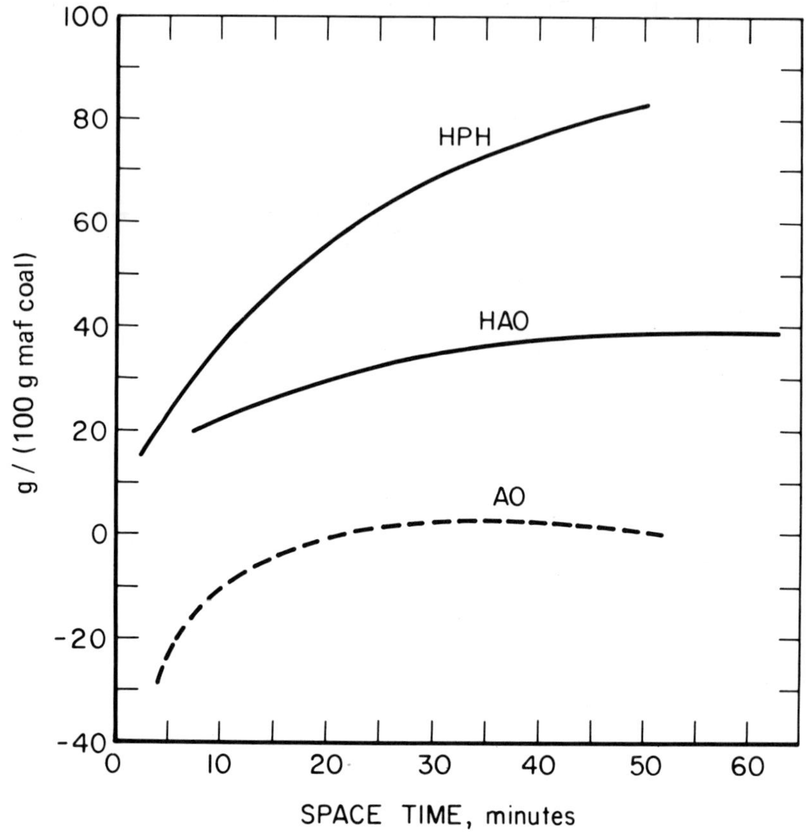

Figure 4-11. Yield of oils using various solvents as a function of space time for the liquefaction of Belle Ayr coal in HAO at 723 K (Cronauer and Ruberto [9]).

The above model, assuming all reactions to be first order and irreversible, was found to fit the data reasonably well for all space times in the temperature range of 673–723 K and for short space times at 733 and 743 K.

The agreement between experimental data and model predictions for the solvation of Belle Ayr coal is shown in Figure 4-15. Equally good agreement was observed for the various lumped fractions in the reaction network [9]. The activation energies and the frequency factors for various rate constants obtained in this study are provided in Table 4-16. The activation energies reported by these authors are lower than those reported by earlier workers. The reason for this is unclear, although there is speculation that it is due to grouping of several reaction intermediates into four distinct groups. It may also be partly due to the differences in analytic procedures employed by the various investigators [4, 24].

The results shown in Figure 4-15 indicate that the soluble portion of coal varies with the nature of the solvent. For a given coal-solvent system (at a fixed

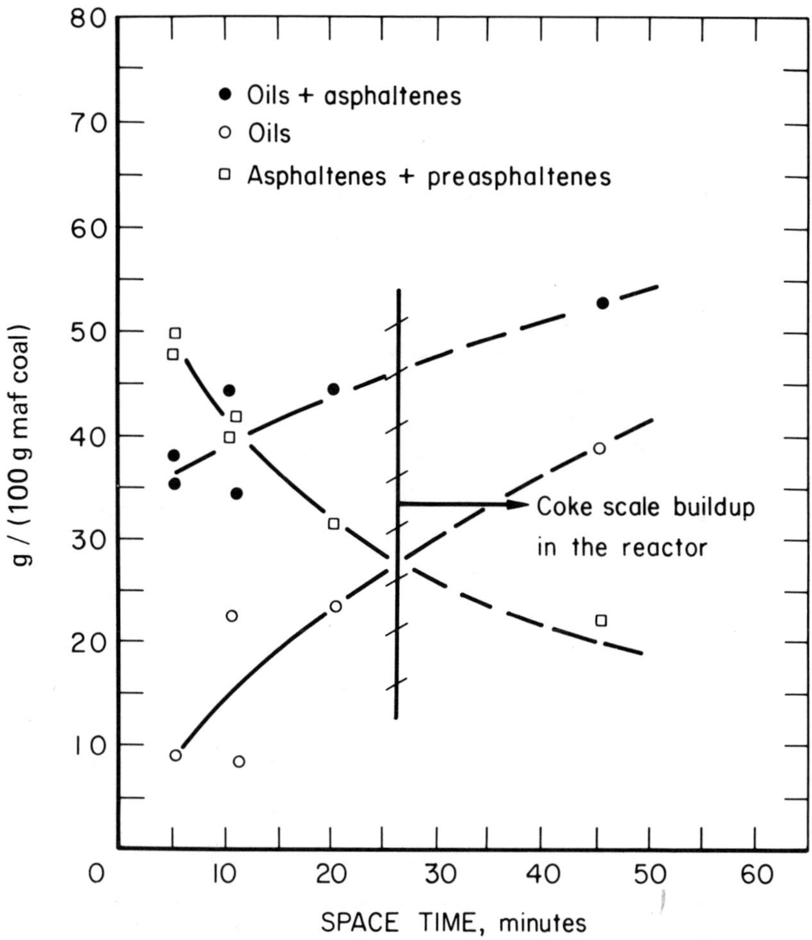

Figure 4-12. Yield of various fractions as a function of space time for the liquefaction of Belle Ayr coal using HAO at 743 K (Cronauer and Ruberto [9]).

temperature and pressure) there appears to exist an equilibrium concentration of insoluble organic matter in coal which varies with the solvent type.

Imbibition of the solvent by the coal during liquefaction has also been reported for some Japanese, Australian, and Indonesian coals by Morita et al [32]. Though the nature of the solvent and the ranks of the various coals used are not known, the above authors found that during reaction of a coal in the presence of a H_2MoO_4 catalyst at a temperature of 723 K the amount of oil present dropped initially to half its initial value, after which it began to increase. The formation of coke and resin-type material exhibited a maximum during this time period. The following generalized model has been proposed by the above authors:

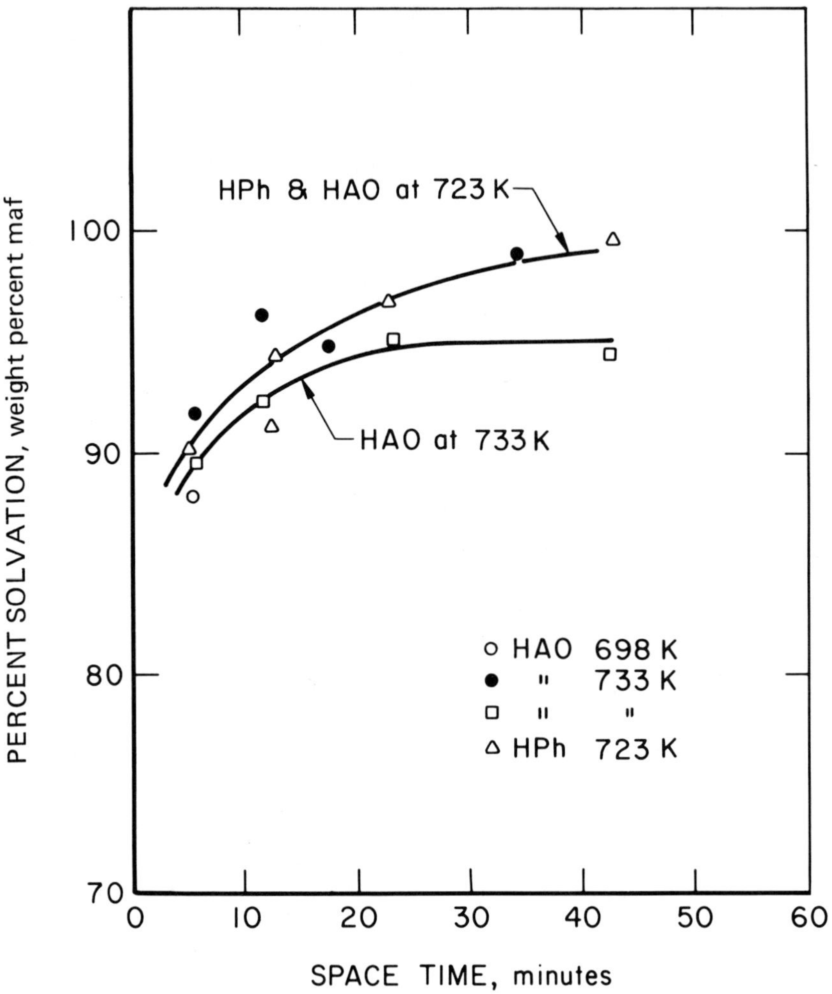

Figure 4-13. Percentage solvation of Burning Star coal using various solvents as a function of space time and temperature (Cronauer and Ruberto [9]).

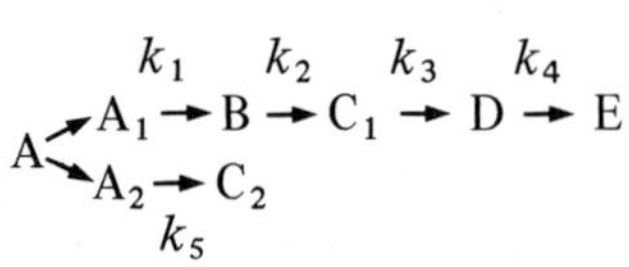

where A, B, C, D, and E represent the coal, asphaltene, oil, resin, and coke fractions respectively. Using benzene and *n*-hexane as solvents for Soxhlet extraction, they defined the following reactive fractions:

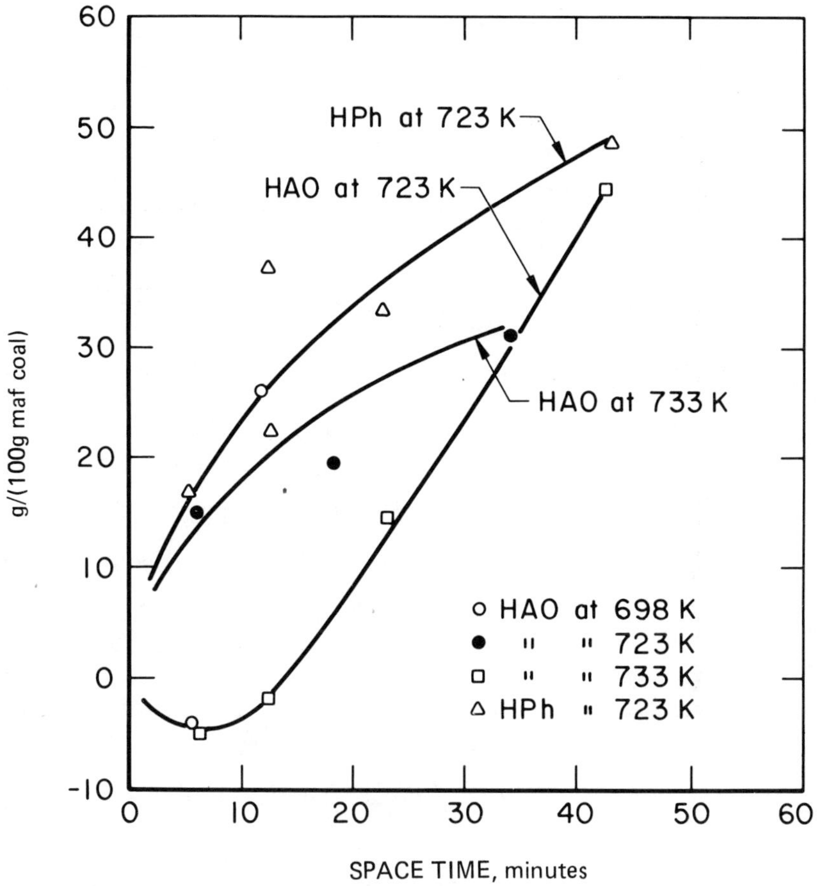

Figure 4-14. Yield of oils as a function of space time and temperature for the liquefaction of Burning Star coal using various solvents (Cronauer and Ruberto [9]).

asphaltene = benzene-solubles, *n*-hexane-insolubles,

oils = *n*-hexane-solubles.

The above mechanism is an extension of the model proposed by Cronauer et al. [8] in that the reactions oils → resins → coke are included in the reaction network. When $A_2 \rightarrow C_2$ is negligible, the model reduces to an extension of the model of Weller et al. [49].

The model of Brunson [5] is based on the following mechanism for liquefaction:

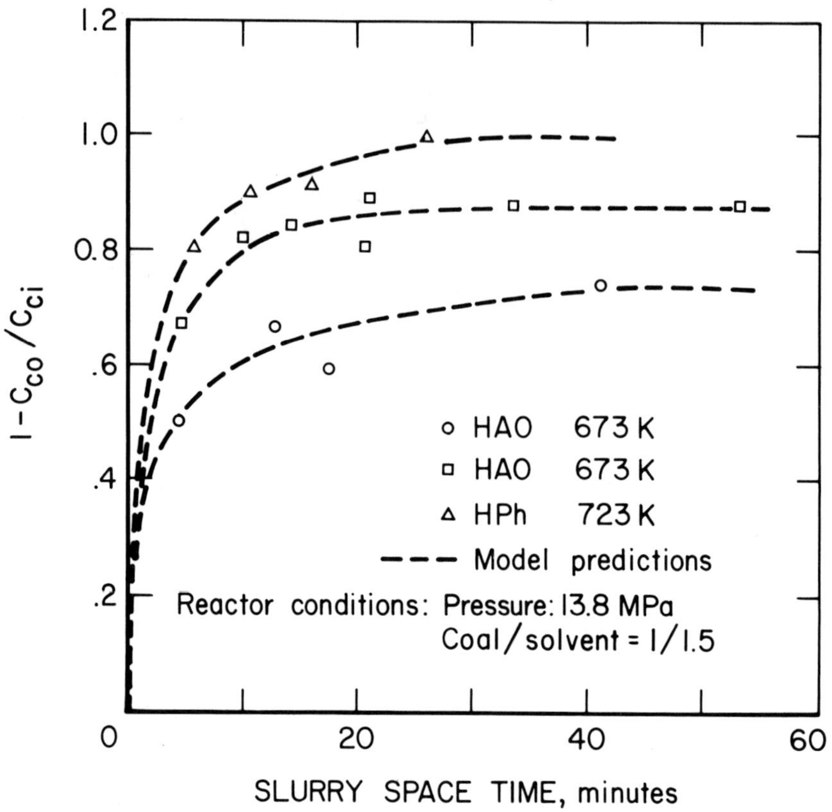

Figure 4-15. Coal solvation as a function of space time (Cronauer et al. [8]).

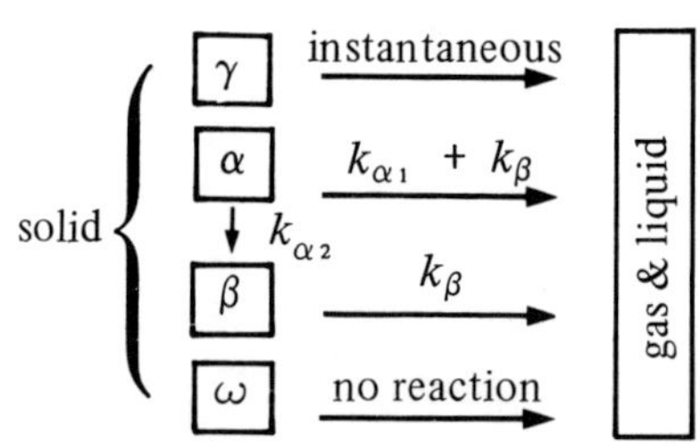

The feed coal is lumped into four classes — γ, α, β and ω — based on their reactivity. The portion labeled γ is volatile and, upon entering the reactor, is converted instantaneously. This fraction was determined by carbonizing the feed coal in a high-vacuum distillation scheme. The second fraction, α, can react by one of two paths. The first is the conversion to fluids via a rate constant $k_{\alpha 1}$ and the second to another type of reactive species, β, via a rate constant $k_{\alpha 2}$. The fraction β is non-

Table 4-16

Activation Energies and Frequency Factors for the Liquefaction of Belle Ayr Coal Using HAO and HPh Solvents*

Reaction	Activation energy, kJ/mol		Frequency factor, min^{-1}	
	HAO	HPh	HAO	HPh
Coal → oils	59.0	121	3.11×10^3	2.1×10^8
Coal → preasphaltenes	57.76	18	2.81×10^3	4.94
Coal → asphaltenes	65.3	36	1.12×10^4	9.63×10^1
Coal → gas	90	43.95	8.72×10^5	3.85×10^2
Coal → oils preasphaltenes, asphaltenes, gases	70	85.8	9.15×10^4	1.59×10^6
Preasphaltenes → asphaltenes	53.6	142	9.66×10^2	2.48×10^9
Asphaltenes → oils	66.9	107.2	1.42×10^3	1.53×10^7

*Cronauer et al. [8].

distillable but can be converted to liquids and gases by reaction via a rate constant k_β. The fraction α is also supposed to be converted to gases and liquids along the path of β. The fractions α and β essentially represent the preasphaltene and asphaltene fractions defined earlier (Cronauer et al. [8]). These fractions were not individually monitored, and the model and calculations described herein represent the combined conversion of the α and β fractions. Finally, the fraction ω is labeled unreactive under the experimental conditions. This fraction is determined from petrographic analysis of coal and liquefaction residues.

The mass-balance equations for the conversion of α and β assuming first-order irreversible kinetics are given as

$$\frac{d\alpha}{dt} = -(k_{\alpha_1} + k_{\alpha_2} + k_\beta)\,\alpha, \tag{4-15}$$

$$\frac{d\beta}{dt} = -k_{\alpha 2}\,\alpha - k_\beta\,\beta, \tag{4-16}$$

with

$$\alpha(0) = \alpha_o \tag{4-17a}$$

and

$$\beta\,(0) = \beta_o\ . \tag{4-17b}$$

Solving the above equations and combining the results, the final solution is given as

$$\frac{\alpha+\beta}{\alpha_o+\beta_o} = g \exp\,[-k_{\alpha_1} + k_{\alpha_2} + k_\beta\,] + (1-g)\exp\,[-k_\beta t]\,, \tag{4-18}$$

where

$$g = \frac{k_{\alpha_1}}{k_{\alpha_1} + k_{\alpha_2}}\left(\frac{\alpha_o}{\alpha_o + \beta_o}\right).$$

The rate constant $k_{\alpha 1}$ includes the combined effects of reaction pressure and mass transfer on the true chemical rate constant. The true rate constant is therefore defined as

$$k_\alpha = \frac{1}{\dfrac{p}{k_{\alpha 1}} - \dfrac{1}{k_d}}\ , \tag{4-19}$$

where p is the reaction pressure in MPa, and k_d, the rate of coal dissolution, is given as

$$k_d = 2 \times 10^{-7} T^2\,, \tag{4-20}$$

where T is the temperature in K. The model was found to fit the data reasonably well. The activation energies for the various reactions were reported as follows:

$$E_{\alpha 1} = 16.1 \text{ kJ/mol}, \tag{4-21}$$

$$E_{\alpha 2} = 79.1 \text{ kJ/mol}, \tag{4-22}$$

$$E_\beta = 66.64 \text{ kJ/mol}. \tag{4-23}$$

The activation energy for the primary reaction $\alpha \rightarrow$ products is comparable to that of Curran et al. [10]. The activation energies for the other two reactions, which are similar to the reactions preasphaltenes → asphaltenes and asphaltenes → oils in the model of Cronauer et al. [8], are low and comparable to those of the latter. Although external mass-transfer resistances have been ruled out, other effects such as intraphase resistance, as well as lumping of several reactions, may be responsible for such low values of the activation energies in both cases.

The use of molten halide catalysts for the hydrogenation of coal and coal extract effluents has been investigated by Zielke et al. [57]. These authors found that zinc halides acted as Lewis acids and were capable of hydrogenating coal under mild operating conditions in the absence of a solvent vehicle. Shinn et al. [41] and Shinn and Vermuelen [42] examined the use of water and methanol respectively as a solvent vehicle for $ZnCl_2$ melts during coal liquefaction. On the basis of their study they proposed the following mechanism for liquefaction using the $ZnCl_2$-methanol system:

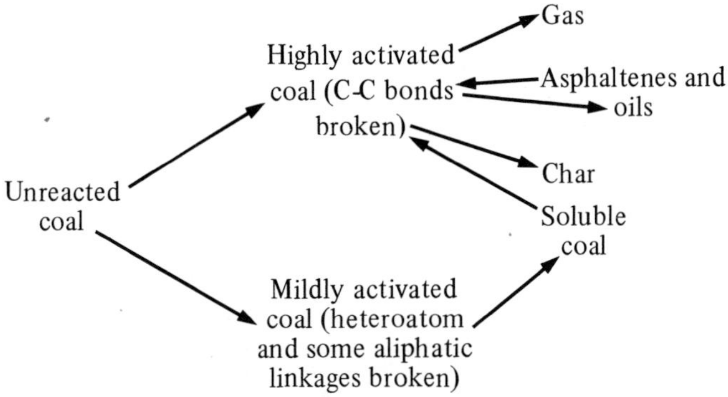

The above model is similar to the models proposed earlier in that the coal is initially brought to an activated state, whereupon further reactions depend upon the nature of the solvent vehicle. Good donor solvents enhance the oil-forming reactions, whereas the formation of char is favored in a hydrogen-poor system. The initial activation, however, is hypothesized to occur in two ways depending on the temperature and type of catalyst used. One interesting feature of this type of liquefaction is that the free radicals formed during liquefaction could be capped by alkyl transfer reactions from methanol resulting in stable alkylated molecules.

A modification of the mechanism of Cronauer et al. [8] has been proposed by Shalabi et al. [40], who included the reaction of preasphaltenes to oils and gases:

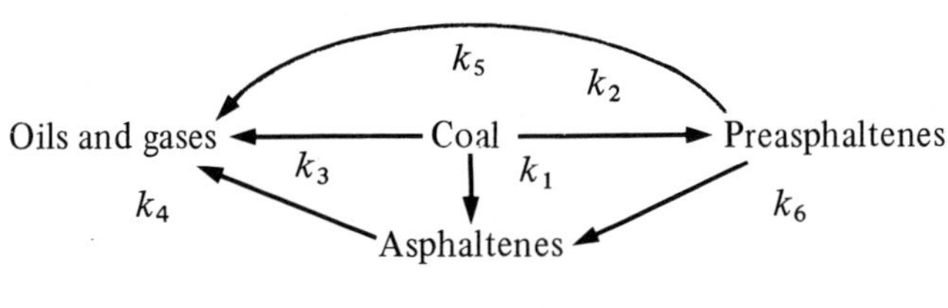

The various reacting fractions were defined as follows:

oils = pentane-solubles,

asphaltenes = benzene-solubles, pentane-insolubles,

preasphaltenes = benzene-insolubles, tetrahydrofuran (THF) solubles,

mineral matter and unreacted macerals = THF-insolubles.

The model was found to fit the experimental data reasonably well in the temperature range of 623–673 K for a highly volatile type-A bituminous coal from the Madisonville #9 seam, Fies Mine, Kentucky. The activation energies for the primary reactions of coal → preasphaltenes, coal → asphaltenes, and coal → oils were found to be 125.6, 121.4, and 167.4 kJ/mol, which is in good agreement with the results of Curran et al. [10] and Brunson [5], and considerably higher than those of Cronauer et al. [8] for the same reactions.

A somewhat different approach to modeling the liquefaction reaction has been taken by Shah et al. [39]. In this model the products of coal liquefaction are classified according to boiling range as shown below:

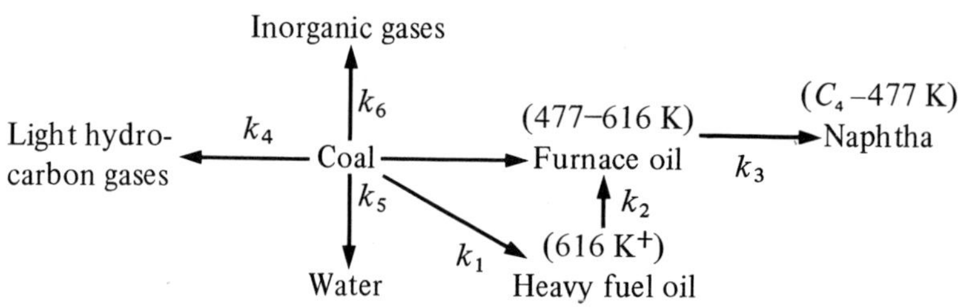

Assuming first-order irreversible kinetics for the above reactions and either zero- or first-order irreversible kinetics respectively for the hydrocracking reactions

(heavy fuel oil) → (furnace oil) → naphtha,

the model was shown to correlate the data reasonably well for the liquefaction of a subbituminous coal from the Big Horn Mine, Wyoming, using a segmented catalytic reactor patented by Gulf Research and Development Company. The various liquid fractions were obtained from the boiling-point distillation curves. The specific gravities of the various boiling-range liquid products were estimated to convert the volume fractions to mass fractions. This model assumes that inorganic gases, light hydrocarbons (C_1–C_4), and water come directly from the coal. This is not a true kinetic model in the sense that the various reactive species are classified according to a physical property, thus ignoring the chemical identity of the molecules. However, such models have proven to be successful in modeling the cracking of petroleum fractions [48]. The activation energies of the various primary reactions obtained by Shah et al [39] were high, implying that they are primarily thermal in nature.

In actual coal-liquefaction reactors the primary reactions commence in the preheater and then proceed to completion, along with secondary reactions, in the reactor. Under these circumstances the following conceptual model, based in part on the model of Liebenberg and Potgieter [27], might be useful. The preasphaltenes and the asphaltenes can be separated into two fractions each, PA* and PA, and A* and A, where the * represents the reactive part of each fraction. On this basis the following reactions may be visualized as occurring in the preheater:

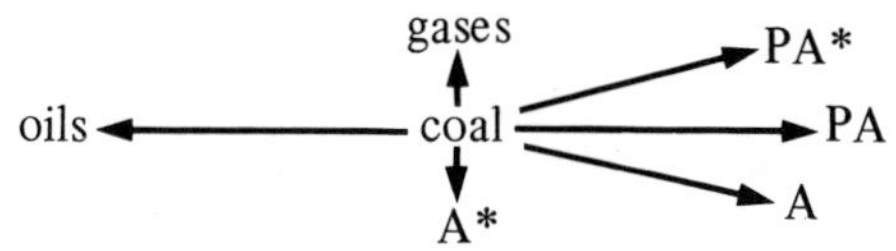

Subsequently, the following secondary reactions may be expected to occur in the reactor:

$$PA^* \rightarrow A^* + A, \tag{4-24a}$$

$$A^* \rightarrow \text{oils} + \text{gases}, \tag{4-24b}$$

$$PA^* \rightarrow \text{oils} + \text{gases}, \tag{4-24c}$$

$$\text{oils} \rightarrow \text{gases}. \tag{4-24d}$$

The reactions (4-24a)–(4-24d) are basically hydrogenation/hydrocracking reactions and will therefore result in a high hydrogen consumption. It should be noted that the lumped parameters in the above model can be replaced by various boiling-range fractions (Shah et al. [39]) of oils. The kinetic constants for various reaction paths with these new lumped parameters would, of course, be different. The implementation of this model requires considerable data at all residence times.

A similar concept in the kinetics of coal liquefaction considers the coal to be composed of two fractions of different reactivities – a highly reactive fraction which could be composed of loosely held bonds that break very easily under liquefaction conditions and therefore yield products that are observed at short contact times, and a fraction of lower reactivity which may be due to more stable structures that decompose slowly and is responsible for the majority of the product formations at large residence times. Due to their different reactivities, the activation energies for the reactions dealing with these two fractions may also be different. The main problem with this conceptual model is the difficulty in quantitatively identifying the two fractions as a function of reaction conditions.

In the studies reviewed so far, the activation energies for various reactions are expressed as kJ/mol. A gram mole of coal is not clearly defined. Under the circumstances, a basic reactive unit of the coal needs to be carefully chosen in the definition of activation energy. If the overall reaction with coal can be appropriately characterized by a number of different reactive units, each reacting at a different rate, the apparent overall activation energy could then be defined as either a number or a weight average of activation energy for each reacting species. Szladow and Given [44] found that the activation energies (defined as temperature coefficients for the rate) for middle Kittaning seam coal, Kentucky #4 seam coal, and Colorado E2 bed seam coal generally increased with the conversion (in terms of pyridine or tetrahydrofuran solubility) levels of the feed coal.

4.5 Free-Radical Reaction Models

4.5.1 *Model of Attar* [3]

The free-radical mechanism of coal liquefaction has been modeled in a manner similar to that used in classical free-radical chemistry, and although experimental verification is yet forthcoming, it represents a very interesting alternative to some of the lumped kinetics models discussed above. The free-radical reaction mechanism is viewed as consisting of three processes: initiation reactions that form more radicals than they consume; propagation reactions that have no effect on the number of free radicals; and termination reactions that consume free radicals. In one such model, the reaction paths are represented as follows:

$$\mathrm{ln}_1 - \mathrm{ln}_2 \rightarrow \mathrm{ln}_1\cdot + \mathrm{ln}_2\cdot, \tag{4-25}$$

$$\mathrm{ln}_1\cdot + \mathrm{DH}_2 \rightarrow \mathrm{ln}_1\mathrm{H} + \mathrm{DH}\cdot, \tag{4-26}$$

$$\mathrm{DH}_2 \xrightarrow[\text{or minerals}]{\text{catalyst}} \mathrm{DH}\cdot, \tag{4-27}$$

$$\mathrm{DH}\cdot + \mathrm{R}_j\mathrm{H} \rightarrow [\mathrm{R}_j\mathrm{HDH}\cdot] \rightarrow \mathrm{R}_{j+1}\mathrm{H} + \mathrm{R}_{j+1}\cdot + \mathrm{D}, \tag{4-28}$$

$$\mathrm{R}_{j+1}\cdot + \mathrm{DH}_2 \rightarrow \mathrm{R}_{j+1}\mathrm{H} + \mathrm{DH}\cdot, \tag{4-29}$$

$$2\mathrm{DH}\cdot \rightarrow \mathrm{DH}_2 + \mathrm{D}. \tag{4-30}$$

Here $\mathrm{ln}_1\cdot$, $\mathrm{ln}_2\cdot$, etc. are the various free radicals formed by the defragmentation of the coal particles.

In the initiation reaction (4-25), very weak bonds in the coal macromolecular network $\mathrm{ln}_1 - \mathrm{ln}_2$ are decomposed, yielding free-radical molecules $\mathrm{ln}_1\cdot$ and $\mathrm{ln}_2\cdot$. The major initiation reactions probably involve the weak crosslinks between the macromolecules consisting of ether bonds, methylene bridges at benzyl positions, and sulfidic sulfur. A hydrogen-donor molecule DH_2 in the solvent can interact (4-26) by capping the initial radicals and forming a donor radical DH·. The donor molecule can also initiate a radical DH· through interaction with a catalyst (4-27) that is either added or simply indigeneous to the coal (mineral matter). The donor radical is capable of breaking carbon-oxygen and carbon-carbon bonds and continues the decomposition of the weak crosslinks and the weakest bonds within each macromolecular cluster (4-28). The coal molecular units $\mathrm{R}_j\mathrm{H}$ and the radicals $\mathrm{R}_j\cdot$ contain several aromatic rings and have molecular weights in the range 300–3000. The index j is used to indicate the number of times a particular molecular unit has participated in previous reactions. The decomposition proceeds according to the hierarchy of bond strengths or activation energies, so that the weakest bond in $\mathrm{R}_{j+1}\mathrm{H}$ is stronger than the corresponding bonds in the parent molecule $\mathrm{R}_j\mathrm{H}$. The dehydrogenated donor molecule D is also a product of this reaction. The hydrogen donation reaction (4-29) proceeds quickly, capping the radicals $\mathrm{R}_j\cdot$, and completes the propagation reaction by maintaining the supply of DH· radicals. Since the radicals $\mathrm{R}_j\cdot$ are much larger molecular units than DH·, it is reasoned that they are much less mobile, so that the dominant termination reaction must be the combination of two donor radicals to form hydrogenated and unhydrogenated donor molecules [4-30].

Based upon the above free-radical chemistry, the detailed kinetic equations have been derived by Attar [3]. While this approach still needs experimental verification, it is the first model of its type and deserves further attention.

4.5.2 *Mechanistic Studies Using Free-Radical Scavengers*

A very fruitful approach has recently been developed for examining the individual roles of temperature, hydrogen solvent, and catalyst in the mechanism of coal liquefaction [15, 16]. The effect of each of these variables was studied separately with and without the presence of a free-radical scavenger such as *p*-benzoquinone or iodine. The effect of temperature was studied by simple pyrolysis in the absence of the other parameters. The free-radical scavengers trapped the initial radicals formed from the coal, as shown by a substantial lowering of the benzene-solubles yield and the presence of chemically bound iodine in the benzene-insoluble residue. Addition of catalyst had no effect under pyrolytic conditions. In the presence of the *p*-benzoquinone scavenger, the yield of hydrogen gas substantially increased, since it was not needed for stabilization of radicals. The scavenger in effect inhibited the hydrogen abstraction and intermolecular rearrangement among the radicals generated from the coal. The following pyrolytic mechanism was proposed:

Initiation (thermal cleavage, molecular dimensions $m < n-m \leq n$):

$$C(\text{coal}) \rightarrow R_n\cdot + R_m\cdot . \tag{4-31}$$

Propagation (intermolecular rearrangement):

$$R_n\cdot \rightarrow R_{n-m}\cdot + P_m \tag{4-32}$$

$$R_n\cdot \rightarrow R_m\cdot + C_1 . \tag{4-33}$$

Termination (intermolecular rearrangement):

$$R_{n-m}\cdot + R_m\cdot \rightarrow C_1 + P_m \text{ (or } C_1\text{)}, \tag{4-34}$$

$$R_{n-m}\cdot + R_{n-m}\cdot \rightarrow C_1 + P_m \text{ (or } C_1\text{)} \tag{4-35}$$

$$R_m\cdot + R_m\cdot \rightarrow P_m . \tag{4-36}$$

In this scheme P_m represents any material in the benzene solubility class, C_1 is material in the insoluble residue class, and $R_m\cdot$ represents small secondary radicals including H•.

In a pressurized hydrogen atmosphere, molecular hydrogen apparently reacts directly with the initial coal radicals. The conversion to benzene-solubles in hydrogen was three times that obtained in a nitrogen atmosphere, but the increase was nearly eliminated by the addition of benzoquinone, a radical scavenger. Addition of catalyst nearly doubled the conversion obtained with hydrogen pressure alone and appeared to activate the hydrogen by dissociative adsorption. Temperatures above 673 K only marginally improved conversion, indicating that reaction of $R_{n-m}\cdot$ radicals with coal radicals becomes a predominant reaction where many more radicals are present than the available hydrogen can handle. The role of hydrogen and catalyst was summarized by the following reactions:

non-catalytic hydrogenation,

$$R_m\cdot + H_2 \rightarrow P_m + R_m\cdot, \tag{4-37}$$

$$R_{n-m}\cdot + H_2 \rightarrow C_1 + R_m\cdot; \tag{4-38}$$

catalytic and thermal hydrogenation,

$$C_1 + H_2 \rightarrow C_1H. \tag{4-39}$$

In the latter reaction, C_1H is hydrogenated benzene-insoluble material that is amenable to further conversion, in contrast to other more stable benzene-insoluble residue, C_1.

The free-radical nature of the reaction between coal and solvent (tetralin) was confirmed in separate solvent and catalyst interaction experiments. The presence of tetralin greatly increased the benzene-solubles yield obtainable from the pyrolysis reactions, apparently by stabilization of the coal-derived radicals and formation of tetralyl radicals. The tetralyl radicals are believed to have participated further in stabilization reactions with coal-derived or secondary-product-derived radicals, and also to have yielded hydrogen which was utilized through catalytic reactions. The hydrogenated intermediate C_1H, although benzene-insoluble can still yield radicals due to thermal cleavage and then react with the donor solvent. The mechanism of coal-solvent interaction is summarized as follows:

hydrogen abstraction,

$$R_m\cdot + TH \rightarrow P_m + R_m\cdot, \tag{4-40}$$

$$R_{n-m}\cdot + TH \rightarrow R_m\cdot + C_1; \tag{4-41}$$

hydrogen transfer,

$$C_1 + TH \rightarrow C_1H + P_m, \tag{4-42}$$

where TH is a coal-derived or externally added hydrogen donor.

The overall reactions occurring in donor-solvent liquefaction of coal can be viewed as proceeding by two parallel paths. One path yields gas and benzene-soluble products directly from coal, and the other path involves the formation of intermediates which then degrade to gas and benzene-soluble products. By combination of all the previous results relating to the roles of temperature, hydrogen, solvent and catalyst in coal conversion, the mechanistic scheme shown in Table 4-17 has been constructed to represent the overall reactions [15, 16]. Rate equations were derived, solved, and shown to fit the available experimental data and account for the variations in reaction order. Based on this analysis, it was also concluded that activation energy for the overall liquefaction reactions has little theoretical significance. Many of the reactions proposed in Table 4-17 have been confirmed in model compound studies, especially the participation of molecular hydrogen directly in the free-radical reactions [46].

4.6 Future Work

An increasing degree of sophistication has been observed in the formulation of lumped models for coal liquefaction since the model of Weller et al. [49]. This has in part been due to the development of improved analytical techniques for product analysis and classification. It is expected that with the further improvement of such techniques, a greater degree of flexibility will be available in formulating lumped kinetic models.

Table 4-17

Proposed Free-Radical Mechanistic Scheme for Donor-Solvent Liquefaction of Coal [16]

Type of reaction	Mechanism
(1) Pyrolysis	$C \rightarrow R_n\cdot + R_m\cdot$
(2) Intramolecular rearrangement	$R_n\cdot \rightarrow C_1 + R_m\cdot$
	$R_n\cdot \rightarrow P_m + R_{n\text{-}m}\cdot$
(3) Biradical coupling and Intermolecular rearrangement	$2R_m\cdot \rightarrow P_m$
	$R_m\cdot + R_{n\text{-}m}\cdot \rightarrow C_1 + P_m$
(4) Hydrogen abstraction from solvent and soluble products	$R_m\cdot + TH \rightarrow P_m + R_m\cdot$
	$R_{n\text{-}m}\cdot + TH \rightarrow R_m\cdot + C_1$
(5) Hydrogen transfer	$C_1 + TH \rightarrow C_1H + P_m$
(6) Catalytic hydrogenation	$C_1 + H_2 \rightarrow C_1H$
(7) Stabilization by hydrogen	$R_m\cdot + H_2 \rightarrow P_m + R_m\cdot$
	$R_{n\text{-}m}\cdot + H_2 \rightarrow C_1 + R_m\cdot$
(8) Hydrogen abstraction from intermediate hydrogenated species	$C_1H + R_m\cdot \rightarrow R_n\cdot + P_m$
	$C_1H + R_{n\text{-}m}\cdot \rightarrow R_n\cdot + C_1$

Very little is understood at the present time about the retrogressive reactions, which could play an important role due to the preponderance of hydrogen-scavenging free radicals. It is an understanding of the retrogressive reactions that is currently needed to obviate their occurrence. The work reported so far clearly indicates that the hydrogen-donor capability of the solvent plays a very important role in various reaction paths. Future work should include a quantitative evaluation of this role. Finally, free-radical reaction models should be experimentally verified with the techniques outlined in Chapter 2.

REFERENCES

1. Aczel, T., Williams, R. B., Pancirov, R. J., and Karchmer, J. H., "Chemical Properties of SYNTHOIL Products and Feeds," Final Report to ERDA under Contract No. E(46-1)-8007, September 1976.
2. Anderson, L. L. and Kang, D., 176th ACS National Meeting (Miami Beach 9/10–15/78), *ACS Div. Fuel Chem. Preprints,* **23** (4), 118 (1978).
3. Attar, A., "The Kinetics of Coal Liquefaction in a Hydrogen Donor Solvent," *ACS Div. Fuel Chem. Preprints,* **23** (4), 169 (1978).
4. Bellman, R. and Kalaba, R., *Quasilinearization and Nonlinear Boundary Value Problems,* American Elsevier, New York 1965.
5. Brunson, R. J., *Fuel,* **58**, 203 (1979).
6. Catalytic, Inc., "SRC Process-Operation of Solvent Refined Coal Pilot Plant at Wilsonville, AL," 1976 Annual Report, FE-2270-15, June 1977.
7. Catalytic, Inc., "SRC Process Operation of Solvent Refined Coal Pilot Plant at Wilsonville, AL," 1977 Annual Report, FE-2270-31, May 1978.
8. Cronauer, D. C., Shah, Y. T., and Ruberto, R. G., *I&EC Proc. Des. & Dev.,* **17,** 281 (1978).
9. Cronauer, D. C. and Ruberto, R. G., "Investigation of Mechanism of Reactions Involving Oxygen-Containing Compounds in Coal-Hydrogenation," EPRI Report AF 913, Final Report under Project 713, March 1979.
10. Curran, A. P., Struck, R. T., and Gorin, E., *I&EC Proc. Des. & Dev.,* **6**, 166 (1967).
11. Curran, G. P., Struck, R. T., and Gorin, E., "The Mechanism of the Hydrogen Transfer Process of Coal and Coal Extract," *ACS Div. Petroleum Chem. Preprints,* **C**, 130 (1966).
12. Farasciu, M., Mitchell, T. O., and Whitehurst, D. D., *ACS Div. Fuel Chem. Preprints,* **21**, 11 (August–September 1976).
13. Francis, W., *Coal,* Edward Arnold Publishers, London, 1961.
14. Guin, J., Tarrer, A., Taylor, L., Jr., Prather, J., and Green, S., Jr., *I&EC Proc. Des. Dev.,* **15**, 490 (1976).
15. Gun, S. R., Sama, J. K., Chowdhury, P. B., Mukherjee, S. K., and Mukherjee, D. K., "A Mechanistic Study of Hydrogenation of Coal. 1," *Fuel,* **58,** 171 (1979).
16. Gun, S. R., Sama, J. K., Chowdhury, P. B., Mukherjee, S. K., and Mukherjee, D. K., "A Mechanistic Study of Hydrogenation of Coal. 2," *Fuel,* **58,** 176 (1979).

17. Han, K. W., and Wen, C. Y., *Fuel,* **58**, 779 (1979).
18. Heredy, L. A. and Fugassi, P., *Coal Science,* Advances in Chemistry Series No. 55, ACS Washington, DC, 1966, Chapter 28.
19. Hill, G. R., Anderson, L. L., and Shifai, M. Y., *Fuel,* **53**, (1), 32–37 (January 1974).
20. Hill, G. R., Hariri, H., Reed, R. I., and Anderson, L. L., *Coal Science,* Adv. in Chem. Series No. 55, Am. Chem. Soc., Washington, DC, 1966, pp 427–447.
21. Hill, G. R. and Lyon, L. B., *I&EC,* **54**, 636 (1962).
22. Kang, D., Anderson, L. L., and Wiser, W. H., "Elucidation of Coal Structural Components by Short Residence-Time Extractive Liquefaction," Technical Report to the U.S. DOE, University of Utah, Contract E(49-18)-2006, August 1979.
23. Kang, C. C., Nongbri, G., and Steward, N., *ACS Div. Fuel Chem. Preprints,* **21**, 19 (August–September 1976).
24. Kittrell, J. R., Mezaki, R., and Watson, C. C., *I&EC,* **58**, 51 (1966).
25. Langanbach, J. R., Droege, J. W., and Chauhan, S. P., *ACS Div. Fuel Chem. Preprints,* **24**, (2), 52 (1976).
26. Lee, E. S., "Coal Liquefaction," Chapter 5 in *Coal Conversion Technology,* C. Y Wen and E. S. Lee, Eds., Addison-Wesley Publishing Co., Reading, MA, 1979.
27. Liebenberg, B. J. and Potgieter, H. G. J., *Fuel,* **52**, 130 (1973).
28. Lowry, H. H., Ed., *The Chemistry of Coal Utilization,* Vols. I, II, John Wiley and Sons, New York, 1945.
29. Lowry, H. H., Ed., *Chemistry of Coal Utilization,* Supplementary Volume, John Wiley and Sons, New York, 1963: Dryden, T. G. C., "Chemical Constitution and Reactions of Coal," Chapter 6; Tschamler, H. and DeRuiter, E., "Physical Properties of Coals," Chapter 2.
30. Mitchell, G. D., Davis, A., and Spackman, W., *ACS Div. Fuel Chem. Preprints,* **21**, 189 (August–September 1976).
31. Morita, M., Sato, S., and Hashimoto, R., *ACS Div. Fuel Chem. Preprints,* **24**, (2), 62 (1976).
32. Morita, M., Sato, S., and Hashimoto, R., *ACS Div. Fuel Chem. Preprints,* **24**, (2), 270 (1976).
33. Neavel, R. C., *Fuel,* **55**, 237 (1976).
34. Oele, A. P., Waterman, H. I., Goldkoop, M. L., and Van Krevelen, D. W., "Extractive Distintegration of Bituminous Coals," *Fuel,* **30**, 169 (1951).
35. Orchin, M., Golumbic, C., and Anderson, J. E., "Studies on the Extraction and Coking of Coal and Their Significance in Relation to Its Structure," U. S. Bureau of Mines Bulletin 505, 1951.
36. Ratto, J. J., Heredy, L. A., and Skowronski, R. P., *ACS Div. Fuel Chem. Preprints* **24** (2), 155 (1979).
37. Ruberto, R. G., and Cronauer, D. C., *Fuel Processing Technology,* **2**, 215–219 (1979).
38. Ruberto, R. G., Jewell, D. M., Jenser, R. K., and Cronauer, D. C., ACS Advances in Chemistry Series, No. 151, 1976, Chapter 3.
39. Shah, Y. T., Cronauer, D. C. McIlvried, H. G., and Paraskos, J. A., *I&EC Proc. Des. and Dev.* **17**, 288 (1978).
40. Shalabi, M. A., Baldwin, R. M., Bain, R. L., Gary, J. H., and Golden, J. O., *I&EC Prod. Des. and Dev.,* **18**, 474 (1979).

41. Shinn, J. H., Hershkowitz, F., Holten, R. R., Vermeulen, T., and Greus, E. A., paper presented at the Annual Meeting of the AIChE, Miami, FL, November 1977.

42. Shinn, J. H. and Vermeulen, T., *ACS Div. Fuel Chem. Preprints,* **24**, (2), 80 (1976).

43. Squires, A. M., paper presented at the Conference on Coal Gasification/Liquefaction, Sponsored by US-USSR Trade and Economic Council, Inc., Moscow, October 1976.

44. Szladow, A. J. and Given, P. H., "Some Aspects of Mechanism and Kinetics of Coal Liquefaction," Technical Report No. 3, submitted to U. S. Department of Energy under Contract No. EX76CO12494, Rept. No. FE-2494-TR-3, 1979.

45. Van Krevelen, D. W., *Coal,* Elsevier, Amsterdam, 1961, p. 393.

46. Vernon, L. W., "Free Radical Chemistry of Coal Liquefaction: Role of Molecular Hydrogen," *Fuel,* **59**, 102 (1980).

47. Walker, P. H., Jr., Spackman, W., Given, P. H., White, E. W., Davis, A., and Jenkins, R. G., "Characterization of Mineral Matter in Coals and Coal Liquefaction Residues," EPRI Report AF-417, Annual Report under Project 366-1, June 1977.

48. Weekman, V. W. and Nace, D. M., *AIChE J.,* **16**, 397 (1970).

49. Weller, S., Pelipetz, M. G., and Friedman, S., *I&EC,* **43**, 1572 (1951).

50. Whitehurst, D. D., Farasciu, M., and Mitchell, T. O., "The Nature and Origin of Asphaltenes in Processed Coals," EPRI Report AF-252, First Annual Report Under Project RP-410, February 1976.

51. Whitehurst, D. D., Farasciu, M., Mitchell, T. O., and Dickert, J. J., Jr., "The Nature and Origin of Asphaltenes in Processed Coals," EPRI Report AF-480, Second Annual Report under Project RP-410, July 1977.

52. Whitehurst, D. D. and Mitchell, T. O., "Short Contact Time Coal Liquefaction," *ACS Div. Fuel Chem. Preprints,* **21** (5), 127 (1976).

53. Winans, R. E., Hayatu, R., McBeth, R. L., Scott, R. G., Moore, L. P., and Studier, M. H., *ACS Div. Fuel Chem. Preprints,* **24**, (2), 196 (1976).

54. Wiser, W. H., *Fuel,* **47**, 475 (1968).

55. Wiser, W. H., Anderson, L. L., Qader, S. A., and Hill, G. R., *J. Appl. Chem. Biotechnol.,* **21**, 82 (1971).

56. Wiser, W. H. and Hill, G. R., "A Kinetic Study of the Thermal Dissolution of a High-Volatile Bituminous Coal," in *Proceedings of the Symposium on the Science and Technology of Coal,* Ottawa, Mines Branch, Dept. of Energy, Mines & Resources (Canada), 1976, pp. 162–167.

57. Zielke, C. W., Struck, R. T., Evans, J. M., Costanza, C. P., and Gorin, E., *I&EC Proc. Des. & Dev.,* **5**, (2), 158, (1966).

5. Design of Coal-Liquefaction Reactors

Y.T. Shah, S.J. Parulekar, and N.L. Carr

INTRODUCTION

In this chapter, some design considerations for coal-liquefaction preheaters and reactors are presented. Pressure-drop and heat-transfer considerations for the preheater, as well as models for the product-yield structure in an isothermal reactor and the thermal behavior of an adiabatic reactor (with and without an intrareactor quench), are evaluated. In view of the uncertainty involved in the development of a definitive lumped kinetic scheme, the reactor models use a generalized and yet simple set of reaction pathways. Wherever possible, use of information obtained from existing reactors is made to check the validity of the models. Physical, transport, and kinetic parameters required for the models are either estimated or measured experimentally. Problems and concepts involved in the reactor scale-up are also briefly discussed.

5.1 Reactor Types

A reactor unit in a direct coal-liquefaction process consists of a slurry preheater and a reactor. The preheater is usually an open coiled tube wherein coal-oil slurry is heated from approximately 340 K to the desired reactor inlet temperature. Simultaneously, a large fraction of the coal is dissolved in the vehicle solvent by the time the slurry reaches the reactor inlet. In the reactor, coal dissolution is completed and hydrogenation and hydrocracking of heavy coal liquids occur. In all liquefaction processes (e.g., SRC, H-coal or Exxon donor solvent) that are currently being investigated, an open-tube, three-phase slurry reactor is in use. In Fort Lewis (Tacoma) SRC-II operation the coal-oil slurry and hydrogen flow cocurrently

Yatish T. Shah, Reaction Engineering in Direct Coal Liquefaction, ISBN 0-201-07381-1

upwards through such a reactor. Some details of the reactor, which is 0.61 m in diameter and approximately 10.1 m long, are shown in Figure 5-1. The diameter of

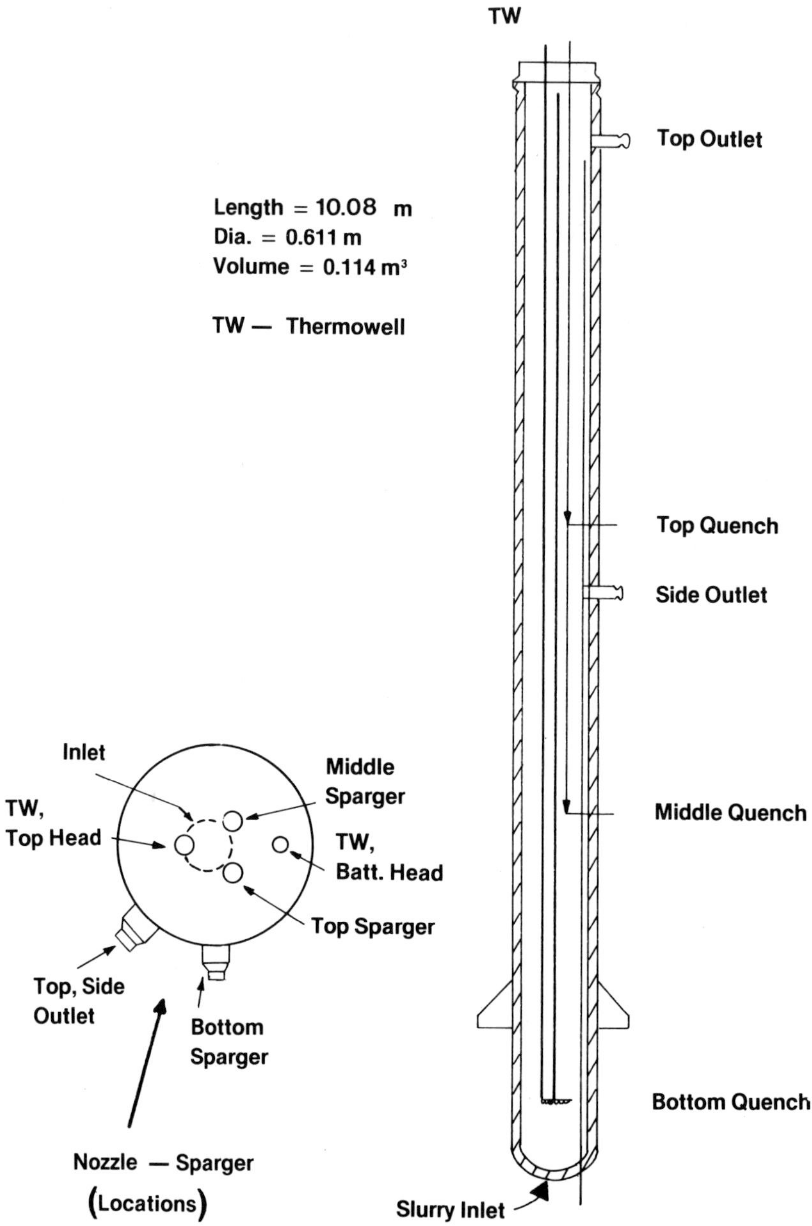

Figure 5-1. A schematic of the Fort Lewis SRC-II reactor.

the 6000-ton/day demonstration plant SRC-II reactor is planned to be approximately 3.96 m. The reactor is essentially adiabatic and contains several quench locations. The Exxon donor-solvent coal-liquefaction process uses a similar reactor configuration. The length-to-diameter ratio of the reactor is, however, kept large in order to improve the plug flow for both gas and slurry phases. This can be achieved by using a number of reactors in series as shown in Figure 5-2. The EDS process at Baytown uses four reactors in series, each 0.61 m in diameter and 18.3 m long. The H-coal process uses an expanded bed reactor, and the reactor contains internal cup as shown in Figure 5-3, which is used for an internal recirculation of the catalyst and coal-oil slurry. This recirculation provides a high degree of mixing. The diameters of the demonstration and commercial reactors are planned to be approximately 1.83 m and 4.26 m respectively. The ranges of temperature, pressure, space time, and solvent:coal ratio examined in the SRC-II, EDS, and H-coal reactors are approximately 673–735 K, 6.9–27.6 MPa, and 15–60 min, and 5–1.5:1 respectively. H-coal uses higher pressure because it improves catalyst aging. At present, the H-coal reactor uses 1.6-mm (1/16-in.) extrudate catalyst at a rate of approximately $2.5 \times 10^{-4} - 3.0 \times 10^{-4}$ (kg catalyst)/(kg coal).

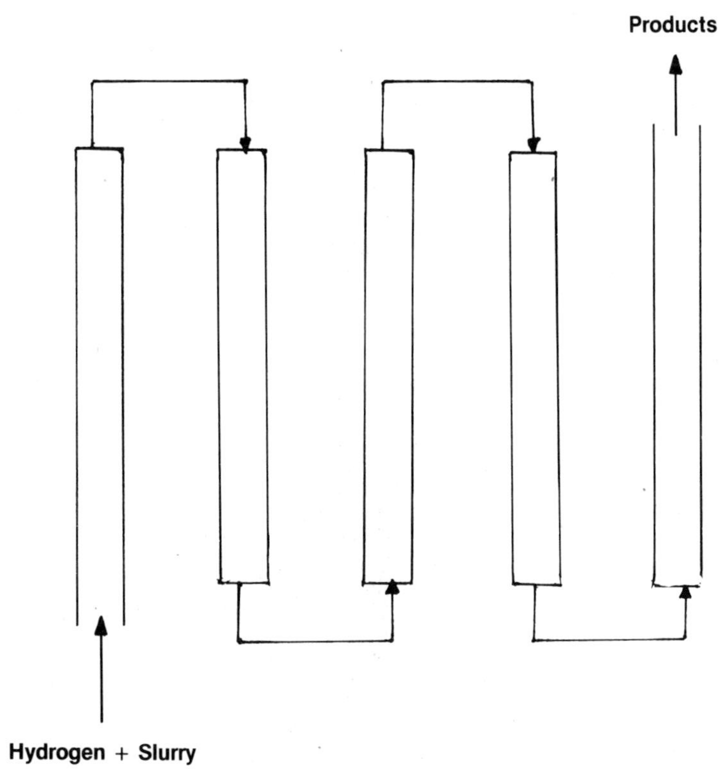

Figure 5-2. Typical coal-liquefaction reactors in series (EDS approach).

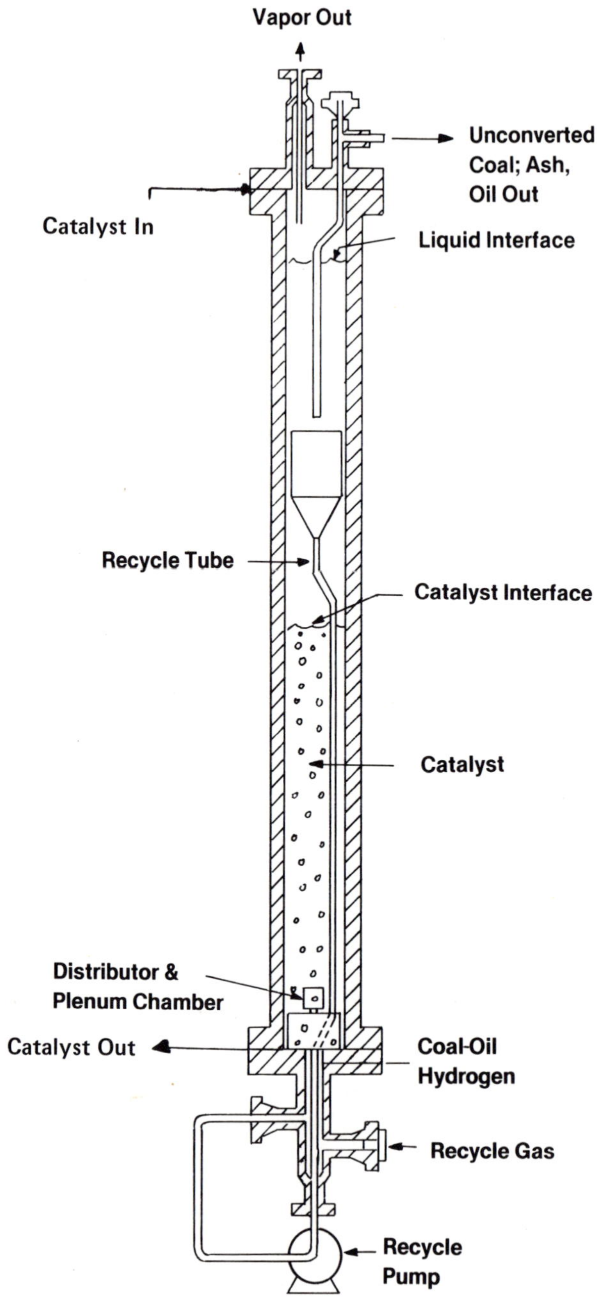

Figure 5-3. A schematic of the H-coal reactor.

5.2 Important Factors in Reactor Design

As mentioned earlier, a typical coal-liquefaction reactor unit consists of a transfer-line slurry preheater and a reactor. The preheater raises the temperature of the coal-oil slurry to the desired reactor inlet temperature. During this heatup, simultaneous dissolution of a large fraction of coal occurs producing coal liquids and gaseous products. The rate of coal dissolution depends upon the coal type, temperature, pressure, and nature of solvent, but it is generally found to be independent of particle size (as long as particle size is below approximately 100–200 μm). Some typical results illustrating the particle-size effect on the coal dissolution rate are shown in Figure 5-4. As shown, neither product particle-size distribution nor conversions (pyridine solubles) significantly change with initial particle size.

An important factor in the preheater design is the flow behavior of the solid-gas-liquid slurry. In the portion close to the preheater inlet, the weight fraction of solids is high. This is followed by a preheater section which contains a very viscous (gel-like) non-Newtonian mixture, particularly for swelling coals. Since the main function of the preheater is to heat the coal-oil slurry, the effects of this non-Newtonian fluid on slurry-wall heat-transfer characteristics as well as flow behavior are important factors in the preheater design. Axial holdup variations due to dissolution of coal can also be very important in the design of the preheater. Normally the combined liquid and solid holdup is much more than the gas holdup, and the amount of solid in the solid-liquid slurry is appreciable, especially in the SRC-II process.

It has been experimentally observed that the fragmentation and dissolution of coal in solvent are extremely rapid. Further reactions of the coal in liquid form are much slower than its dissolution. The overall coal-liquefaction reaction scheme can thus be divided into two parts based on the magnitude of the reaction rates. Fragmentation of coal into gases and intermediate liquid products is assumed to be complete in the preheater, and the associated reactions can be termed "fast" reactions. These intermediates undergo further "slow" reactions, mainly in the reactor, to yield the final products. This approach indicates that the models for the reactor and the preheater can be separated as a slow-reaction zone and a fast-reaction zone, respectively.

The assumption that the major part of the coal dissolution is complete in the preheater can be substantiated by examining the particle-size distribution observed in actual reactors. McGinnis [41] studied the solids sampled from the Fort Lewis SRC-II reactor for the liquefaction of Robinson Run and Powhatan source coals. The particle-size distribution of reactor solids separated from sampling at various reactor locations (bottom, midpoint, top or equivalent) were found to be fairly constant. The median size of the particles was about 3–4 μm, whereas the feed-coal particle size was greater than 100 μm and 95% were smaller than 20 μm. The size of the reactor solids was surprisingly unaffected by continuous long-run operations, when lined-out conditions had been reached. For Powhatan coal, the solids contained 64.8% ash. The typical particle-size distribution obtained in this study is illustrated in Figure 5-5.

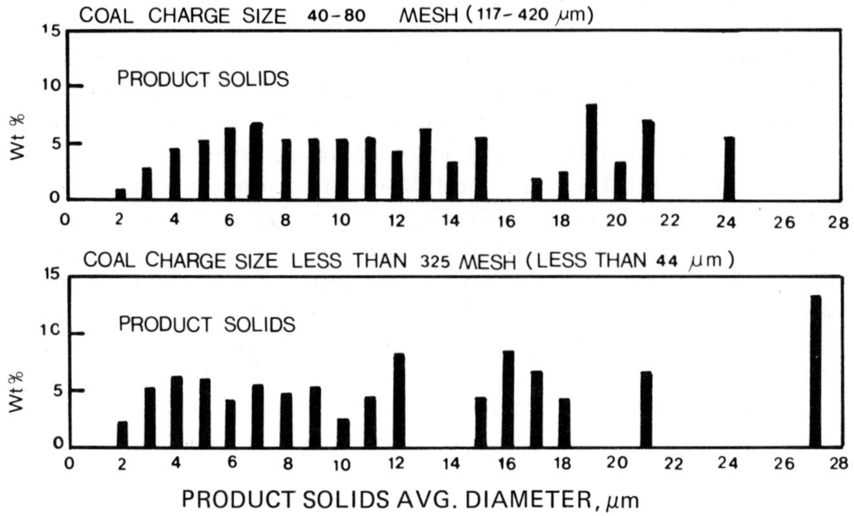

Figure 5-4(a). Effect of coal-charge particle size on product-solids particle size (Gulf data). Autoclave charge: 33 wt% Big Horn coal, 67 wt% anthracene oil. Operating conditions: 700 K, 24.1 MPa, 1.0 hr.

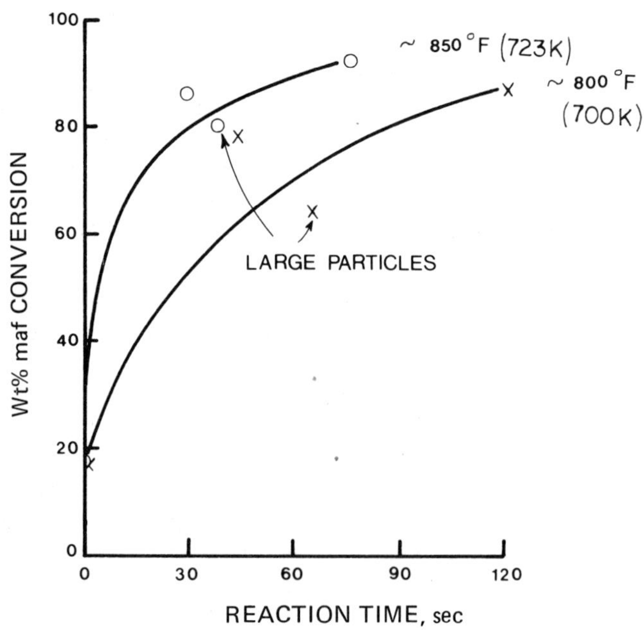

Figure 5-4(b). Burning Star coal conversions: particle-size effect (Whitehurst et al. [66]).

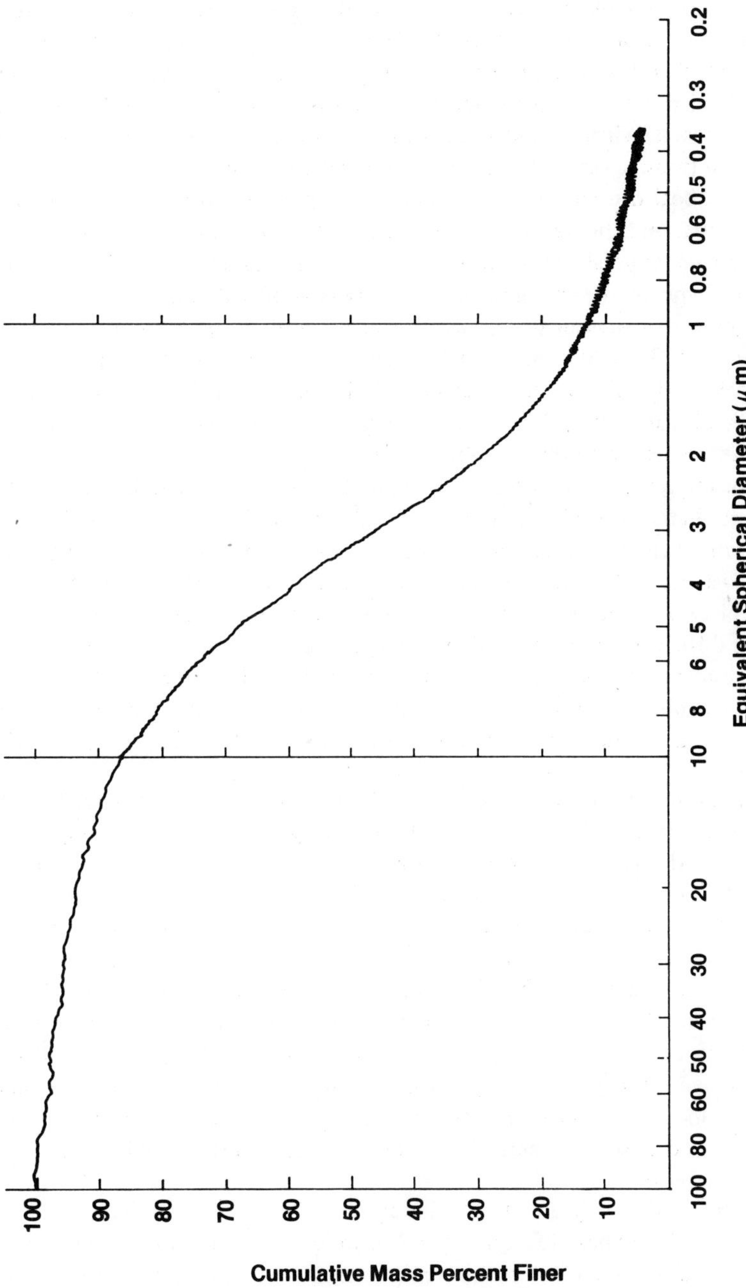

Figure 5-5. Typical particle-size distribution obtained in the Fort Lewis Tacoma SRC-II reactor [41].

The petrographic data for these solids showed a significant amount of pyrrhotite and partially reduced pyrite. Quartz, clay, and shale (or bone) were also identified. Semifusinite, fusinite, and micrinite-macrinite constituted coal-derived inerts. Evidence of some semi-inert material was also obtained in photomicrographs. Partially reacted vitrinite and semicoked vitrinite also appeared with other materials. Vitroplast (mosaic carbon) was found to be relatively scarce.

The coal dissolution rate is dependent upon the coal type, the temperature, the pressure, and the nature of the solvent. The solid phase in the reactor consists of inorganic mineral matter, organic matter, and catalyst particles (if externally added). Some of the mineral matters possess good catalytic activity. It has been found by earlier investigators, such as Henley et al. [25], Prather et al. [51], and Guin et al. [23], that coal minerals catalyze the hydrogenation and desulfurization reactions in the SRC process. Han et al. [24] found that the catalyst plays a role in hydrodesulfurization and conversion of heavier liquids (referred to later as heavy distillate) to lighter products (light distillate).

In the reactor, hydrogenation and hydrocracking yield additional gaseous and liquid products, whereas the amount of coal is reduced by further dissolution. Holdup variations resulting from these effects and from feed composition, the settling of solid coal minerals and solid catalyst particles, and partial vaporization of the liquid phase may be very important factors in the reactor design. The major factors causing difficulties in design and modeling of a coal-liquefaction reactor are: (1) unequal residence times of the various phases and (2) different extents of mixing in the various phases. These result from unequal holdups and velocities of the various phases, which in turn are due to the complex flow patterns prevailing within the reactor.

Because of the complexities of three-phase flow accompanied with mass transfer and chemical reaction, it is difficult to predict the performance of the preheater and the reactor from basic principles in a simple way. It has been experimentally found that the absorbed hydrogen does not increase the weight of the liquid phase appreciably, either in the preheater or in the reactor. The desorbed gases (consisting of light hydrocarbons C_1–C_4, some naphtha, water, and byproducts such as CO_2, CO, NH_3, and H_2S) have usually much higher molecular weight than the gas being absorbed (mainly hydrogen). The formation of these volatile products results in significant loss of the liquid products. Both gas and liquid velocities may vary axially in the reactor depending upon the relative magnitudes of local absorption and desorption rates, and average molecular weights of absorbed and desorbed gases. The extent of this change would depend upon the reaction severity.

The work reported in the literature indicates that the holdups of various phases in a three-phase reactor depend on velocities of the various phases, column size, configuration and internals, coal (or catalyst) particle size, particle concentration, and various physical properties (such as density, surface tension, viscosity) of the fluid phases. Hence, if the velocities of the respective phases as well as the solid concentration vary axially in the reactor. the holdups of various phases will also vary. All these factors further add to the complexities in design and modeling of such reactors.

Finally, in a catalytic coal liquefaction reactor, (such as the H-coal reactor), the availability of hydrogen at the catalyst surface is quite important. The catalyst concentration can play an important part in the kinetics of various reactions (in particular, the hydrogen consumption rate), and hence the proper prediction of the axial catalyst concentration distribution is also required.

5.3 Preheater Design

The main purpose of the preheater is to raise the temperature of coal-oil slurry and hydrogen from approximately 340 K to the desired reactor inlet temperature. During this heating a significant amount of coal dissolution occurs. The preheater in coal liquefaction thus also functions as an extension of the reactor.

Much work has been performed in tubular preheaters in order to develop an understanding of the dissolution process and to obtain design data for larger-scale liquefaction-process preheaters [61, 62]. Although much has been learned from these studies, the rate data are very difficult to interpret because the flow regimes in the preheater are not well understood, nor are the physical-property changes fully defined. Consequently, it is difficult to calculate the actual slurry residence time. Figure 5-6 shows some typical conversion data for a four-section upflow coiled preheater having an internal diameter of 3.1 mm and a total length of 18.6 m. The space time is at the isothermal temperature indicated and does not include the

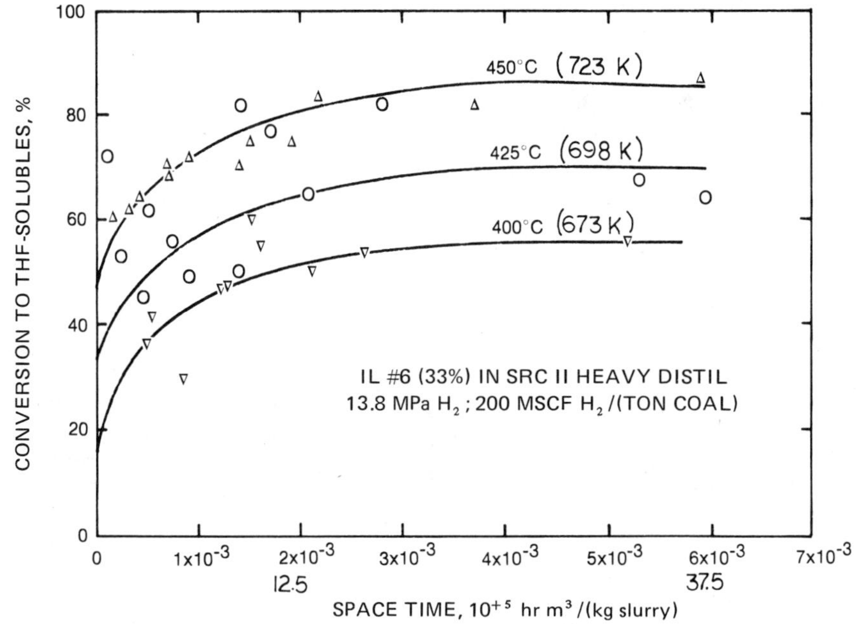

Figure 5-6. Conversion to THF-solubles in preheaters [61].

preheater length required to heat the slurry to that temperature (about 0.61 m). The slurry rate was varied from 0.41 to 1.77 kg/hr with 33% Illinois #6 Burning Star Coal in 67% SRC-II heavy distillate. The hydrogen rate was 200 MSCF/(ton of coal) at a pressure of 2000 psi (13.8 MPa). If plug flow conditions are assumed to prevail in the preheater, the flow times can be estimated to be about 30 sec at a space time of 3.12×10^{-5} hr m^3/kg and about 240 sec at a space time of 1.87×10^{-4} hr m^3/kg. The range of data extends from about 20 sec to 8 min, based on these estimates. Conversion was measured in terms of tetrahydrofuran (THF) solubility as shown in Figure 5-6. Conversion to THF-soluble products was essentially completed within 3 to 4 min, while the conversion to benzene-soluble products occurred at a much slower rate. Product distributions were generally the same as obtained in short-residence-time autoclave studies, and the maximum in the preasphaltene yield shifted to shorter residence times as the temperature was increased from 673 to 723 K.

For thermal hydrodynamic modeling the preheater can be divided into three axial regions as illustrated in Figure 5-7:

Region I. The coal-slurry-gas mixture is considered to be a Newtonian two-phase two-component mixture if the solvent is recycled process fluid devoid of SRC. (SRC is essentially non-Newtonian.) The viscosity decreases smoothly with increasing temperature. There is virtually no change in the flow rates of the components. Region I ends when the viscosity reaches its local minimum value. The flow regime can be turbulent-laminar or laminar-laminar. For horizontal preheaters the flow pattern is probably slug or stratified.

Region II. This region is characterized by a sharp increase in viscosity as agglomeration and swelling becomes the dominant mechanism. The slurry becomes non-Newtonian. The fluid flow regime changes to laminar flow, and the flow pattern may change. The viscosity reaches a maximum due to swelling and solvation and levels off. Region II ends when significant chemical reaction and dissolution occur.

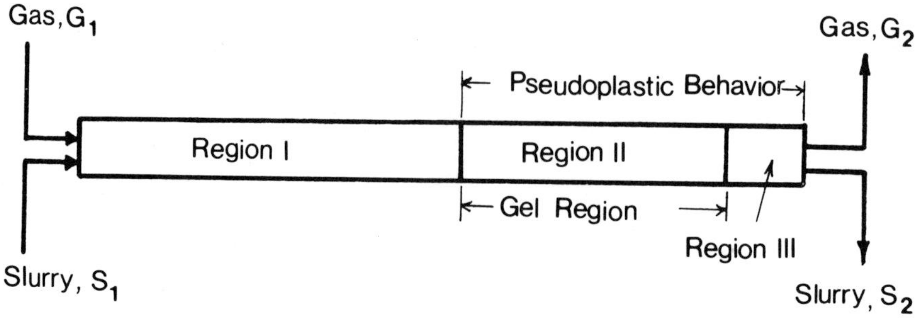

Figure 5-7. Coal-liquefaction preheater: hydrodynamic and heat-transfer model.

Region III. Depolymerization and dissolution of the gel take place due to chemical reaction. The viscosity decreases rapidly. Subsequently the viscosity decreases smoothly as the temperature increases. The mixture remains non-Newtonian [59]. There is a considerable decrease in the average density. Turbulent two-phase non-Newtonian flow may occur.

For a reliable preheater design, boundaries for the three regions need to be established. With certain low-grade subbituminous coals and lignites which are non-swelling, the gel region is eliminated. The three regions characterize slurry heating, swelling, chemical reaction and dissolution, and initial hydrogenation; the typical viscosity behavior in these three regions is described in Figure 5-8.

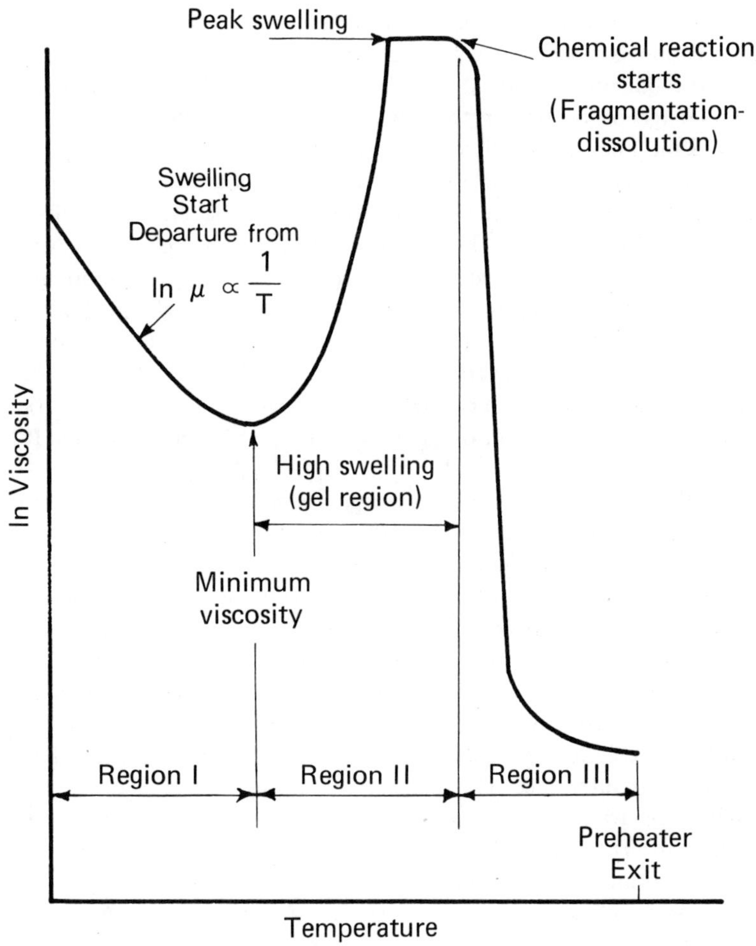

Figure 5-8. Viscosity behavior in preheater.

In the next few pages we briefly discuss some aspects of the viscosity of the coal-oil slurry, the pressure drop, and the heat transfer in the preheater. A successful knowledge of these three design parameters plays an important role in the preheater design. Finally, a simple kinetic model, which can form the basis for evaluating kinetic performance of the preheater, is briefly described.

5.3.1 *Coal-Oil Slurry Viscosity*

Accurate knowledge of the coal-slurry viscosity is important because the viscosity is used in pressure-drop, heat-transfer, holdup, and flow-pattern calculations and model development. Viscosity is also thought to be an important variable in maintaining stable slurry mixing and pumping [49].

As the coal-oil slurry passes through the preheater the coal swells and partially dissolves, and hydrogenation is initiated. This region of swelling and the initial chemical reaction is associated with large changes in viscosity and highly pseudoplastic (shear-thinning) behavior: the "gel" region. To obtain an optimized preheater design, the range of parameters over which coal slurry exhibits non-Newtonian behavior and the value of the viscosity must be known and be predictable. It is not surprising that coal slurries exhibit non-Newtonian pseudoplastic characteristics, because coal is commonly accepted as a macromolecular material not unlike a polymer [52, 65]. Moreover, it is known that asphalts and asphaltenes, which are related to coal, show non-Newtonian behavior [4, 36, 65]. Several important parameters which influence the viscosity of coal slurries are: (1) coal and solids concentration, (2) solvent viscosity, (3) type of coal, (4) shear rate, (5) time, (6) temperature, and (7) degree of coal conversion.

Viscosity data from reference [67] are shown in Figure 5-9 for 20%, 25%, 28.5%, and 33% weight fraction of Kentucky #11 coal in process solvent (SRC-I). Three important facts are revealed by the data. As the weight percent coal increases: (1) the peak viscosity in the gel region increases significantly, (2) the gel region begins at a lower temperature, but ends at same temperature, and (3) the temperature span of the gel region is increased.

The nature of solvent has been found to have some effects on coal-oil slurry viscosity. Recent data obtained by Castillo and Williams [8] for the viscosity of coal-oil slurries using Aroclor and glycerin as solvent at 293 K show such an effect. The low-strain-rate viscosity at 40% volume fraction coal was found to be an order of magnitude higher using Aroclor. Differences are also caused by different solvents such as SRC-I, SRC-I process solvent, etc. Use of filtered or unfiltered feed produces significant viscosity differences [19].

Swelling behavior, coking, and the formation of a gel region are observed in bituminous coals. The temperature and the extent of the gel region depend on the coal type. Subbituminous coal and lignites do not exhibit a gel region. Typical viscosity-temperature relationships for swelling and nonswelling coals are illustrated in Figure 5-10. Increasing shear rate reduces the viscosity increase in the gel region. Because of its non-Newtonian rheology, the coal-oil slurry viscosity is time dependent (viscosity increases with time). This dependence is stronger at higher temperature.

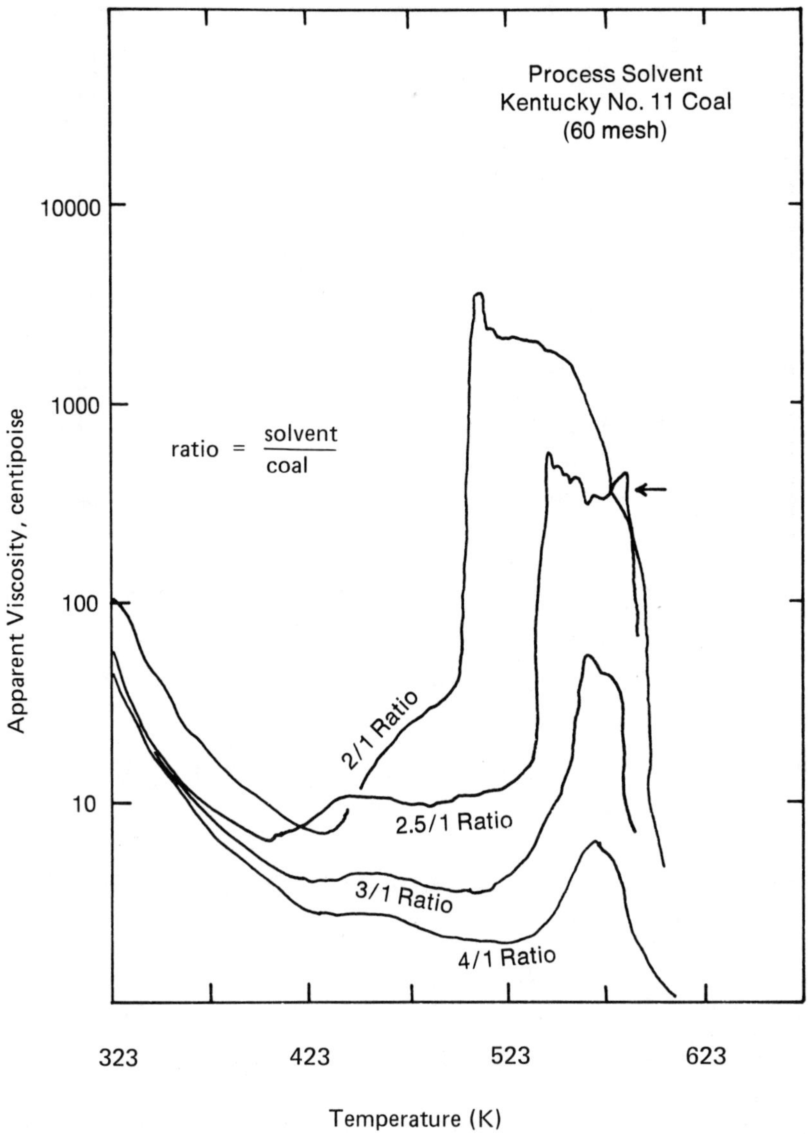

Figure 5-9. Viscosity change during coal dissolution as a function of coal concentration [67].

Temperature affects the viscosity of coal-oil slurry in three important ways: (1) normal behavior (viscosity decreasing as temperature increases), (2) swelling-plastic behavior, and (3) chemical reaction and dissolution. Outside the gel region,

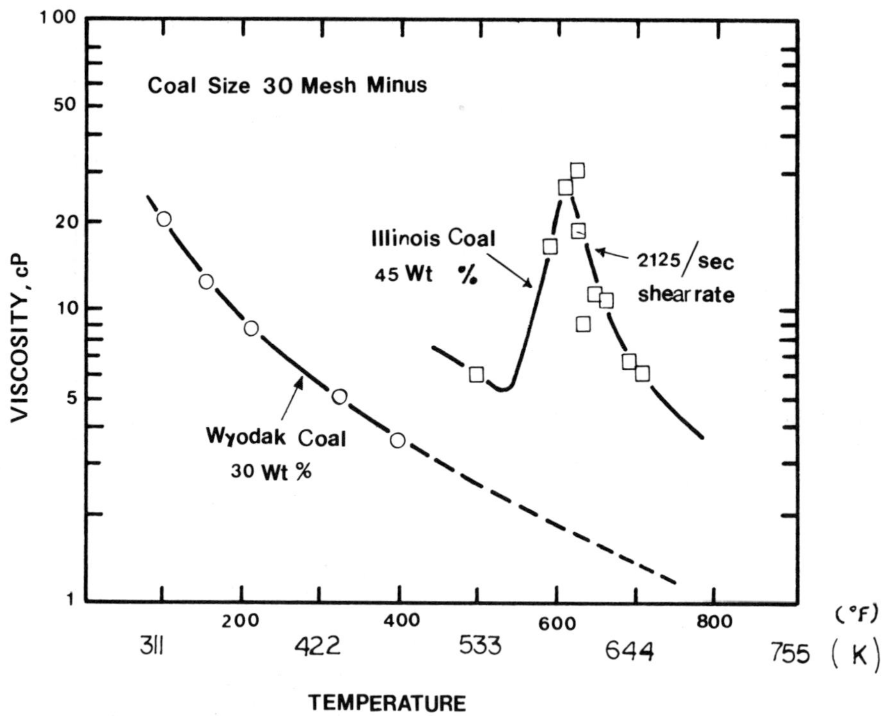

Figure 5-10. Feed slurry viscosity for swelling and nonswelling coal [20].

the viscosity of a coal-oil slurry decreases smoothly as temperature increases; the normal behavior is expressed by

$$\ln \mu = A + \frac{B}{T^n} . \tag{5-1}$$

or a related form. Note that when $\ln \mu$ versus $1/T$ is graphed for coil-oil slurry a linear relationship is not obtained.

Viscosity data obtained at Sandia Laboratory for 33% Illinois #6 coal-oil slurry flowing through a 3.1-mm-diameter tube at 723 K are shown in Figure 5-11. The viscosity decreased as the conversion increased; an increase in conversion from 60 to 67% decreased the viscosity by a factor of 7. This behavior gives additional evidence that the role of coupled chemical reaction and dissolution to decrease the viscosity following the increase caused by swelling, agglomeration, and/or gel formation.

Swelling and plasticity could increase the slurry viscosity in several ways. Particles could agglomerate, thereby trapping solvent; swelling would increase the

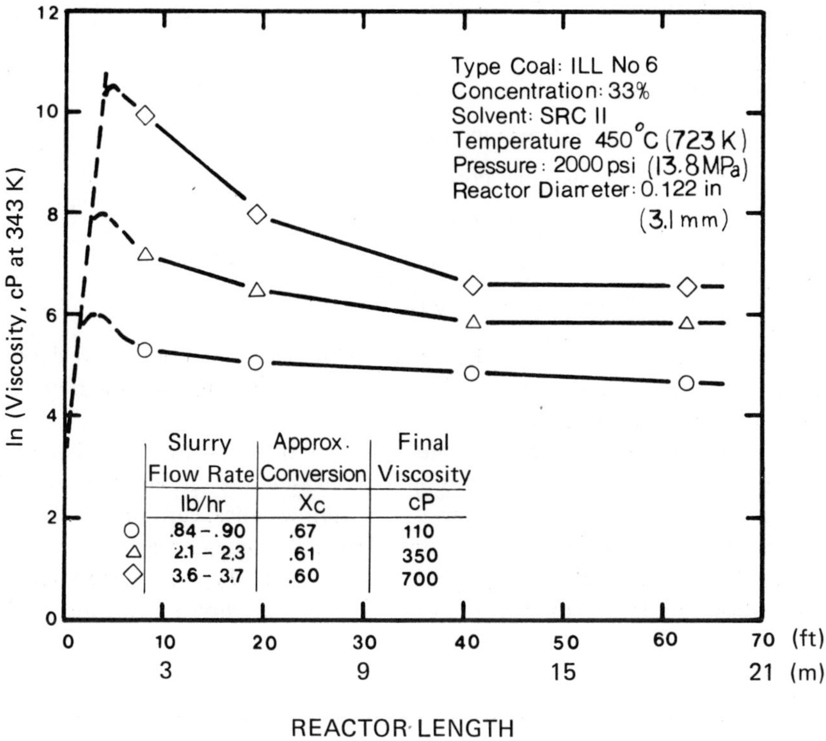

Figure 5-11. Viscosity versus degree of reaction and slurry flow rate [60].

effective volume fraction of the coal particles; and solvent could be adsorbed or absorbed, which again would cause a solvent deficiency insofar as viscosity is concerned. In fact, extensive swelling and agglomeration can cause transformation from a suspension to a gel.

When a caking coal is heated, it generally softens, becomes plastic, and swells. Swelling is influenced significantly by coal type, temperature, heating rate, pressure, and gaseous environment. Berkowitz [2, 3] suggests that swelling is primarily an intra- rather than an interparticle phenomenon, governed by the internal pressure due to the difference between the rates of formation and release of volatile material.

5.3.2 *Pressure Drop and Temperature Distribution*

A widely used empirical correlation for the pressure drop in preheater is expressed as

$$\Delta P = 5.2712 \times 10^{-15} C^{2.444} P^{2.307} R^{1.034} V^{0.953} L^{0.431}, \qquad (5\text{-}2)$$

having a standard deviation of 9.5 psi (65.5 kPa) and a correlation coefficient equal to 0.8616,

where ΔP = preheater pressure drop (psi),

C = coal concentration in feed slurry (wt%),

P = preheater inlet pressure (psia),

V = superficial fluid velocity, ft/hr,

R = ratio of feed slurry volume to feed-gas volume,

L = length of preheater (ft).

The above relation was derived from Wilsonville and Fort Lewis data for Illinois coal [50]. At present, no theoretical correlation for the pressure drop under preheater conditions is available.

In experiments performed on the preheater of the Wilsonville pilot plant, [9, 10, 62] temperature and pressure drop were measured at various points over the length of the preheater. The preheater was an upflow helical coil constructed from 184.6 m of 31.75-mm, schedule-160 pipe. The coil diameter was 1.52 m with a pitch of 114.3 mm and an overall height of 4.36 m. Typical temperature and pressure-drop profiles for Indiana V bituminous coal are shown in Figures 5-12(a)

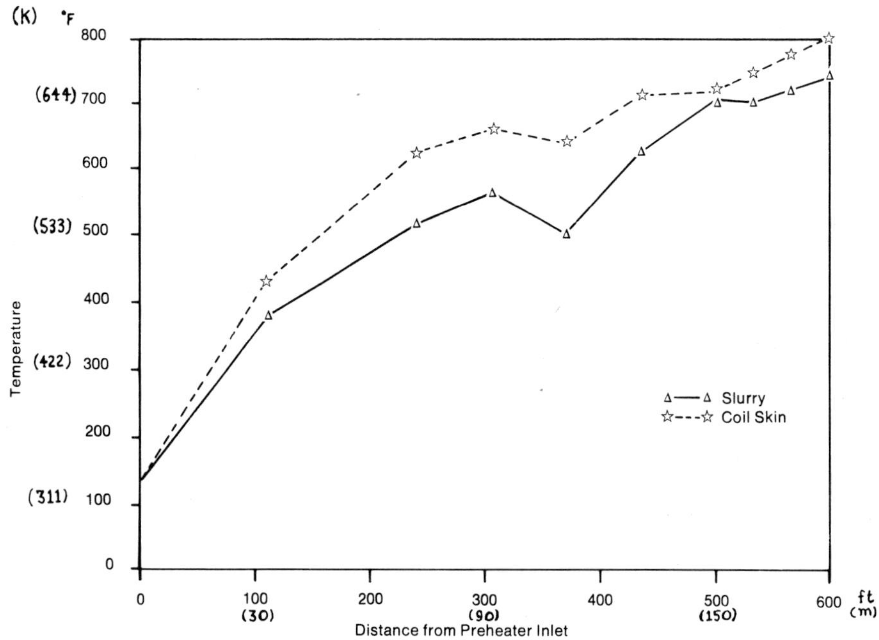

Figure 5-12(a). Temperature profile for preheater [64].

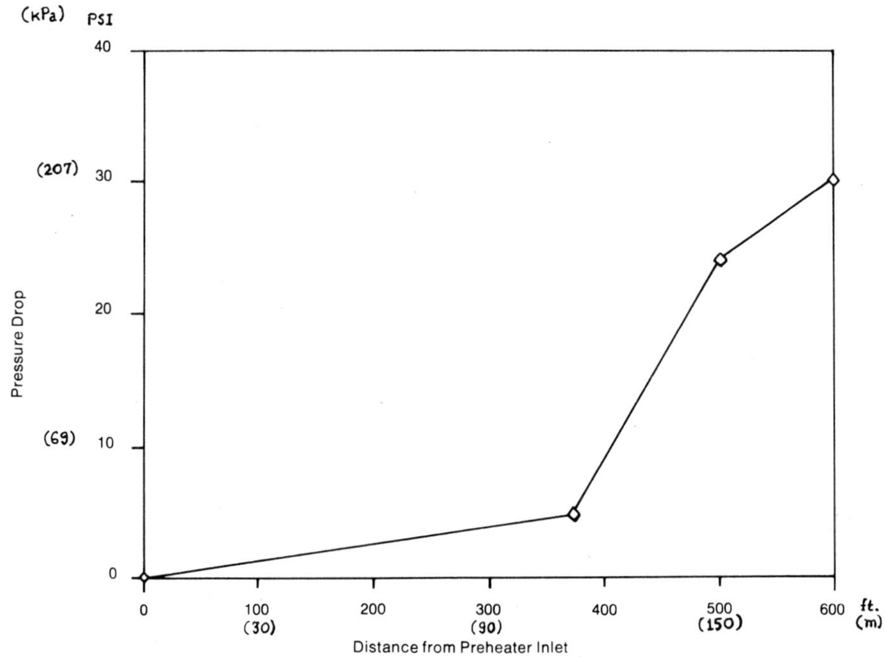

Figure 5-12(b). Pressure drop through preheater [64].

and 5-12(b) respectively [62]. A decrease in the slurry and preheater skin temperature occurs once the slurry reaches a temperature of 555 K at 91.41 m. This indicates that rapid thermal depolymerization of the coal has initiated and is probably an endothermic reaction for this coal. The temperature attains a minimum at about 114.2 m, and at this point the pressure drop begins to increase exponentially. Most of the pressure drop across the preheater occurs over the 114- to 183-m section, and is believed to be a result of the "gel" formation as the coal particles swell and disintegrate in the solvent. Very similar results were obtained in early runs in the Fort Lewis pilot plant. More recently, somewhat more linear variations in the pressure drop have been obtained, the reasons for which are not very clear. More experimental data under a wider range of operating conditions are currently being obtained.

5.3.3 *Heat-Transfer Coefficient*

For the calculation of the heat-transfer rate in a preheater, an estimation of the heat-transfer coefficient is required. Various literature correlations that have been used to correlate the experimental data are illustrated in Table 5-1. None of the literature correlations have been found to correlate the experimental data very

Table 5-1

Heat-Transfer Correlations Currently Used in Coal-Liquefaction Preheater Heat-Transfer Calculations*

Source	Correlation†	Comments
Sieder & Tate [58]	$\dfrac{h}{C_pG} = -.023\left(\dfrac{DG}{\mu}\right)_b^{-0.2} (\mathrm{Pr}_b)^{-2/3}\left(\dfrac{\mu_b}{\mu_w}\right)^{0.14}$	(1) Reynolds number $> 10^4$. (2) All properties at bulk temperature except μ_w. (3) Single-phase, fully developed turbulent flow.
Crewe & Simons [15] (modified form)	$\dfrac{h}{C_pG} = 0.0285\left(\dfrac{DG}{\mu}\right)_b (\mathrm{Pr}_b)^{-2/3}\left(\dfrac{\mu_L}{\mu_s}\right)_b^{-0.25}$	(1) μ_L solvent viscosity. (2) Single-phase slurry flow.
Collier & Pulling [12, 13]	$\dfrac{hD}{k} = 0.175\left(\dfrac{\rho_L}{\rho_G}\right)^{0.28} (\mathrm{Re}')^{0.7}(\mathrm{Pr}_L)^{0.4}$ Re' = Total-flow Reynolds number, slurry viscosity	(1) Saturated forced-convection boiling. (2) One-component flow. (3) Primarily annular flow.
Chen [11]	$\dfrac{hD}{k} = 0.023 \left[G\dfrac{1-X}{\mu}D\right]^{0.8} (\mathrm{Pr}_L)^{0.4}F$ $F = \phi(X_{tt})$	(1) Saturated forced-convection boiling. (2) One-component flow. (3) Primarily annular flow.

*Personal communication with Dr. A. A. Bishop.

†X_{tt} = Lockhart-Martinelli factor, G = mass flow rate of slurry, X = conversion.

well. A typical comparison between experimental heat-transfer coefficient and the ones calculated from Sieder-Tate and Crewe-Simons equations is shown in Figure 5-13. More work is required in this area.

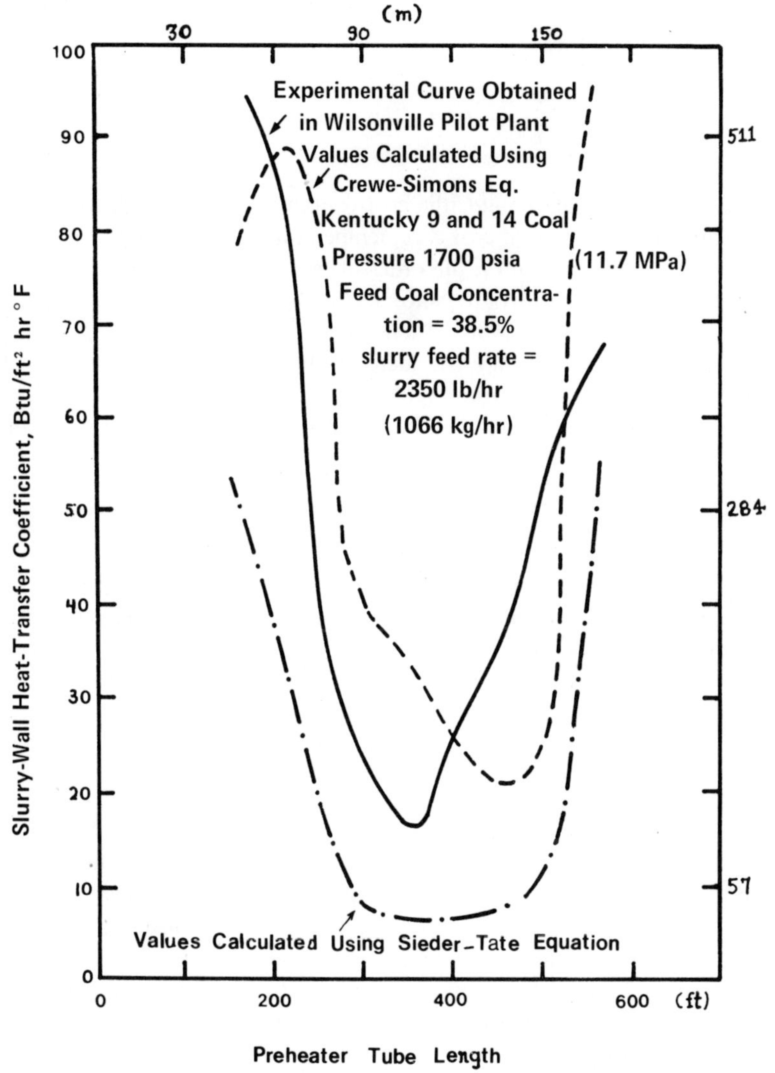

Figure 5-13. Comparison of experimental heat-transfer coefficients with those calculated using Sieder-Tate and Crewe-Simons Equations [30].

5.3.4 *Simple Kinematic Model*

Since the length-to-diameter ratio in most preheaters would be very large, it is reasonable to assume that both gas and slurry phases are in plug flow in the preheater. The most important reaction in the preheater is the coal dissolution, producing a variety of liquid and gaseous products. A simple differential coal material balance can be expressed as

$$\frac{dC}{d\tau} = -RC^n \epsilon_{s\ell}, \tag{5-3}$$

where τ is the slurry space time; C the coal concentration; R the dissolution rate constant, which is a function of coal type, temperature, pressure, nature of solvent, etc.; $\epsilon_{s\ell}$ the slurry holdup; and n the order of the dissolution reaction. The literature indicates n to be either 1 or 2. Since all the products come from coal, similar expressions for each product can be written easily.

Coal dissolution is generally considered to be an endothermic reaction. The total heat of the reaction is, however, generally considered to be small compared to the heat transfer from the walls. The heat balance on the preheater can thus be expressed as

$$\frac{dT}{d\tau} = u\ (T-T_s), \tag{5-4}$$

where u is the dimensionless overall heat-transfer coefficient between coal-oil slurry and hydrogen and the surrounding, and T_s is the surrounding temperature. As shown in the previous sections, u is a complex function of the slurry rheology, the temperature, and the thermal resistance of the preheater walls which needs to be determined experimentally. Equations (5-3) and (5-4) are coupled and together describes the kinetic and thermal behavior of the preheater.

5.3.5 *Summary*

The most important factor in the preheater design for a commercial coal-liquefaction reactor is the size of the preheater. This size can be accurately minimized only with better theoretical understanding of heat transfer and fluid rheology under preheater conditions. Future work should be concentrated in this area. A kinetic model of the preheater would require better understanding of the fast-reaction zone, i.e., short-contact-time coal-dissolution kinetics under the entire temperature range of interest. An accurate description of fluid rheology is also important for this purpose.

5.4 Reactor Models

Several lumped kinetic models for donor-solvent coal liquefaction have been reviewed in the previous chapter. The complexity of the kinetic model depends upon the number of the lumped species used and the chosen reaction paths. As indicated in the previous chapter, our knowledge in the area of lumped kinetic modeling is in the development stage. Since our objective here is to underline the fundamentals of reactor modeling, we illustrate it by choosing a simple kinetic model which carries the basic features of the detailed reaction paths.

Let A denote the heavy distillate (which contains asphaltenes, preasphaltenes, and the reactive organic matter in the coal [16], B denote the light distillate, and C denote the volatile products. The reaction mechanism used for the reactor model can thus be described as shown below.

Reaction 1:

$$\mathrm{A}\,(\ell) \rightarrow \mathrm{B}\,(\ell). \tag{5-5}$$

Reaction 2:

$$\mathrm{A}\,(\ell) \rightarrow \mathrm{C}\,(g), \tag{5-6}$$

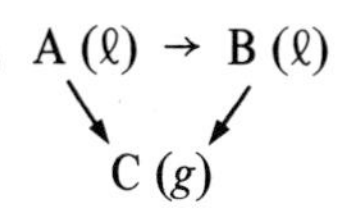

Reaction 3:

$$\mathrm{B}\,(\ell) \rightarrow \mathrm{C}\,(g). \tag{5-7}$$

Reaction 4:

$$\mathrm{H_2}\text{+nonhydrogenated coal liquids} \rightarrow \text{hydrogenated coal liquids.} \tag{5-8}$$

Here ℓ and g denote liquid and gas phases respectively. In general, all reactions are partly thermal and partly catalytic. Species A and B are the mixtures of hydrogenated and nonhydrogenated coal liquids.

5.4.1 *Axial-Dispersion Model for an Isothermal Reactor*

Parulekar and Shah [44] described a simple model for an isothermal reactor. The model assumed: (1) the gas phase follows the ideal gas law, (2) the reactor operates under isobaric conditions (since the total pressure drop is not significant as compared to the total pressure), (3) the axial dispersion model is applicable to nonvolatile liquid and solid components, (4) for simplicity, all reaction rates are directly proportional to the catalyst concentration, (5) the gas phase moves as plug flow, (6) the reactor is the cocurrent upflow type, (7) all the global reaction rates can be expressed in terms of gas phase concentration of hydrogen. (This means that the gas-liquid interface mass-transfer resistance is negligible.) r_1 represents rate of the reaction (5-5) in kg/m³ liq. sec, and r_2, r_3, r_4 represent the rates of the reactions (5-6), (5-7), and (5-8) in kg-mol/m³ liq. sec, so that

$$r_1 = k_i^* e^{-E_i/R_gT} C_{A\ell} C_{H_2} C_s, \qquad i=1,2 \tag{5-9}$$

$$r_3 = k_3^* e^{-E_3/R_gT} C_{B\ell} C_{H_2} C_s, \tag{5-10}$$

$$r_4 = k_4^* e^{-E_4/R_gT} C_{H_2} C_s. \tag{5-11}$$

Finally, (8) there are no radial variations in phase holdup and the concentrations of the reacting species.

A. SOLID DISTRIBUTION

For constant liquid- and gas-phase superficial velocities, low catalyst loading, and small particle size, the model proposed by Cova [14] can be used to predict the catalyst concentration profile in the reactor. The model assumes that the particle motion follows Stokes's law and that the solid mass flux is much smaller than the liquid mass flux. The model also assumes free-settling conditions (settling of individual particles is not influenced by the presence of surrounding particles). The model is not applicable when the particles are large, when the solid loading is high, or when gas and liquid velocities as well as the gas and liquid phase holdups change in the direction of flow.

The model proposed by Parulekar [44] assumes hindered-settling conditions (settling of an individual particle is influenced and hindered by the presence of surrounding particles) and is applicable over a large range of particle size and solid loading. According to his model the mass flux n_s of solid particles may be represented by a modification of Fick's law, the modification being necessary to take into account the effect of settling of the catalyst particles. In the case of a cocurrent upflow reactor the flux n_s can be represented as

$$n_s = \frac{C_s(n_s + n_\ell)}{\rho_{s\ell}} - C_s U_t' - \frac{E_{zs}}{(1-\epsilon_g)} \frac{dC_s}{dx} , \tag{5-12}$$

where n_ℓ is the mass flux of liquid. Both n_s and n_ℓ are based on the cross-sectional area occupied by slurry.

As has been shown by Brown [5] and McCabe and Smith, [40] the settling velocity with respect to the column is related to that with respect to the liquid by the relation

$$U_t' = U_t \frac{\epsilon_\ell}{1-\epsilon_g} , \tag{5-13}$$

where U_t is the terminal settling velocity of the particles with respect to liquid. Information available in the literature on the terminal settling velocity as a function of various variables mentioned earlier can be used to calculate the local value of U_t.

The terminal settling velocity of catalyst particles, U_t', is a function of slurry density, volume fraction of liquid in slurry, viscosity of slurry, catalyst density, and particle diameter. Since the weight fraction of solid in feed slurry encountered in industrial practice is appreciable, the settling of particles may be hindered by the presence of surrounding particles.

The parameters slurry velocity $U_{s\ell'}$ slurry density $\rho_{s\ell'}$ and liquid holdup ϵ_ℓ in the above equations are defined as

$$U_{s\ell} = \frac{n_s + n_\ell}{\rho_{s\ell}} = U_\ell - \frac{\rho_s \epsilon_s U_t}{\rho_{s\ell}}, \tag{5-14}$$

$$\rho_{s\ell} = \frac{\rho_s \epsilon_s + \rho_\ell \epsilon_\ell}{1 - \epsilon_g}, \tag{5-15}$$

and

$$\epsilon_\ell = (1 - \epsilon_g)\left(1 - \frac{C_s}{\rho_s}\right). \tag{5-16}$$

B. MODEL EQUATIONS

We use the following dimensionless variables:

$$z = x/L,$$

$$\bar{U}_g = \frac{U_{0g}}{(U_{0g})_0}, \qquad \bar{U}_\ell = \frac{U_{0\ell}}{(U_{0\ell})_0},$$

$$H' = C_{H_2}/C_T, \quad C_T = P/R_gT, \quad C = 1 - H',$$

$$A = C_{A\ell}/\rho_\ell, \quad B = C_{B\ell}/\rho_\ell,$$

$$\mathrm{Pe}_\ell = \frac{(U_{0\ell})_0 L}{E_{z\ell}}, \qquad \mathrm{Pe}_s = \frac{(U_{0\ell})_0 L}{E_{zs}}, \tag{5-17}$$

$$R_1 = \frac{k_1 \rho_\ell L}{(U_{0\ell})_0}, \quad R_2 = \frac{k_2 \rho_\ell L}{(U_{0\ell})_0}, \quad R_3 = \frac{k_3 \rho_\ell L}{(U_{0\ell})_0}, \quad R_4 = \frac{k_4 L}{(U_{0\ell})_0},$$

$$r = \frac{(U_{0\ell})_0}{(U_{0g})_0},$$

$$f = \frac{U_t}{(U_{0\ell})_0}, \qquad f_1 = \frac{U_t'}{(U_\ell)_0},$$

$$\bar{e}_s = \frac{E_{zs}}{(E_{zs})_0}, \qquad \bar{e}_\ell = \frac{E_{z\ell}}{(E_{z\ell})_0}.$$

Here $k_i = k_i^* \, e^{-E_i/R_gT}$ for $i = 1,2,3,4$.

The relevant equations for this model can be represented as follows:

Gas-phase hydrogen.

$$\frac{d(\overline{U}_g H')}{dz} = - R_4 \, r H' C_s \, \epsilon_\ell \ . \tag{5-18}$$

Gas velocity.

$$\frac{d\overline{U}_g}{dz} = (R_2 A + R_3 B - R_4) \, r \, H' C_s \, \epsilon_\ell \ . \tag{5-19}$$

Liquid phase A.

$$\frac{1}{\mathrm{Pe}_\ell} \frac{d}{dz} \left(\overline{\epsilon}_\ell \frac{dA}{dz}\right) - \frac{d(\overline{U}_\ell A)}{dz} - R_1 \, A \, C_s \, \epsilon_\ell - \frac{R_2 A H' C_s M_\mathrm{C} \, \epsilon_\ell C_T}{\rho_\ell} = 0 \ . \tag{5-20}$$

Liquid phase B.

$$\frac{1}{\mathrm{Pe}_\ell} \frac{d}{dz} \left(\overline{\epsilon}_\ell \frac{dB}{dz}\right) - \frac{d(\overline{U}_\ell B)}{dz} + R_1 A C_s \, \epsilon_\ell - \frac{R_3 \, B H' \, C_s \, \epsilon_\ell \, M_\mathrm{C} \, C_T}{\rho_\ell} = 0 \ . \tag{5-21}$$

Liquid velocity.

$$\frac{d\overline{U}_\ell}{dz} = \frac{C_T}{\rho_\ell} \left(R_4 \, M_{H_2} - (R_2 A + R_3 B) M_\mathrm{C}\right) H' C_s \, \epsilon_\ell \ . \tag{5-22}$$

Solid balance.

$$\frac{1}{\mathrm{Pe}_s} \frac{d}{dz} \left(\overline{\epsilon}_s \frac{dC_s}{dz}\right) - \frac{d}{dz} \left[(\overline{U}_{s\ell} - f_1)(1 - \epsilon_g) \, C_s\right] = 0 \ , \tag{5-23}$$

where

$$\overline{U}_{s\ell} = \overline{U}_\ell - \frac{\rho_s \epsilon_s}{\rho_{s\ell}} f \ .$$

Depending upon the relative magnitudes of the absorption of hydrogen and production of volatile products (C) [note that the rate of production of C is proportional to the rates of the reactions (5-6) and (5-7)], the gas-phase velocity may either increase or decrease. Any such change in the gas-phase velocity is obtained by considering a total gas balance over a differential section of the reactor. Equation (5-19) results from the total gas balance. The principle of the conservation of mass implies that total mass of the three-phase mixture is the same everywhere in the direction of flow. Since it is assumed that almost all the chemical changes that occur in the reactor are in the liquid phase (where they are catalyzed by solid particles), it is obvious that the mass of the gas-liquid mixture should remain axially constant (solid particles do not get depleted in the reactions).

The change in liquid velocity is thus related to the change in the gas velocity and subsequently to the rates of individual reactions. The equation for the change in liquid velocity in dimensionless form is represented by (5-22).

The initial conditions for Equations (5-18), (5-19), and (5-22) are

$$H' = H'_0, \quad \bar{U}_g = 1, \quad \bar{U}_\ell = 1 \quad \text{at} \quad z = 0^+ . \tag{5-24}$$

The dimensionless dispersion coefficients $\bar{e}$ and $\bar{e}_s$ can be obtained from the equations [44]

$$\bar{e}_\ell = \bar{U}_g^{0.3} , \tag{5-25}$$

$$\mathrm{Fr} = \mathrm{Fr}_o \bar{U}_g , \tag{5-26}$$

and

$$E_{zs} = \frac{\bar{U}_g (U_{og})_o D (1+8\,\mathrm{Fr}^{0.85})}{13\,\mathrm{Fr}\,(1+0.009\,\mathrm{Re}_p\,\mathrm{Fr}^{-0.8})} , \tag{5-27}$$

where

$$\left.\begin{aligned} \mathrm{Fr} &= U_{og}/\sqrt{gD} , \\ \mathrm{Re}_p &= \frac{d_p U_t \rho_\ell}{\mu_\ell} . \end{aligned}\right\} \tag{5-28}$$

The boundary conditions for Equations (5-20), (5-21), and (5-23) are

at $z = 0^+$:

$$\frac{dA}{dz} = \mathrm{Pe}_\ell (A - A_o) , \tag{5-29a}$$

$$\frac{dB}{dz} = \mathrm{Pe}_\ell (B - B_o) , \tag{5-29b}$$

$$\frac{dC_s}{dz} = \frac{\mathrm{Pe}_s}{\bar{e}_s} [(\bar{U}_{s\ell} - f_1)(1-\epsilon_g) C_s - C_{so}] ; \tag{5-29c}$$

at $z = 1^-$:

$$\frac{dA}{dz} = \frac{dB}{dz} = \frac{dC_s}{dz} = 0\,, \tag{5-30a}$$

$$C_s \overline{U}_\ell = C_{so}\,. \tag{5-30b}$$

The boundary condition (5-30b) is obtained from an overall solid-phase mass balance (flux of solid entering the reactor must equal that leaving the reactor).

In Equations (5-17) through (5-30), subscript 0 denotes values corresponding to reactor feed conditions. C_{s0} is related to the weight fraction of solid in the feed slurry, w_s, as

$$C_{so} = \frac{W_s\ \rho_\ell}{1 - \dfrac{\rho_s - \rho_\ell}{\rho_s}\ W_s}\,. \tag{5-31}$$

The gas holdup can be estimated using (5-97) below.

C. OVERALL YIELDS AND CONSUMPTIONS

Overall yields and consumptions of various components can be based on the amount of slurry entering the reactor (expressed as amount produced or consumed per unit of slurry fed). The yields and consumptions can be expressed in terms of various dimensional and dimensionless variables as follows:

Consumptions.

$$Y_\mathrm{A} = A_o - (A\,\overline{U}_\ell)_1\,, \tag{5-32}$$

$$Y_{\mathrm{H}_2} = [H_o - (H\overline{U}_g)_1]\ C_\mathrm{T}\ M_{\mathrm{H}_2}/r\rho_\ell\,, \tag{5-33}$$

where Y_A and Y_{H_2} are the consumptions of A and H_2, respectively.

Yields.

$$Y_\mathrm{B} = (B\overline{U}_\ell)_1 - B_o \tag{5-34}$$

$$Y_\mathrm{C} = [(C\overline{U}_g)_1 - C_o]\ C_\mathrm{T}\ M_\mathrm{C}/r\,\rho_\ell \tag{5-35}$$

where Y_B and Y_C are the yields of B and C, respectively. Subscript 1 denotes the values of various variables at the reactor outlet.

D. RESULTS AND DISCUSSION

The above equations were solved by Parulekar [44] using the values of physical properties of the three phases, the kinetic parameters for the reactions listed in Table 5-2, and the operating conditions listed in Table 5-3. The values of the

Table 5-2

Physical Properties and Kinetic Parameters [44]

(1)	Molecular weight, kg/kg-mol:	
	hydrogen	2
	volatile products (C)	40
(2)	Liquid density ρ_ℓ	$0.9 \times 10^{+3}$ kg/m^3
(3)	Solid density ρ_s	$1.3 \times 10^{+3}$ kg/m^3
(4)	Liquid viscosity μ_ℓ	0.9×10^{-2} Pa sec
(5)	Interfacial tension σ	8×10^{-3} N/m
(6)	Universal gas constant (R_g)	$8.315 \times 10^3 \dfrac{\text{m}^3\,\text{Pa}}{\text{kg-mol K}}$
		$8.29 \dfrac{\text{kJ}}{\text{kg-mol K}}$
(7)	Frequency factors:	
	k_1*	$20 \dfrac{\text{m}^3\ \text{slurry}}{(\text{kg cat.})\,(\text{sec})}$
	k_2*	$10 \dfrac{(\text{m}^3\ \text{gas})\,(\text{m}^3\ \text{slurry})}{(\text{sec})\,(\text{kg cat.})\,(\text{kg A})}$
	k_3*	$50 \dfrac{(\text{m}^3\ \text{gas})\,(\text{m}^3\ \text{slurry})}{(\text{sec})\,(\text{kg cat.})\,(\text{kg B})}$
	k_4*	$0.4 \dfrac{(\text{m}^3\ \text{gas})\,(\text{m}^3\ \text{slurry})}{(\text{m}^3\ \text{liq.})\,(\text{kg cat.})\,(\text{sec})}$
(8)	Activation energies, kJ/kg-mol:	
	E_1	8.29×10^4
	E_2	1.16×10^5
	E_3	5.86×10^4
	E_4	1.24×10^5
(9)	Weight fraction of heavy oil in feed liquid, $A_o = 0.7$.	
	Weight fraction of light oil in feed liquid, $B_o = 0.3$.	
	Mole fraction of hydrogen in feed gas, $H'_o = 1$.	

Table 5-3

Operating Conditions and Reactor Dimensions [44]

(1)	Superficial liquid velocity $(U_{o\ell})_o$	1.5×10^{-3} m/sec
(2)	Superficial gas velocity $(U_{og})_o$	1.5×10^{-2} m/sec
(3)	Weight fraction of catalyst in feed slurry, W_s	0.15
(4)	Reactor feed temperature	623 K
(5)	Total pressure	10.13 MPa
(6)	Particle size	150 μm
(7)	Column (reactor) diameter	0.08 m
(8)	Reactor length	5.4 m

kinetic parameters [(7) and (8) and Table 5-2] for the reaction (5-4) were chosen so that hydrogen consumption was around 3–5% of the heavy-distillate consumption. While some of the assumed values in Tables 5-2 and 5-3 may not be completely correct for the coal-liquefaction reactor, they represent the basic features of H-coal, EDS, and SRC-II reactors.

Parulekar [44] found that in the range of reactor temperature 573–673 K, Y_A and Y_C increased and $\overline{U}_\ell$ decreased with an increase in temperature, whereas Y_B exhibited a maximum with respect to temperature. The catalyst concentration within the reactor increased with the temperature. The hydrogen consumption increased almost linearly with temperature. Figure 5-14 shows a plot for hydrogen consumption, as a function of reactor temperature, obtained from several pilot plants (e.g. Fort Lewis, Wilsonville, Gulf R&D) for the liquefaction of Kentucky coal [33]. This plot is in qualitative agreement with the hydrogen-consumption–temperature plot obtained by Parulekar. The predictions were also found to be in qualitative agreement with the experimental results of Shah et al [55]. An increase in operating pressure enhances the hydrogen consumption, and the formation of C due to an increase in hydrogen concentration or its partial pressure. These predictions are in qualitative agreement with the experimental observations.

An increase in liquid velocity $(U_{0\ell})_0$ decreases backmixing in both the liquid and solid phases and also reduces the importance of settling of catalyst particles. The catalyst concentration gradient throughout the reactor decreases with the increase in liquid velocity. Figure 5-15 illustrates the typical effect of liquid velocity on the yields and consumptions of various reacting species. Both Y_C and Y_{H_2} decrease as liquid velocity is increased. Y_B exhibited a maximum because of the

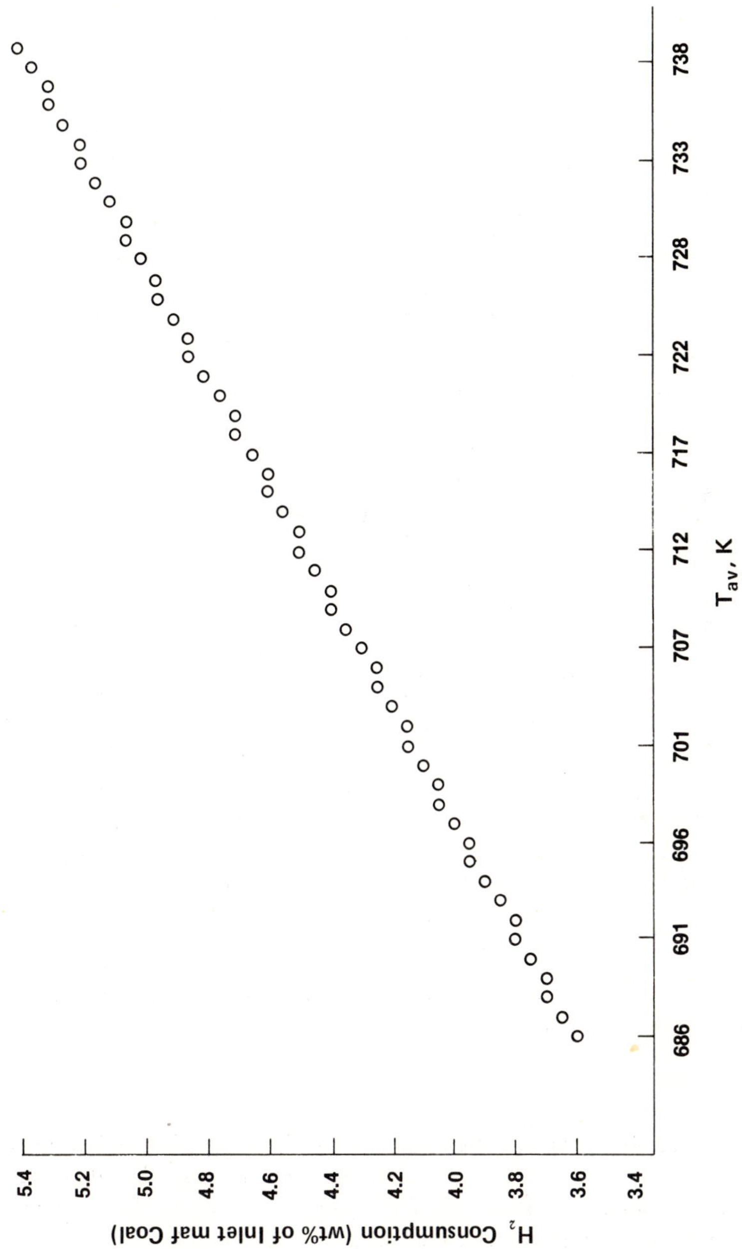

Figure 5-14. Hydrogen consumption versus average reactor temperature for Kentucky coal (SRC-II process) [33]. Basis values: pressure P_{H_2} = (13.79 at a gas purity of 91%) = 12.55 MPA; coal concentration (CIN) = 30 kg/(100 kg slurry feed); Slurry residence time (SRT) = 1 hr.

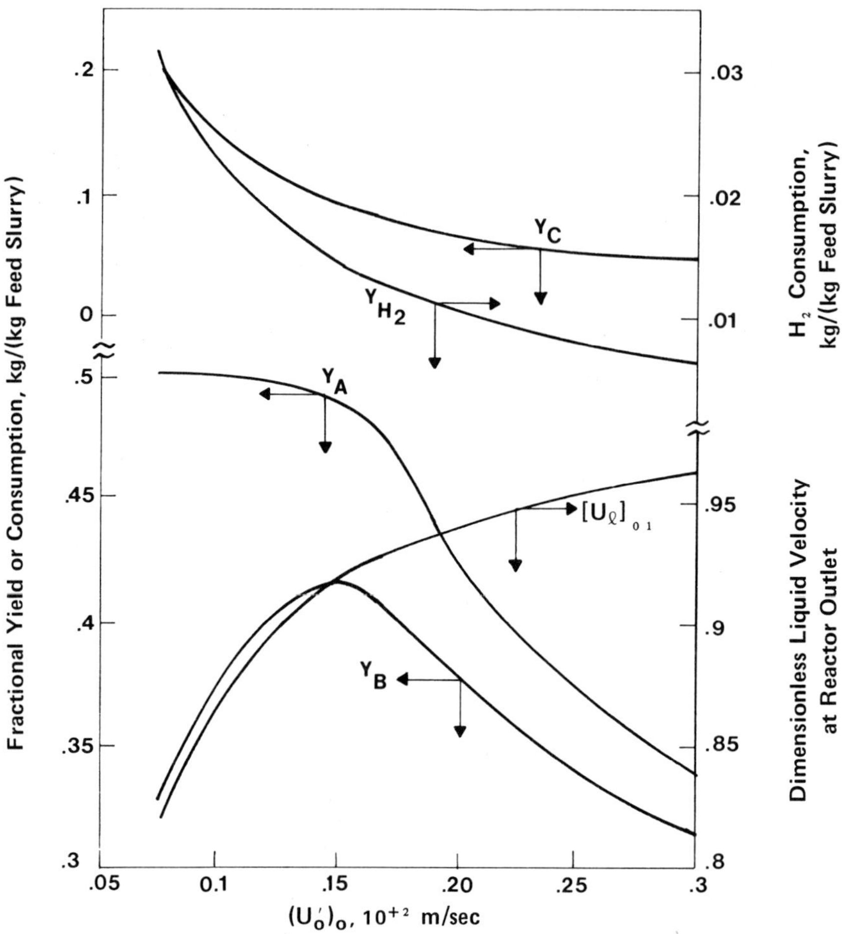

Figure 5-15. Effect of feed liquid velocity on various yields and consumptions and degree of vaporization (after Parulekar [44]).

counteracting effects of increased liquid-phase Peclet number and decreased residence time and local catalyst concentration with increasing liquid velocity. The effect of decreased catalyst settling and liquid-phase residence time dominates the effect of increased liquid Peclet number on the consumption of heavy distillate (A), and hence consumption of A decreases with an increase in liquid velocity.

An increase in the superficial gas velocity results in increased backmixing in both the liquid and the solid phases, thereby causing a decrease in local catalyst concentration. The typical effects of gas velocity on individual yields and consumption are outlined in Figure 5-16. The consumption of A (conversion) and the

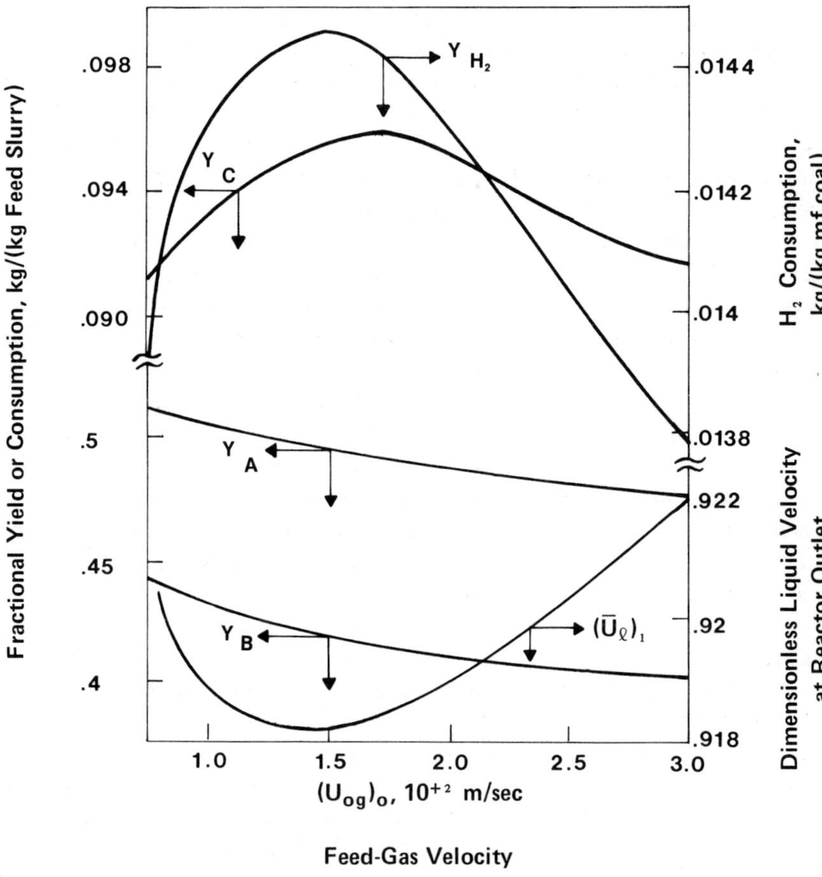

Figure 5-16. Effect of feed-gas velocity on various yields and consumptions and degree of vaporization (after Parulekar [44]).

yield of B decrease with an increase in gas velocity, because of increasing backmixing in both liquid and solid phases. The mole fractions of volatile products (C) in the exit stream decreases as the superifical gas velocity is increased. This results in more moles of gas throughout the reactor as well as in less conversion (as indicated by the mole fractions of H_2 and C, respectively). These two counteracting effects cause the observed maxima in the yield of C and hydrogen consumption, as shown in Figure 5-16. The results are in qualitative agreement with the experimental results reported by Shah et al. [55].

An increase in column diameter results in a somewhat decrease in the gas holdup and an increase in backmixing of the liquid and solid phases. The increased backmixing of the solid phase results in a decrease in catalyst concentration and its gradient throughout the reactor and a significant increase in the liquid-phase holdup. Parulekar [44] found that for his range of column diameter (0.08–0.16 m) all yields and consumptions decreased with an increase in column diameter.

An increase in length results in an increase in residence time as well as in the Peclet numbers Pe_ℓ and Pe_s. Parulekar [44] found that for the range 4.2–6.0 m these factors gave better yields of B and C and consumptions of hydrogen and A with increasing length. His results were in very good qualitative agreement with the regression plots for the yields of various products as a function of reactor residence time, for Kentucky coal obtained in pilot-plant (Fort Lewis, Wilsonville, Gulf, etc.) operations [33]. These regression plots, which are shown in Figures 5-17 to 5-19, give good estimates of the experimental facts. More sophisticated nonlinear regression analysis of the experimental data is also given by Jones [33].

A change in the weight fraction of solid in the feed slurry causes a change in the local solid concentration, as well as the solid and liquid holdups throughout the reactor. From Figure 5-20 it is seen that in the range of W_s between 0.15–0.5 the effect of an increase in the local catalyst concentration on all the reaction rates is dominant and thus the yields and consumptions of the various species increase as W_s is increased. As W_s exceeds 0.5, the effect of a decrease in the liquid holdup on all the reaction rates becomes dominant, and hence the yields and consumptions of all the species decrease with any further increase in W_s.

An increase in particle size of the catalyst increases the effect of settling and thus increases the catalyst concentration gradient and local catalyst concentration throughout the reactor. Simultaneously, it also causes a decrease in the local liquid holdup. Parulekar [44] found that up to a particle diameter of 275 μm, the effect of an increase in the local catalyst concentration on the reaction rates is dominant over the effect of decrease in liquid-phase holdup, and thus the yields and consumptions of all the species increase as particle size increases in the range 50–275 μm. Any further increase in particle size beyond 275 μm results in a reduction of the yields and consumptions of all the species due to increasing importance of the effect of reduction in liquid holdup on the reaction rates.

Typical axial variations of the dimensionless concentrations of A, B, C, and H_2, the dimensionless velocities of gas, liquid, and slurry, the holdup of the gas, liquid, and solid phases, and the catalyst concentration C_s, obtained by Parulekar [44], are shown in Figures 5-21 and 5-22. These results indicate that depending upon the operating conditions, there could be significant axial variations in the liquid and gas velocities and the holdups of various phases. The validity of these predictions needs to be checked with actual experimental data.

5.4.2 *Model of Lee et al.* [38]

An isothermal reaction model, which considers the gas-liquid interface mass-transfer resistance, has been analyzed by Lee et al. [38]. They applied an axial dispersion model to the preheater-reactor assembly of the SRC liquefaction process, and simulated the Wilsonville SRC-I pilot plant. The preheater and transfer-line section were assumed to be essentially a plug-flow reactor, since the length-to-diameter ratio for this section was as high as 1000. The plug-flow reactor (pre-

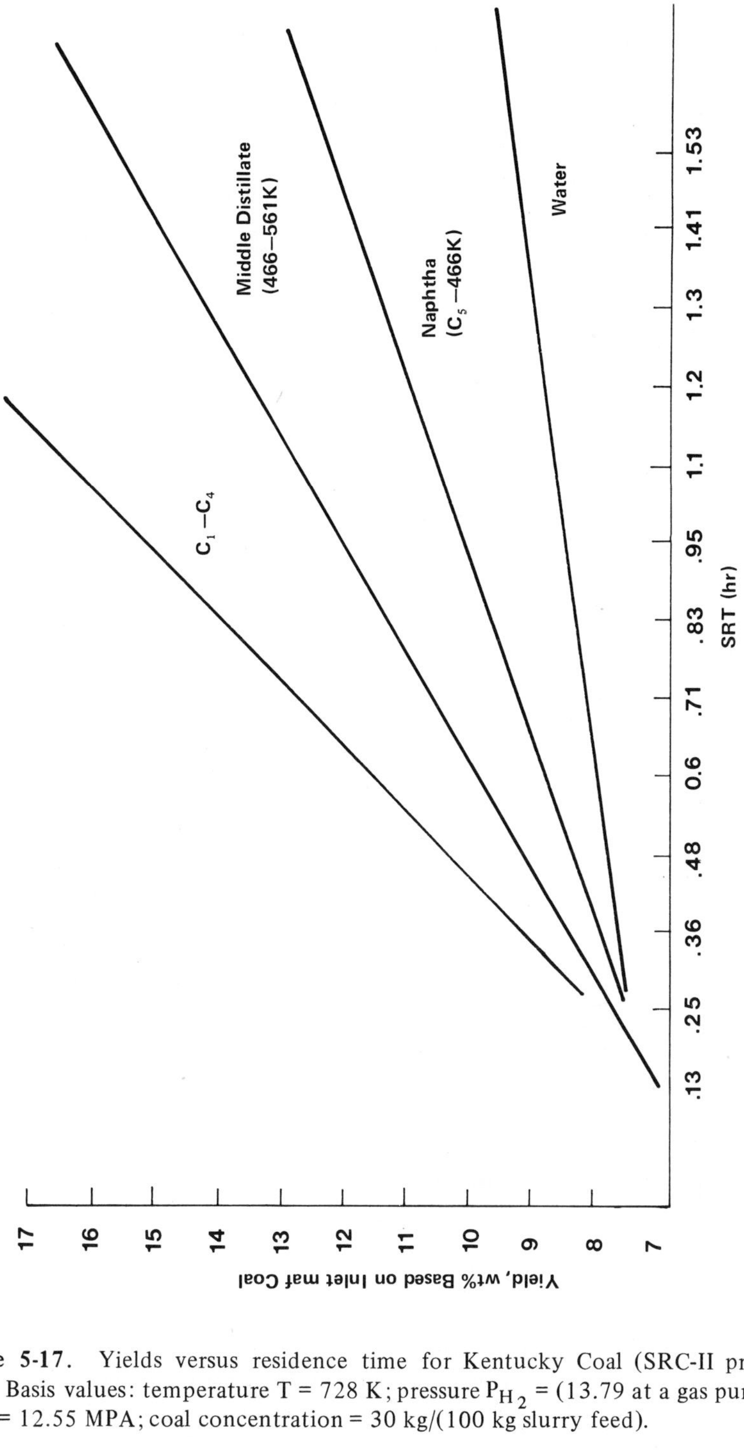

Figure 5-17. Yields versus residence time for Kentucky Coal (SRC-II process) [33]. Basis values: temperature T = 728 K; pressure P_{H_2} = (13.79 at a gas purity of 91%) = 12.55 MPA; coal concentration = 30 kg/(100 kg slurry feed).

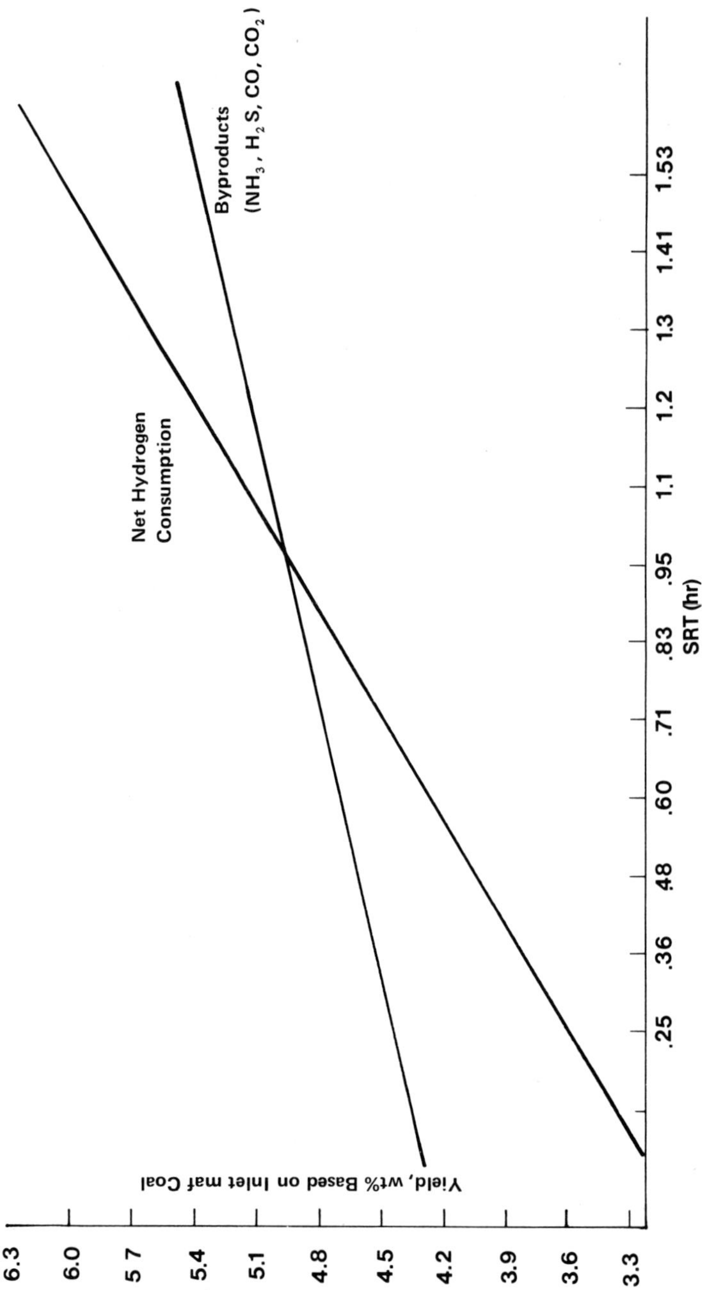

Figure 5-18. Yields versus residence time for Kentucky coal [33]. Basis values: temperature T_{avg} = 728 K; Pressure P_{H_2} = (13.79 at a gas purity of 91%) = 12.55 MPA; coal concentration = 30 kg/(100 kg slurry feed).

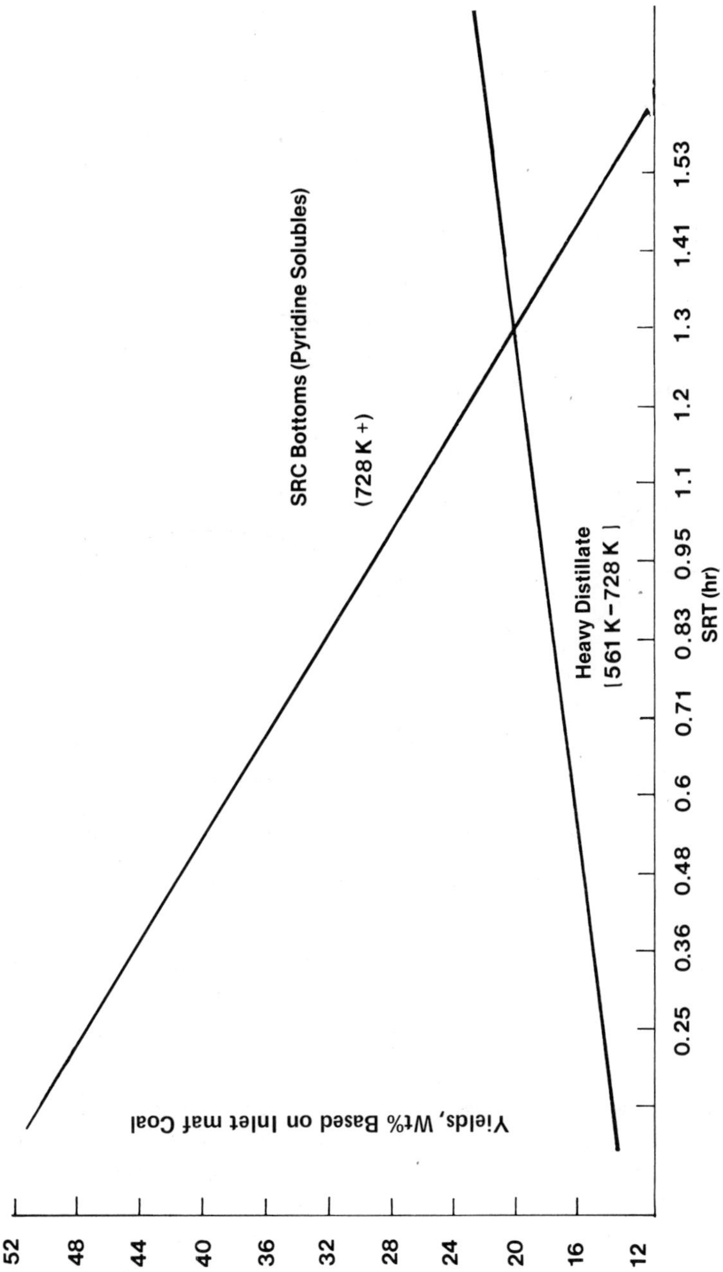

Figure 5-19. Yields versus residence time for Kentucky coal [33]. Basic values: temperature T_{avg} = 728 K; Pressure P_{H_2} = (13.79 at a gas purity of 91%) = 12.55 MPA; coal concentration = 30 kg(100 kg slurry feed).

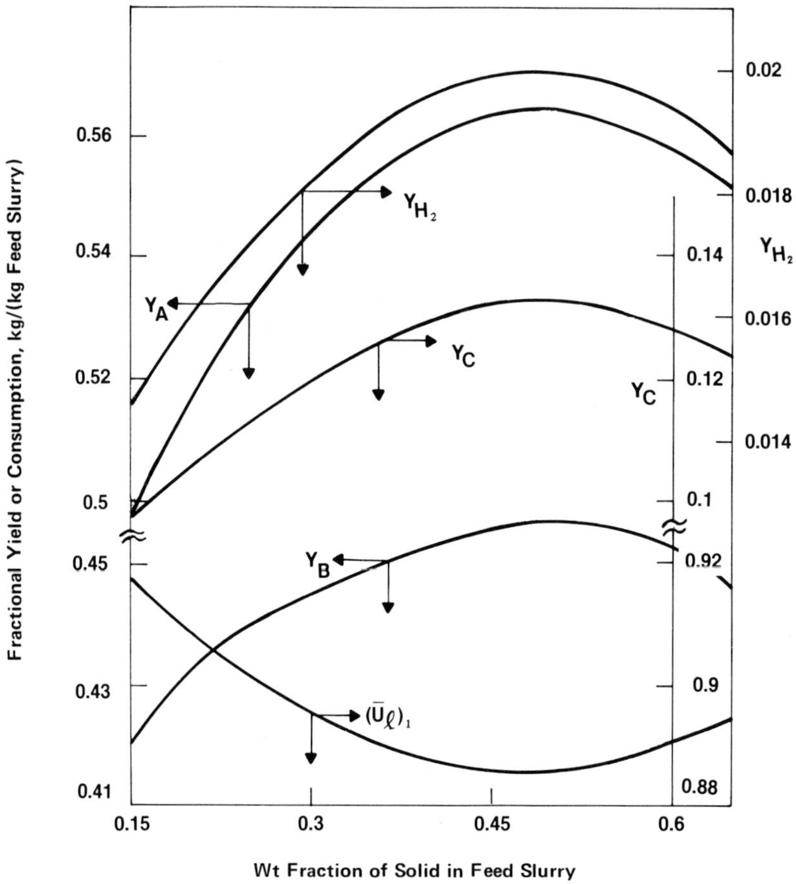

Figure 5-20. Effect of weight fraction of solid in feed slurry on various yields and consumptions and degree of vaporization (after Parulekar [44]).

heater–transfer-line section) was assumed to be operating at a temperature which was the average of preheater inlet and outlet temperatures. Since the length-to-diameter ratio for the reactor was small, they assumed a dispersed plug-flow model for the reactor.

Lee et al. [38] assumed that the coal dissolution, hydrodesulfurization, and hydrogenation reactions are the key reactions, and hence considered only these in the design equations. Rate expressions developed by Pitts [48], describing the hydrogen consumption, coal dissolution, and hydrodesulfurization, were used in the analysis. The velocities and holdups of all the three phases were assumed to remain unchanged in both the reactor and the preheater–transfer-line sections.

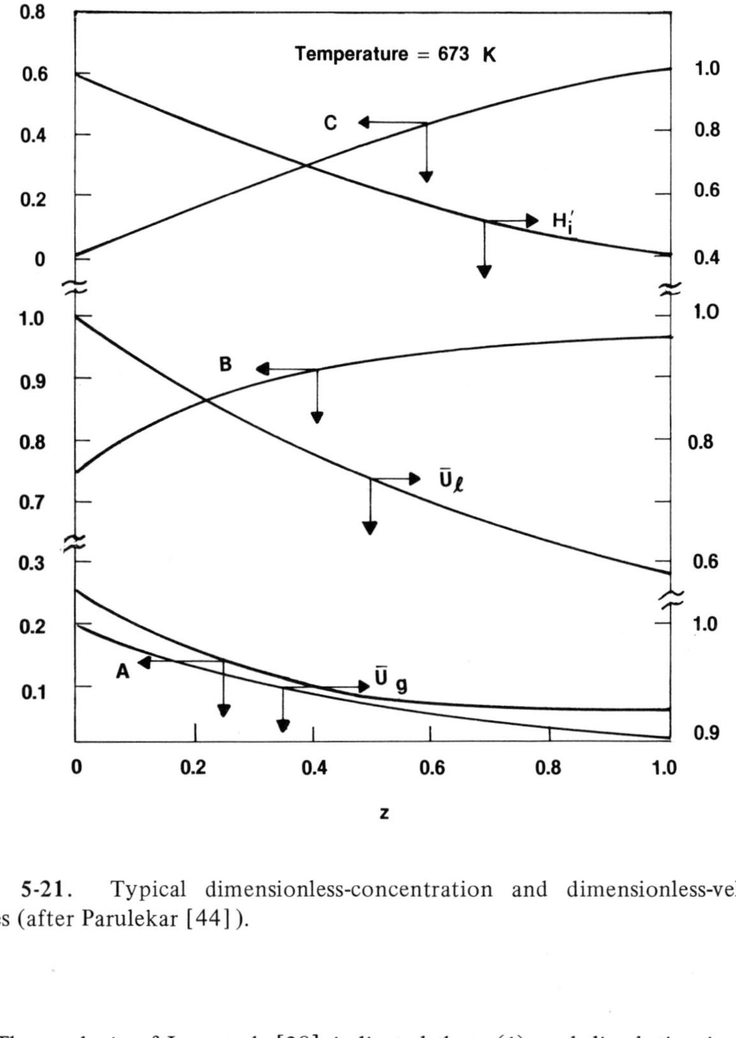

Figure 5-21. Typical dimensionless-concentration and dimensionless-velocity profiles (after Parulekar [44]).

The analysis of Lee et al. [38] indicated that: (1) coal dissolution is nearly complete in the preheater, (2) the effect of gas and liquid Peclet numbers on coal dissolution becomes more significant when the hydrogen mass-transfer coefficient $(k_\ell a_\ell)_{H_2}$ is low, and (3) the Wilsonville SRC pilot plant operates in a kinetically controlled reaction regime.

5.4.3 *Thermal Behavior of an Adiabatic Reactor*

Commercial reactors will be operated under adiabatic conditions, i.e., there will be no significant heat lost relative to that generated. The predictions of the performance of these reactors require an energy balance along with the mass balances introduced earlier. In this section, we briefly examine the thermal behavior

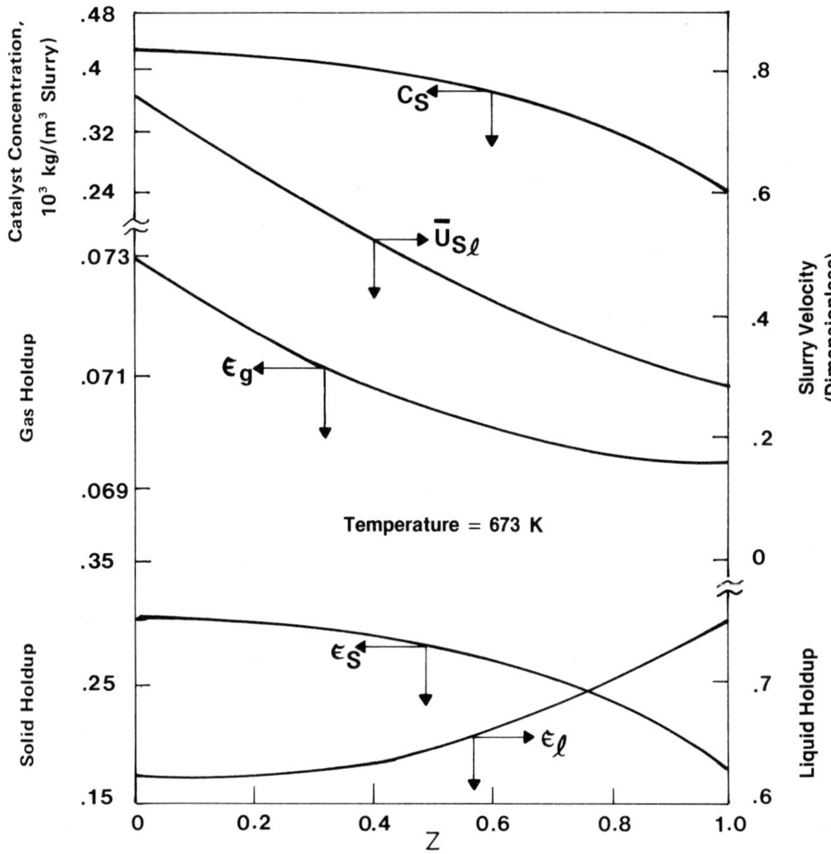

Figure 5-22. Typical holdup, solid-concentration, and dimensionless-slurry-velocity profiles (after Parulekar [44]).

(i.e., temperature distribution) of such reactors under various operating conditions, based on the axial-dispersion model.

Although, as mentioned earlier, the dissolution process occurring in the preheater is endothermic, hydrogenation and hydrocracking reactions taking place in the reactor are exothermic. While a number of reactions generate or absorb heat, in practice the overall heat generation is expressed in terms of hydrogen consumption alone. This is done because the reactions involving hydrogen are the most energetic. This consideration allows an evaluation of the thermal behavior of the reactor by using only a hydrogen mass balance and an energy balance. In the following discussion we follow this assumption and briefly describe the reported analysis [45, 46] for two cases: (1) when the gas-liquid interface mass-transfer resistance is insignificant, and (2) when this mass-transfer resistance is important. We restrict our discussion to the cocurrent upflow reactor. The analysis made the following additional assumptions:

1. Various physcial properties, such as the liquid-side mass-transfer coefficient for hydrogen and product gases, the specific interfacial area, a_ℓ, the densities and specific heats of the solid and liquid phases, and the axial dispersion coefficients (E_{zg}, $E_{z\ell}$, and E_{zs}) of various phases, are independent of temperature. The assumption is valid as long as the temperature variations in the reactor are not excessively large.

2. The gas, liquid, and solid velocities, as well as the holdups for the three phases, are constant within a given reactor. This assumption is reasonable when the extent of vaporization is low.

3. The density of the gas phase is assumed to be approximately constant. The assumption may be questionable, since hydrogen absorption from the gas to the liquid phase results in simultaneous desorption of volatile products (C) which have much higher molecular weight than hydrogen. The increase in density due to this may be somewhat counterbalanced by the increase in temperature and the slight decrease in the pressure along the length of the reactor. Since the gas phase carries only a small fraction of the total sensible heat, a variation in the gas-phase density need not significantly affect the temperature distribution in the reactor.

4. The size of the particles and their concentration in the slurry are assumed to be small, so that the model proposed by Cova [14] can be used for predicting the solid-concentration profile.

5. The gas phase moves as plug flow. This assumption is valid in the pilot-scale and prototype reactors. There may be some gas backmixing in the commercial reactors, but its importance on the hydrogen consumption is assumed to be negligible for the dynamic models.

A. CASE (a): NO GAS-LIQUID INTERFACE RESISTANCE [45]

For this case the required hydrogen-balance and heat-balance equations can be expressed in dimensionless form as follows:

$$\frac{dH'}{dz} = -R_4' r H' \epsilon_\ell \qquad (5\text{-}36)$$

and

$$\frac{1}{\mathrm{Pe}_\theta}\frac{d^2\theta}{dz^2} - \frac{d\theta}{dz} + \Gamma_R H' \epsilon_\ell = 0\,, \qquad (5\text{-}37)$$

where

$$z = x/L,$$

$$\theta = \frac{T-T_o}{T_o},$$

$$r = \frac{(U_{o\ell})_o}{(U_{og})_o}$$

$$R_4' = K_4' \,(\mathrm{SRT}) \exp\left(\frac{-E_4'}{R_g T_o\,(1+\theta)}\right),$$

$$\mathrm{Pe}_\theta = \frac{[\rho_g C_{pg}\,(U_{og})_o + \rho_\ell C_{p\ell}\,(U_{o\ell})_o + \rho_s C_{ps}\,(U_{os})_o]\,L}{D_{Hg} + D_{H\ell} + D_{Hs}},$$

$$\Gamma_R = \frac{R_4'\,\Delta H_R\,(U_{o\ell})_o\,C_{To}}{T_o\,[\rho_g C_{pg}\,(U_{og})_o + \rho_\ell C_{p\ell}\,(U_{o\ell})_o + \rho_s C_{ps}\,(U_{os})_o]},$$

and

$$\mathrm{SRT} = L/(U_{o\ell})_o\ .$$

The above equations assume that the catalyst concentration is uniform across the reactor, so it can be lumped with the reaction rate constant.

Equations [5-36] and [5-37] are subjected to the following boundary conditions:

at $z = 0^+$,

$$H' = H_o', \tag{5-38}$$

$$\frac{1}{\mathrm{Pe}_\theta}\frac{d\theta}{dz} = (\theta - \theta_o)\,; \tag{5-39}$$

at $z = 1^-$,

$$\frac{d\theta}{dz} = 0\,. \tag{5-40}$$

B. CASE (b): GAS-LIQUID INTERFACE MASS TRANSFER RESISTANCE IMPORTANT

This case was recently studied by Parulekar and Shah [46]. The necessary mass- and heat-balance equations along with the boundary conditions can be expressed in dimensionless form by defining the following dimensionless variables:

$$z = x/L,$$

$$r = \frac{(U_{o\ell})_o}{(U_{og})_o}, \quad R_\ell = \frac{(k_\ell a_\ell)_{H_2} L}{(U_{og})_o}, \quad R_o = \frac{Lk_4^*}{(U_{og})_o},$$

$$\Gamma_R = \frac{\Delta H_R \, C_{To} \, (U_{og})_o}{[(U_{o\ell})_o \, \rho_\ell C_{p\ell} + (U_{og})_o \rho_g C_{pg} + \rho_s C_{ps} \, (U_{os})_o] \, T_o},$$

$$\Gamma_s = \frac{(\Delta H_s)_{H_2} \, C_{To} \, (U_{og})_o}{[\rho_g C_{pg} \, (U_{og})_o + \rho_\ell C_{p\ell} \, (U_{o\ell})_o + \rho_s C_{ps} \, (U_{os})_o] \, T_o},$$

$$H'_g = \frac{C_{H_2 g}}{C_{To}}, \quad H'_\ell = \frac{C_{H_2 \ell}}{C_{To}}, \tag{5-41}$$

$$\mathrm{Pe}_\ell = \frac{(U_{o\ell})_o \, L}{E_{z\ell}}, \quad \mathrm{Pe}_s = \frac{(U_{os})_o \, L}{E_{zs}},$$

$$\delta_s = -\frac{(\Delta H_s)_{H_2}}{R_g T_o}, \quad \delta_R = -\frac{E_4}{R_g T_o},$$

$$\theta = \frac{T - T_o}{T_o},$$

$$\mathrm{Pe}_\theta = \frac{(\rho_g C_{pg} \, (U_{og})_o + \rho_\ell C_{p\ell} (U_{o\ell})_o + \rho_s C_{ps} \, (U_{os})_o) L}{(D_{Hg} + D_{H\ell} + D_{Hs})},$$

$$H_{H_2} = H^*_{H_2} \exp\left(\frac{\delta_s}{1+\theta}\right).$$

The governing equations are:

Gas-phase hydrogen.

$$\frac{dH'_g}{dz} = -R_\ell \, (H_{H_2} H'_g - H'_\ell). \tag{5-42}$$

Liquid-phase hydrogen.

$$r\left(\frac{1}{\mathrm{Pe}_\ell} \frac{d^2 H'_\ell}{dz^2} - \frac{dH'_\ell}{dz}\right) + R_\ell \, (H_{H_2} H'_g - H'_\ell) - R_o \exp\left(\frac{\delta_R}{1+\theta}\right) H'_\ell \, \epsilon_\ell = 0. \tag{5-43}$$

Heat balance.

$$\frac{1}{\mathrm{Pe}_\theta}\frac{d^2\theta}{dz^2} - \frac{d\theta}{dz} + R_\ell\,(H_{\mathrm{H}_2}H_g' - H_\ell')\,\Gamma_s$$
$$+R_\rho \exp\left(\frac{\delta_R}{1+\theta}\right) H_\ell'\,\epsilon_\ell\,\Gamma_R = 0\,. \tag{5-44}$$

Solid phase.

$$\frac{1}{\mathrm{Pe}_s}\frac{d^2C_s}{dz^2} - \frac{dC_s}{dz}\frac{V_s}{V_\ell} = 0\,, \tag{5-45}$$

where $V_\ell = U_\ell/\epsilon_\ell$, and

$$V_s = \frac{U_\ell}{\epsilon_\ell} - V_t\,. \tag{5-46}$$

The boundary conditions for Equations (5-42) through (5-45) can be expressed as follows:

at $z = 0^+$,

$$H_g' = H_{go}'\,, \tag{5-47}$$

$$\frac{1}{\mathrm{Pe}_\ell}\frac{dH_\ell'}{dz} = (H_\ell' - H_{\ell o})\,, \tag{5-47b}$$

$$\frac{1}{\mathrm{Pe}_\theta}\frac{d\theta}{dz} = \theta - \theta_o\,, \tag{5-47c}$$

$$\frac{1}{\mathrm{Pe}_s}\frac{dC_s}{dz} = \left(\frac{V_s}{V_\ell}\right) C_s - C_{so}\,. \tag{5-47d}$$

at $z = 1^-$,

$$\frac{dH_\ell}{dz} = \frac{d\theta}{dz} = 0 \tag{5-48}$$

and

$$\frac{1}{\mathrm{Pe}_s}\frac{dC_s}{dx} = -\frac{V_t}{V_\ell}\,C_{so}\,. \tag{5-49}$$

The subscript 0 in all these equations refers to the feed conditions. Equation (5-45) along with the accompanying boundary conditions (5-47d) and (5-49) can be solved analytically, and the solution is given by Parulekar and Shah [46].

The correlation proposed by Kato [34] was used to estimate the Peclet number for the liquid phase [see Equation (5-75) below]. Another correlation proposed by the same authors was used to estimate the Peclet number for the solid phase [see Equation (5-78)]. A correlation proposed by Akita and Yoshida [1] for estimation of the gas-phase holdup in three-phase reactors was used to estimate the gas holdup [Equation (5-92)].

The liquid-phase holdup was evaluated by using the relation (5-99). For the liquid-phase mass-transfer coefficient for hydrogen absorption $[(k_\ell a_\ell)_{H_2}]$, the correlation shown below was used:

$$(k_\ell a_\ell)_{H_2} = \mathscr{K}_o D^{0.17} \epsilon_g^{1.1}, \qquad 5\text{-}50)$$

where D is the diameter of the reactor and $\mathscr{K}_0$ is a parametric constant.

The dependence of $(k_\ell a_\ell)_{H_2}$ on reactor diameter and gas holdup in the above relation has been taken to be of the same form as that proposed by Akita and Yoshida [1]; [see Equation (5-101)]. The heat of the reaction $\triangle H_R$ and the constant Γ_0 were chosen so that the hydrogen consumption lies between 2 and 3% in most of the cases [32]. The heat-dispersion coefficient was estimated using the analogy between heat and mass-transfer coefficients.

Henry's constant H_{H_2} was expressed as a function of the reactor temperature T by the standard exponential relation. The frequency factor k_4^*, the activation energy E_4, the constant $H_{H_2}^*$, and the heat of solution were obtained from the work of Guin et al. [23] and Lee et al. [38]. The terminal settling velocity of the catalyst particles, U_t, was evaluated using the criteria proposed by McCabe and Smith [40]. The consumption of hydrogen based on moisture-free coal can be expressed as

$$Y_{H_2} = (H_{go} - (H_g)_1)\ C_{To} M_{H_2}/rC_i, \qquad (5\text{-}51)$$

where $(H_g)_1$ denotes the dimensionless hydrogen concentration at the dissolver outlet and C_i represents the concentration of moisture-free coal in (kg mf coal)/ m³ slurry).

C. RESULTS AND DISCUSSION

Case (a) has been analyzed by Shah and Parulekar [45]. For the experimental conditions described in Table 5-4, the experimentally measured temperature distribution in the Fort Lewis SRC-II reactor is illustrated in Figure 5-23. As shown in the figure, the experimental data were well correlated by the case (a) model.

The effects on the hydrogen consumption and the overall temperature rise of reaction variables such as feed temperature, total pressure, superficial gas

Table 5-4

Typical Operating Conditions for an Adiabatic Reactor With and Without Intrareactor Quench

	Run 78SR-19*	Test run*
Furnace outlet temp., K	740	703
Reactor outlet temp. T_o,† K	733	733
Temp. rise in reactor, K	30	42.2
"0"-position temperature, K	662.44	698.22
Quench ratio, m	0.51	(no quench)
H_2 Consumption, % mf coal	3.28	3.4
H_2 Consumption (model), % mf coal	3.35	3.5
Quench location η	0.05	–
C_i, kg/(m^3 slurry)	328	328
H_2 flow rate (preheater), kg/hr	112.27	–
H_2 flow rate (quench), kg/hr	69.55	–
Total gas rate, kg/hr		
(1) Feed	407.77	–
(2) Quench	206.66	–
Diameter, m	0.6	0.6
Length, m	9	9

*IL coal.
†Temperature sensor near quench injector questionable.

velocity, superficial liquid velocity, reactor diameter, reactor length, and gas-liquid mass-transfer coefficient for both cases (a) and (b) were examined by Parulekar and Shah [45, 46]. The results obtained in these studies are briefly summarized below.

An increase in reactor feed temperature increases the hydrogen consumption and hence the heat generation. As one would expect, the temperature effect on the hydrogen consumption was found to be more pronounced for case (a) than for the case (b), which includes diffusional resistances. For both cases, an increase in total pressure increased the total gas-phase concentration C_{T0}, which increased the hydrogen consumption. The increase in pressure also increased $(U_{0g})_0$, which in turn increased the thermal Peclet number and hence the temperature rise.

The reactor diameter affects the extent of liquid-phase backmixing and the gas-liquid mass-transfer coefficient. For case (a), an increase in reactor diameter did not affect the hydrogen consumption, but thermal backmixing was increased and

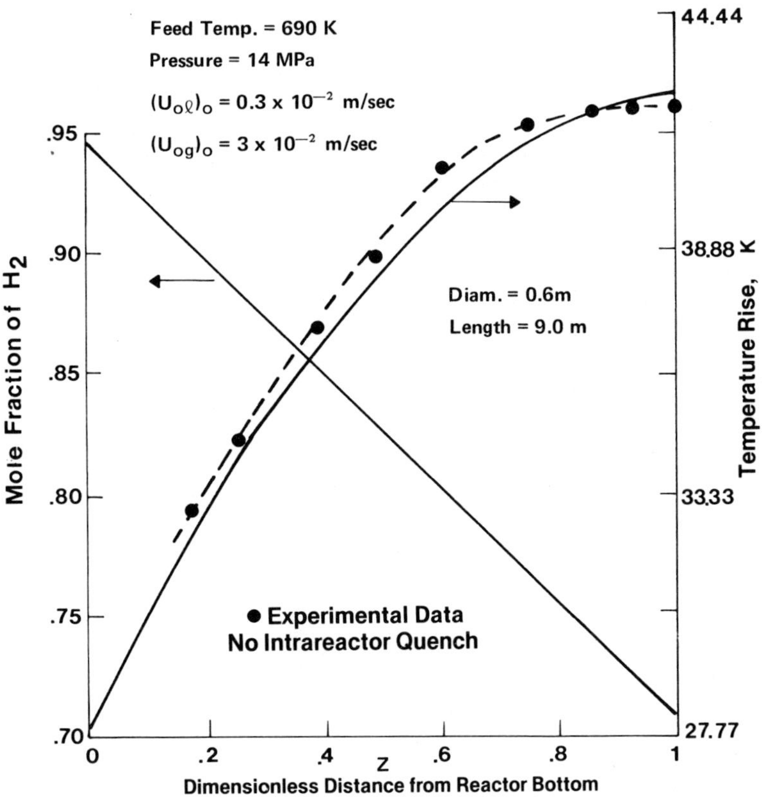

Figure 5-23. Comparison between predicted and experimental temperature profiles (after Parulekar and Shah [45]). H_2 consumption: (1) model prediction = 3.7%; (2) experimental = 3.6%.

hence a decrease in temperature rise was observed. The results were somewhat different when the gas-liquid mass-transfer resistance was significant. In this case, an increase in reactor diameter increased gas-liquid mass-transfer coefficient and hence the hydrogen consumption rate. This, in turn, increased the temperature rise in spite of the counteracting effect of increased thermal backmixing.

An increase in liquid velocity resulted in a decrease in the extent of thermal backmixing and also a decrease in slurry space time. While these two changes generally have counteracting effects on the hydrogen consumption and the temperature rise, Parulekar and Shah [45, 46] found that in most practical cases, both hydrogen consumption and overall temperature rise were lowered with an increase in liquid velocity. An increase in superficial gas velocity decreased the time spent by a molecule of hydrogen in the reactor and increased the thermal backmixing, and in most practical conditions this results in reduced hydrogen consumption and temperature rise [45, 46]. An increase in the reactor length increases the hydrogen consumption and also increases the temperature rise due to decreased thermal backmixing and increased slurry space time. Finally, for case (b) an increase in the gas-liquid mass-transfer coefficient increases the hydrogen consumption and hence the temperature rise. A summary of hydrogen-consumption and temperature-rise responses to the reactor variables reported by Parulekar and Shah [45, 46] is given in Table 5-5.

Table 5-5

Summarized Responses to Reactor Variables [45, 46]

Variable studied	Basis value*	Responses due to an increase in variable: Hydrogen consumption†	Responses due to an increase in variable: Temperature Rise
Mf coal conc.	30% [cases (a) and (b)]	Decreases	Decreases slightly
Feed temp.	693 K	Increases	Increases
Reactor pressure	14 MPa	Increases	Increases
Reactor diam.	0.60 m	No effect [case (a)] Increases [case (b)]	Decreases [case (a)] Increases [case (b)]
Length	9 m	Increases	Increases
Velocity of: liq.	3×10^{-3} m/sec	Decreases	Decreases
gas	3×10^{-2} m/sec	Decreases	Decreases

*Approximate Ft. Lewis conditions. Variables changed one at a time.
†MF-coal basis.

The results discussed above were also obtained with the use of a stagewise backmixing model (interstage-recirculation model). It was found that the temperature-rise predictions of the axial-dispersion model and the stagewise interstage-recirculation model agreed well [46]. Han et al. [24] and Parulekar and Shah [46] found that the rate-determining mechanism changes from the mass-transfer-controlled regime to the chemical-reaction-controlled regime as the flow of mixtures of gas and slurry becomes more turbulent. For the mass-transfer-controlled regime [case (b)], both hydrogen consumption and the adiabatic temperature rise were lower than the corresponding values for the kinetically controlled regime [case (a)].

5.4.4 *Operation with an Intrareactor Hydrogen Quench*

It is a well-known fact that in an adiabatic reactor, an exothermic reaction may lead to temperature excursions causing difficulty in temperature control. An excessive local temperature can cause rapid aging of catalyst. There is also a maximum allowable temperature set by the metallurgical limits of the reactor vessel. A quench gas is normally used to minimize the temperature rise across the reactor. Shah et al. [57] and Mhaskar et al. [42] examined the effect of quench location on the cycle life of a plug-flow hydrodesulfurization (HDS) reactor.

The use of quench in an adiabatic coal-liquefaction reactor is, however, planned to be carried out for somewhat different reasons than in an HDS reactor. As mentioned earlier, in the liquefaction process, coal fragments and depolymerizes to give various fractions of coal liquids such as asphaltenes, preasphaltenes, and oils of varying boiling range. It has been found by several investigators that at temperatures near 743 K repolymerization of coal liquids can occur, which is obviously not desirable. Just as in other hydroprocessing units, there are some metallurgical limitations on the liquefaction reactors as well. The authors believe that the maximum desirable temperature for coal-liquefaction reactions is around 738 K. Temperatures below this can be controlled by the use of one or more hydrogen quenches. Mathematically, the problem of a quench in an adiabatic backmixed reactor is different from that in a plug flow reactor, because in the former case the temperature has been found to be a continuous function at the quench location.

The basic nature of hydrogen-balance and heat-balance equations in each section of the reactor and the boundary conditions at the reactor inlet and outlet in this case would remain the same. The magnitudes of various reaction parameters may, however, be different in prequench and postquench sections of the reactor. In the following discussion we briefly evaluate the appropriate boundary conditions at the quench location. It is assumed that addition of a quench does not result in significant changes or differences in the flow patterns of various phases in the prequench and the postquench sections of the reactor. The quench is evenly distributed across the cross-section at the point of entry.

Let the quench be added at $z = \eta$. Let m denote the ratio of mass flow rate of quench added to mass flow rate of gas in the feed. The total mass flux balance gives the relation between superficial gas velocities in the prequench and the postquench section:

$$(U_{og})_2 = (m+1)\ (U_{og})_o\ . \qquad (5\text{-}52)$$

Hydrogen flux balance results in the following equation:

$$(H')_{\eta^+} = \frac{(H')_{\eta^-} + H_g'\ U_{og}/(U_{og})_o}{1 + m}\ . \qquad (5\text{-}53)$$

Balance of heat flux gives us the following equation:

$$-\frac{1}{(\mathrm{Pe}_\theta)_1}\left.\frac{d\theta}{dz}\right|_{\eta^-} + P_g\,(1+\theta_g) + (1+\theta)_{\eta^-}$$
$$= \frac{-1}{(\mathrm{Pe}_\theta)_2}\left.\frac{d\theta}{dz}\right|_{\eta^+} + (1+mp_3)\,(1+\theta)_{\eta^+}\ , \qquad (5\text{-}54)$$

$$\theta_{\eta^+} = \theta_{\eta^-}\ , \qquad (5\text{-}55)$$

where

$$(\mathrm{Pe}_\theta)_1 = \frac{(\rho_g C_{pg}(U_{og})_o + \rho_\ell C_{p\ell}(U_{o\ell})_o + \rho_s C_{ps}(U_{os})_o)L}{(D_{Hg} + D_{H\ell} + D_{Hs})_1}\ , \qquad (5\text{-}56)$$

$$(\mathrm{Pe}_\theta)_2 = \frac{(\rho_g C_{pg}(U_{og})_o + \rho_\ell C_{p\ell}(U_{o\ell})_o + \rho_s C_{ps}(U_{os})_o)L}{(D_{Hg} + D_{H\ell} + D_{Hs})_2}\ , \qquad (5\text{-}57)$$

$$p_3 = \frac{\rho_g C_{pg}(U_{og})_o}{(\rho_g C_{pg}(U_{og})_o + \rho_\ell C_{p\ell}(U_{o\ell})_o + \rho_s C_{ps}(U_{os})_o)}\ , \qquad (5\text{-}58)$$

$$P_g = \frac{m\,\rho_g C_{pg}(U_{og})_o}{(\rho_g C_{pg}(U_{og})_o + \rho_\ell C_{p\ell}(U_{o\ell})_o + \rho_s C_{ps}(U_{os})_o)}\ . \qquad (5\text{-}59)$$

The condition (5-55) simply states that there is no discontinuity in temperature at the quench location. In (5-52) through (5-59) subscripts 1 and 2 denote conditions in the prequench and the postquench sections, ρ_g denotes the gas density, M_{H_2} is the molecular weight of hydrogen, θ_q is the dimensionless temperature of the quench, and U_{0q} and H'_q are the quench velocity based on the dissolver cross-section and the dimensionless hydrogen concentration in the quench, respectively. The subscripts η^- and η^+ refer to the conditions just before and just after the quench location, respectively.

The above derivation assumes a constant gas velocity $(U_{og})_{\eta-} = (U_{og})_0$. This is a reasonable assumption when the relative amount of quench is small. It also assumes that the molecular weight of the gas at the quench point is the same as that of the gas at the reactor inlet. This is in general a restricting assumption, because the molecular weight of the gas produced is higher than that of the feed gas, which consists of hydrogen and light hydrocarbons.

For case (b) the values of most of the dimensionless parameters, such as R_ℓ, R_0, Γ_R, Γ_s, Pe_ℓ, Pe_s, and Pe_θ, are significantly different in the prequench $(0\text{–}\eta^-)$ and postquench $(\eta^+\text{–}1)$ sections. Equation (5-49) for the nonreacting-solid-phase balance is defined for each of the two sections as follows:

$$C_s = a+b \exp\left(+\frac{U_{os}}{(U_{o\ell})_o} \mathrm{Pe}_s\, z\right), \tag{5-60}$$

where a and b are constants. a, b, and U_{0s} are defined for the prequench and postquench sections as follows:

for $\eta^+ < z < 1$ (postquench section),

$$a = C_{so}\left[\frac{(U_{o\ell})_o}{U_{os}} + \left(1 - \frac{(U_{o\ell})_o}{U_{os}}\right)\left(\frac{(\mathrm{Pe}_s)_2}{(\mathrm{Pe}_s)_1} - 1\right)\right.$$
$$\left.\exp\left(-\frac{U_{os}}{(U_{o\ell})_o}(\mathrm{Pe}_s)_2(1-\eta)\right)\right], \tag{5-61}$$

$$b = C_{so}\left(1 - \frac{(U_{o\ell})_o}{U_{os}}\right)\exp\left(-\frac{U_{os}}{(U_{o\ell})_o}(\mathrm{Pe}_s)_2\right), \tag{5-62}$$

$$U_{os} = (U_{o\ell})_o - U_t, \tag{5-63}$$

$$\mathrm{Pe}_s = (\mathrm{Pe}_s)_2; \tag{5-64}$$

for $0 < z < \eta^-$ (prequench section),

$$a = C_{so}(U_{o\ell})_o / U_{os}, \tag{5-65}$$

$$b = \frac{(\mathrm{Pe}_s)_2}{(\mathrm{Pe}_s)_1} C_{so}\left(1 - \frac{U_{o\ell}}{U_{os}}\right)$$
$$\times \exp\left\{-\frac{U_{os}}{(U_{o\ell})_o}[\eta(\mathrm{Pe}_s)_1 + (1-\eta)(\mathrm{Pe}_s)_2]\right\}, \tag{5-66}$$

$$U_{os} = (U_{o\ell})_o - U_t , \tag{5-67}$$

$$\mathrm{Pe}_s = (\mathrm{Pe}_s)_1 . \tag{5-68}$$

Earlier it was pointed out that in the Fort Lewis reactor, the solid's median particle size is of the order of 3 μm. For this small particle size, the solid concentration profile within the reactor can be assumed to be flat, and solid and liquid velocities can be considered to be equal. In this case, the solid balance equations can be simplified by taking $U_{0s} = (U_{0\ell})_0$.

The hydrogen consumption [(g/g mf coal)] can be obtained from the following expression:

$$Y_{\mathrm{H}_2} = \frac{M_{\mathrm{H}_2} C_{To}}{rC_i} \left(H_o' + \frac{U_{og}}{(U_{og})_o} H_q' - H_1' (1+m) \right) . \tag{5-69}$$

Shah and Parulekar [45] presented a numerical solution to the equations for the reactor with a quench. In this study the theoretical prediction of case (a) for the reactor temperature distribution was compared with the one measured experimentally in the SRC-II reactor at Fort Lewis for the run 78-SR-19 described in Table 5-4. As shown in Figure 5-24, the predicted temperature distribution agreed

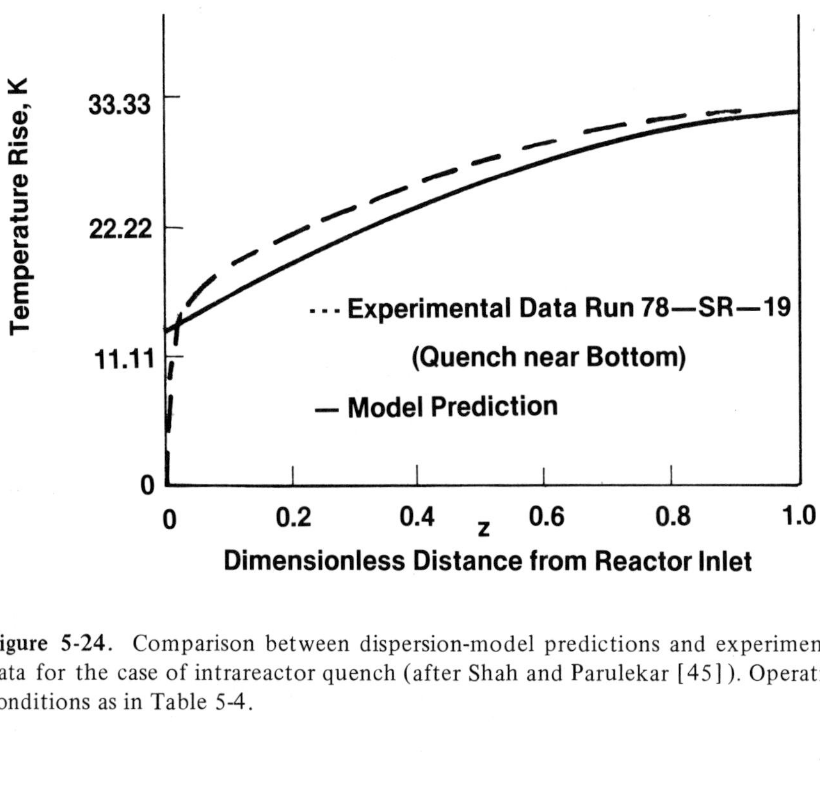

Figure 5-24. Comparison between dispersion-model predictions and experimental data for the case of intrareactor quench (after Shah and Parulekar [45]). Operating conditions as in Table 5-4.

fairly well with the measured one. The comparison indicates that since the temperature increase at the reactor inlet is very high, in the future experimental measurements, more data should be taken near the entrance region of the reactor. Parulekar and Shah [45, 46] also evaluated the quench requirement for the specified reactor temperature distribution under various operating conditions. The results were also checked with the help of the interstage-recirculation model [46].

5.5 Model-Parameter Estimations

As indicated earlier, the reactor models for three-phase slurry reactors to be used in SRC-II, H-coal, or Exxon donor-solvent processes are evaluated here. While the basic models outlined here should be applicable to any of these reactors, the values of the parameters used in the models depend on the reactor internals and the flow regime prevailing within the reactor [22]. A detailed literature search on three-phase slurry column hydrodynamics is given by Vasolos et al. [63] and Shah [56]. The reader is referred to these publications for more details.

In this section, we present a brief review on the following model parameters and how they can be determined:

1. Mass-dispersion coefficients (both axial and radial) for the gas, liquid, and solid phases.
2. Thermal-dispersion coefficients (both axial and radial) for the gas, liquid, and solid phases.
3. Holdups of the gas, liquid, and solid phases.
4. The gas-liquid mass-transfer coefficient.

5.5.1 *Axial Mass-Dispersion Coefficients*

A. GAS PHASE

Pavlica and Olson [47] have reviewed the literature concerning the dispersion coefficient in the liquid and gas (bubble) phase of cocurrent-countercurrent bubble columns. The Peclet number for the gas phase according to them can be obtained by the following relation (for $U_{0g} > 0.1$ m/sec):

$$\mathrm{Pe}_g = \frac{U_{og}\, L}{E_{zg}} = \frac{0.2}{\epsilon_g} \left(\frac{U_{og}}{U_{\mathrm{slip}}}\right) \left(\frac{L}{D}\right), \tag{5-70}$$

where U_{slip} is the slip velocity, which can be obtained from the relation

$$\frac{U_{og}}{\epsilon_g} \pm \frac{U_{o\ell}}{\epsilon_\ell} = U_{\mathrm{slip}}, \tag{5-71}$$

and L and D are the length and diameter of the reactor, respectively. For cocurrent flow, a minus sign is used; for countercurrent flow, a plus sign.

For the three-phase system encountered in coal-liquefaction reactors, we can modify Equation (5-71) as follows:

$$\frac{U_{og}}{\epsilon_g} \pm \frac{U_{s\ell}}{1-\epsilon_g} = U_{\text{slip}} \tag{5-72}$$

where $U_{s\ell}$ is the liquid-solid slurry velocity.

The above correlation has been obtained as an extension of work of Diboun and Schugerl [18]. The reader is cautioned that Equation (5-70) will give a very rough estimate of the gas Peclet number in a three-phase reactor. It is recommended that future experimental work should include the estimation of gas-phase Peclet number in large-diameter three-phase reactors.

B. LIQUID PHASE

Pavlica and Olson [47] have reviewed the literature on the liquid dispersion coefficient and proposed a correlation of the type

$$\text{Pe}_\ell = \frac{U_{o\ell}L}{E_{z\ell}} = \frac{3.0}{\epsilon_\ell}\left(\frac{U_{o\ell}}{U_{\text{slip}}}\right)\left(\frac{L}{D}\right), \tag{5-73}$$

where U_{slip} is the slip velocity, which can be calculated using either Equation (5-71) or Equation (5-72). Deckwer et al. [17], on the other hand, proposed the use of this dispersion-coefficient relationship:

$$E_{z\ell} = 0.00108\ D^{1.4}\ U_{og}^{0.3}, \tag{5-74}$$

where D is in m, U_{og} in m/sec, and $E_{z\ell}$ in m²/sec.

Another study made on the dependence of the liquid-phase dispersion coefficient on the liquid-phase physical properties has been reported by Hikita and Kikukawa [27]. With the help of dimensional analysis, they proposed the following functional form for the liquid-phase axial dispersion coefficient:

$$\frac{E_{z\ell}}{(D^3g)^{1/2}} = \left[0.037 + 0.188\left(\frac{U_{og}}{gD}\right)^{0.72}\right] \times \left(\frac{\mu_\ell{}^4 g}{\rho_\ell \sigma^3}\right)^{-0.03}, \tag{5-75}$$

where μ_ℓ and σ are the viscosity and interfacial tension of liquid phase, and g is the gravitational constant. Thus, the Peclet number can be represented as

$$\text{Pe}_\ell = \left(\frac{\mu_\ell{}^4 g}{\rho_\ell \sigma^3}\right)^{0.03} \frac{U_{o\ell}L}{\sqrt{D^3 g}}\left[0.037 + 0.188\left(\frac{U_{og}}{\sqrt{gD}}\right)^{0.72}\right]^{-1} \tag{5-76}$$

According to these authors the average deviation between the observed values and the calculated values using the above correlation of the liquid-phase dispersion coefficient was around 3%.

Kato et al. [35] have measured dispersion coefficients in 0.666-, 0.122-, and 0.214-m-i.d. columns with particle sizes ranging from 63 through 177 μm and solid concentrations up to 200 kg solid per m^3 of slurry. They also correlated their data for the longitudinal dispersion coefficient of the liquid in the slurry by the following dimensionless relation:

$$\mathrm{Pe}_{\ell} = \frac{U_{o\ell}L}{E_{z\ell}} = \frac{13\ \mathrm{Fr}_g\ (L/D)\ (U_{o\ell}/U_{og})}{1+8\ \mathrm{Fr}_g^{\ 0.85}}, \tag{5-77}$$

where $\mathrm{Fr}_g = U_{og}/\sqrt{gD}$ is the Froude number. The above equation was subsequently modified; however, in our opinion, more experimental data are needed to verify its applicability in coal-liquefaction reactors.

C. SOLID PHASE

The axial dispersion characteristics of solids in a continuous (cocurrent upflow) three-phase flow operation have been studied by Imafuku et al. [29] and Kato et al. [35] It has been found that for small particles and in small-diameter columns, the longitudinal dispersion coefficient for solids in a three-phase fluidized bed is the same as that of liquid and can be given by Equation (5-75). For large-diameter particles and large columns, Kato [34] has proposed the following correlation:

$$\mathrm{Pe}_s = \frac{U_{o\ell}L}{E_{zs}} = 13\left(\frac{L}{D}\right)\left(\frac{U_{o\ell}}{U_{og}}\right)$$

$$\mathrm{Fr}_g = \frac{1+0.009\ \mathrm{Re}_p\ \mathrm{Fr}_g^{\ -0.8}}{1+8\ \mathrm{Fr}_g^{\ +0.85}}, \tag{5-78}$$

where the particle Reynolds number is

$$\mathrm{Re}_p = \frac{d_p U_t \rho_\ell}{\mu_\ell} \tag{5-79}$$

and U_t is the terminal settling velocity of solid particles.

5.5.2 *Thermal-Dispersion Coefficients*

It has been found by earlier investigators that the thermal diffusivity (eddy) for a fluid or a phase is of the same order of magnitude as the mass diffusivity (eddy). Thus, one can correlate the mass-dispersion coefficient and the thermal-dispersion coefficient for a particular fluid or phase.

If we denote the eddy thermal diffusivities for the gas, liquid, and solid phases by E_{Hg}, $E_{H\ell}$, and E_{Hs}, we propose the following correlations by analogy:

$$E_{Hg} \sim E_{zg} \tag{5-80}$$

$$E_{H\ell} \sim E_{z\ell} \tag{5-81}$$

$$E_{Hs} \sim E_{zs} \tag{5-82}$$

The thermal dispersion coefficients D_H can be related to eddy thermal diffusivities by:

$$D_{Hg} = \rho_g C_{pg} E_{Hg} , \tag{5-83}$$

$$D_{H\ell} = \rho_\ell C_{p\ell} E_{H\ell} , \tag{5-84}$$

$$D_{Hs} = \rho_s C_{ps} E_{Hs} . \tag{5-85}$$

To the best of the authors' knowledge there has been no reported literature on the evaluation of thermal-dispersion coefficients for various phases. Future work in this area is necessary and is highly recommended.

5.5.3 *Holdups of the Gas and Liquid Phases*

A. GAS-PHASE HOLDUP

By performing experiments in 0.1-m- and 0.19-m-i.d. bubble columns (1.5 m and 2.40 m in height, respectively) with single-nozzle gas spargers and superficial gas velocities of 0.043–0.338 m/sec, Hikita and Kikukawa [26] proposed this correlation for their experimental data:

$$\epsilon_g = 0.505\, U_{og}^{0.47} \left(\frac{72\times 10^{-3}}{\sigma}\right)^{2/3} \left(\frac{1}{\mu_\ell}\right)^{0.05} \tag{5-86}$$

where σ is the liquid phase surface tension in dyne/cm. The above correlation is valid for gas-phase holdup in aqueous nonelectrolyte solutions. The average deviation between the observed values and the calculated values of ϵ_g was 2.8%.

Akita and Yoshida [1] also proposed a correlation for nonelectrolyte and electrolyte solutions. With the help of dimensional analysis, they obtained the following functional form for the gas-phase holdup:

$$\epsilon_g = f_1 \left(N_{Bo}, N_{Ga}, \mathrm{Fr}_g, d_o/D\right), \tag{5-87}$$

where

$$N_{Bo} = gD^2 \rho_\ell / \sigma = \text{Bond number,} \tag{5-88}$$

$$N_{Ga} = gD^3 \rho_\ell{}^2 / \mu_\ell{}^2 = \text{Galileo number,} \tag{5-89}$$

$$\mathrm{Fr}_g = U_{og} / \sqrt{gD} = \text{Froude number,} \tag{5-90}$$

$$d_o/D = \text{diameter ratio.} \tag{5-91}$$

The effect of the diameter of the single inlet orifice d_o has been shown to be unimportant by Yoshida and Akita [68]. Fair et al. [21] and Yoshida and Akita [68], using several columns of various diameters, showed that the effect of the column diameter on the gas holdup was unimportant for columns larger than 15 cm in diameter. Thus, the dependence of ϵ_g on d_o D in Equation [5-87] can be eliminated. After a series of experiments, the following correlation for ϵ_g was proposed by Akita and Yoshida [1]:

$$\frac{\epsilon_g}{(1-\epsilon_g)^4} = C_1 N_{Bo}^{1/8} N_{Ga}^{1/12} \mathrm{Fr}_g^{1.0} . \tag{5-92}$$

For nonelectrolyte solutions or liquids the value of the constant C_1 is 0.2, while for electrolyte solutions, it is 0.25.

Another study on the gas-phase holdup in three-phase reactors was made by Kito et al. [37], who employed a batch (i.e. no liquid flow) system. They correlated their data for gas phase holdup by a relation

$$\frac{\epsilon_g}{[\epsilon_g (1-\epsilon_g)^2]^{0.44}} = 0.5\, \mathrm{We}_\ell^{0.11} \mathrm{Fr}_g^{0.22} , \tag{5-93}$$

where We_ℓ is the Weber number:

$$\mathrm{We}_\ell = DU_{og}^2 \rho_\ell / \sigma , \tag{5-94}$$

$$\mathrm{Fr}_g = U_g / \sqrt{gD} . \tag{5-95}$$

The data of Kito et al. [37] showed that the gas holdup is independent of the ratio D/d_p. The gas holdup in aqueous solutions of electrolytes is slightly larger than that in nonelectrolyte solutions.

Lee et al. [38] have observed that for $D > 0.15$ m Equation (5-92) gives a better estimate of ϵ_g than Equation (5-86). For $D < 0.15$ m they have recommended the use of Equation (5-86). The reader can verify these comments by referring to Illustration 5-1 below.

Hovmand and Davidson [28] proposed a slug-flow model to correlate the gas holdup, in gas-liquid-solid fluidized beds, at superficial gas velocities in excess of that required for the incipient fluidization of solids. They correlated the gas holdup and superficial gas velocity by the equation

$$\frac{\epsilon_g - \epsilon_g^*}{\epsilon_g^*} = \frac{U_{og} - U_{og}^*}{0.35\,(gD)^{1/2}} \tag{5-96}$$

where U_{og}^* is the superficial gas velocity at the incipient fluidization of solid particles (in cm/sec), and ϵ_g^* is the fractional gas holdup at the incipient fluidization.

Very recently, Shah and Parulekar [45] proposed a correlation for the gas-phase holdup which is valid for low gas velocities and small-diameter columns. They used the data presented by Kato et al. [35] and Reith et al. [53] to obtain the following correlation:

$$\epsilon_g = 0.0866 \left(\frac{1}{\sigma}\right)^{0.1} \frac{U_{og}^{\;0.5}}{D^{0.08}} \tag{5-97}$$

For relatively high gas velocities this equation was modified by Shah and Parulekar as

$$\epsilon_g = \frac{1}{\sigma^{0.1}} D^{-0.08} \,(0.18 \ln U_{og} - 0.05) \tag{5-98}$$

Under identical conditions, larger columns result in lower gas holdups. This has been explained on the basis that in the larger columns, random circulation patterns (eddies) exist. Smaller columns exhibit higher holdups because of the absence of these eddies. However, after a certain diameter, the effect of diameter on gas holdup is negligible, as can be seen from the work of Akita and Yoshida [1].

It is recommended that use of any of the equations (5-86), (5-92), and (5-93) be made to evaluate ϵ_g for $D < 0.15$ m. For $D > 0.15$ m, use of the equations (5-92) and (5-93) is recommended.

B. LIQUID-PHASE HOLDUP

The liquid-phase holdup can be evaluated by measuring the concentration of solid phase experimentally or by computing it from one of the appropriate solid-phase mass-balance equations. It can be easily shown that the liquid-phase holdup is a function of the gas-phase holdup and the solid concentration C_s (expressed as weight of solid per unit volume of slurry). The relation is

$$\epsilon_1 = (1-\epsilon_g)\left(1 - \frac{C_s}{\rho_s}\right), \tag{5-99}$$

where ρ_s is the density of the solid phase.

Illustration 5-1

The solvent-refined coal (SRC-II) process is normally operated at about 460°C (733 K) and 15.2 MPa pressure. Typical dimensions of the pilot-scale reactor are 0.6 m diameter and 10 m height, and the corresponding dimensions for a commercial reactor are 3.2 m and 35 m, respectively. Typical values of superficial gas and liquid velocities used are 0.04 and 0.004 m/sec in the pilot-scale reactor, and 0.12 and 0.012 m/sec in the commercial-scale reactor. Typical values for the densities of the gas, liquid, and solid phases within the reactor are 10, 90, and 130 kg/m^3, respectively. The viscosity and surface tension of the liquid under the reaction conditions are estimated to be 8.0×10^{-4} Pa sec and 0.01 N/m respecitvely. The median particle diameter is about 50 μm. The average solid concentration is 100 kg/m^3 slurry).

Using this information, we can estimate the holdup and the mass-dispersion coefficient (or Peclet number) for the gas, liquid, and solid phases, and also the Peclet number for heat. The reactor is operated in a cocurrent upflow fashion. The results obtained using various correlations can be compared.

Solution

1. *Holdup of gas, liquid, and solid phase.* Here we have used the correlations described in Equations (5-86), (5-92), and (5-93) to estimate the gas holdup.

For the pilot-scale reactor Equation (5-86) gives $\epsilon_g = 0.505(4)^{0.47} (72 \times 10^{-3}/10)^{2/3}(1/0.008)^{0.05} = 0.046$. Use of Equation (5-92) results in the following value of gas holdup:

$$\frac{\epsilon_g}{(1-\epsilon_g)^4} = 0.2\left(\frac{981 \times 80 \times 80 \times 0.9}{10}\right)^{1/8}\left(\frac{981(80)^3 \times (0.9)^2}{(0.008)}\right)^{1/12}\left(\frac{4}{\sqrt{981 \times 80}}\right),$$

from which we obtain $\epsilon_g = 0.109$. Using (5-93), we get $\epsilon_g{}^{0.56}/1-\epsilon_g{}^{0.48} = 0.5 (80 \times 4 \times 4 \times 0.9/10)^{0.11}(4/\sqrt{981 \times 80})^{0.22}$, and thus $\epsilon_g = 0.114$.

The corresponding values of the gas-phase holdup for a large-scale reactor, calculated from Equations (5-86), (5-92), and (5-93), are 0.77, 0.207, and 0.221, respectively. It is observed that Equation (5-86) gives a low estimate of the gas holdup at larger diameters and high values of the gas velocity. The values of ϵ_g predicted by Equations (5-92) and (5-93) are very close, and an average of these has been used in all further calculations.

The average liquid holdup can be calculated from (5-99) as

$$\epsilon_\ell = (1-0.214)\left(1-\frac{0.1}{1.5}\right) = 0.7336 \qquad \text{for a large-scale reactor,}$$

$$\epsilon_\ell = (1-0.1115)\left(1-\frac{0.1}{1.5}\right) = 0.83 \qquad \text{for a pilot-scale reactor.}$$

Thus it can be seen that the average liquid-phase holdup is expected to be less in the large-scale reactor.

The solid holdups in the pilot-scale and large-scale reactor are 0.0585 and 0.0524, respectively.

2. *Mass- and thermal-dispersion coefficients.* Gas-phase Peclet numbers can be evaluated using (5-70). For a large-scale reactor,

$$U_{\text{slip}} = \frac{12}{0.214} - \frac{1.2}{0.73} = 0.544 \text{ m/sec},$$

$$\text{Pe}_g = 0.2\left(\frac{12/0.214}{54.43}\right) \times \frac{3500}{320} = 2.32;$$

for pilot-scale reactors Equation (5-70) cannot be used.

It is seen by the relatively low Pe that significant backmixing of gas in large-scale reactors is expected from Equation (5-70).

A liquid-phase Peclet number can be evaluated using Equations (5-73), (5-76), and (5-77). Using Equation (5-73), for a large-scale reactor we have

$$\text{Pe}_\ell = \frac{3.0}{0.7336}\left(\frac{1.2}{54.43}\right)\frac{3600}{320} = 1.0\ .$$

Using Equation (5-76), we get the following value of Pe_ℓ:

$$\text{Pe}_\ell \left(\frac{(0.008)^4 \times 981}{0.9 \times (10)^3}\right)^{0.03} \left(\frac{1.2 \times 3600}{\sqrt{981 \times (320)^3}}\right)\left[0.037 + 0.188\left(\frac{12}{\sqrt{981 \times 320}}\right)^{0.72}\right]^{-1}$$

$$= 0.46.$$

Using Equation (5-77) we get

$$\text{Pe}_\ell = \frac{13 \times (12/\sqrt{981 \times 320})\,(3600/320)\,(1 - 2/12)}{1 + 8\,(12/\sqrt{981 \times 320})^{0.85}} = 0.24.$$

Thus it can be seen that the three correlations predict different degrees of backmixing for the liquid phase. But in all cases, a high degree of backmixing is predicted.

The value of the liquid-phase Peclet number for the pilot-scale reactor using Equations (5-73), (5-76), and (5-77) is 0.5, 0.22, and 0.19, respectively. It is recommended that correlation proposed by Kato et al. [35] [Equation (5-77)] be used to evaluate axial dispersion coefficient for the liquid phase. The above calculations also indicate that the use of Equation (5-73) may not give correct estimates of backmixing coefficients for the gas and liquid phase.

The Peclet number for the solid phase can be evaluated from (5-78). For the large-scale reactor conditions,

$$U_t = 9.083 \times 10^{-4} \text{ m/sec},$$

$$\mathrm{Pe}_s = 13 \left(\frac{3600}{320}\right) \left(\frac{1.2}{12}\right) \times \frac{12}{\sqrt{981 \times 320}}$$

$$\frac{\left[1+0.009 \left(\frac{0.005 \times 0.0908 \times 0.9}{0.008}\right) \left(\frac{12}{\sqrt{981 \times 320}}\right)^{-0.8}\right]}{1+8 \left(\frac{12}{\sqrt{981 \times 320}}\right)^{0.85}} = 0.242 .$$

The corresponding value of Pe_s for the pilot-scale reactor is 0.193.

The thermal Peclet number can be obtained from Equations (5-80) through (5-85), and it can be related to the mass Peclet number for the three phases as follows:

$$\mathrm{Pe}_\theta = \frac{\rho_g C_{pg} U_{og} + \rho_\ell C_{p\ell} U_{o\ell} + \rho_s C_{ps} U_{os}}{\dfrac{\rho_g C_{pg} U_{og}}{\mathrm{Pe}_g} + \dfrac{\rho_\ell C_{p\ell} U_{o\ell}}{\mathrm{Pe}_\ell} + \dfrac{\rho_s C_{ps} U_{os}}{\mathrm{Pe}_s}} .$$

Assuming $\mathrm{Pe}_g >> \mathrm{Pe}_\ell$ for the pilot-scale reactor, we have

$$\mathrm{Pe}_\theta = \frac{0.01 \times 0.8 \times 4 + 0.9 \times 0.6 \times 0.4 + 1.3 \times 0.3 \times 0.3092}{\dfrac{0.9 \times 0.6 \times 0.4}{0.19} + \dfrac{1.3 \times 0.3 \times 0.3092}{0.193}} = 0.25.$$

The corresponding value for the large-scale reactor is

$$\mathrm{Pe}_\theta = \frac{0.01 \times 0.8 \times 12 + 0.9 \times 0.6 \times 1.2 + 1.3 \times 0.3 \times 1.1092}{\dfrac{0.01 \times 0.8 \times 12}{2.32} + \dfrac{0.9 \times 0.6 \times 1.2}{0.24} + \dfrac{1.3 \times 0.3 \times 1.1092}{0.242}} = 0.26.$$

Thus it is seen that the estimated thermal Peclet numbers are nearly equal for the pilot-scale and large-scale reactors. It is also seen that for small particles of the order of 50 μm, the Peclet numbers for the liquid and solid phases are very close, which implies that the extent of backmixing in the two phases is nearly the same.

5.5.4 *Gas-Liquid Interface Mass-Transfer Coefficient*

Calderbank and Moo-Young [6, 7] correlated their data for the gas-liquid mass-transfer coefficient by the following relation:

$$(k_\ell)_i\,(\mu_\ell/\rho_\ell D_{i\ell})^{2/3} = 0.31\,(\Delta\rho_\ell g\mu_\ell/\rho_\ell{}^2)^{1/3}\,, \tag{5-100}$$

where $\Delta\rho_\ell$ is the density difference between gas bubbles and liquid phase, $D_{i\ell}$ is the molecular diffusivity of component i in the liquid phase, and k_ℓ is the gas-liquid mass-transfer coefficient (in cm/sec). The correlation is applicable to bubbles rising in a liquid by gravitational forces in the absence of any mechanical stirring. Although the correlation was developed for a column with no solids present, it has been applied with some success to bubble columns containing solids by Matsuura [39].

Akita and Yoshida [1], by using dimensional analysis and performing a series of experiments, proposed the following correlation for the prediction of the liquid-phase mass transfer coefficient:

$$(k_\ell a_\ell)_i = 0.6\,D_{i\ell}^{0.5}\,\nu_\ell^{-0.12}\,(\sigma_\ell/\rho_\ell)^{-0.62}\,D^{0.17}\,g^{0.93}\,\epsilon_g^{1.1}\,, \tag{5-101}$$

or

$$\begin{aligned}\mathrm{Sh}\,(a_\ell)_i\,D &= (k_\ell a_\ell)_i\,D^2/D_{i\ell} \\ &= 0.6\,\mathrm{Sc}^{0.5}\,N_{\mathrm{Bo}}^{0.62}\,N_{\mathrm{Ga}}^{0.31}\,\epsilon_g^{1.1}\,,\end{aligned} \tag{5-102}$$

where

$$\mathrm{Sh} = (k_\ell)_i\,D/D_{i\ell} \quad = \text{Sherwood number}, \tag{5-103}$$

$$\mathrm{Sc} = \mu_\ell/\rho_\ell D_{i\ell} \quad = \text{Schmidt number}, \tag{5-104}$$

and N_{BO} and N_{Ga} have already been defined in Equations (5-88) and (5-89), respectively. $(k_\ell a_\ell)_i$ refers to liquid-phase mass-transfer coefficient for component i, in sec^{-1}.

The assumption that $k_\ell a_\ell$ increased with the column diameter to the 0.17 power was verified by experiments only up to a diameter of 0.6 m. Lee et al. [38] have recommended that for design purposes, it might be safer to use the $k_\ell a_\ell$-values for the 0.6-m column for columns of larger diameter as well.

Although the use of Equations (5-101) and (5-102) in the calculation of $k_\ell a_\ell$ is valid only for systems where no chemical reaction occurs, it can also be considered a reasonable approximation for systems with very slow reactions.

Use of Equation (5-101) is recommended for estimating the gas-liquid mass-transfer coefficient.

Illustration 5-2

Using the values of various parameters supplied in Illustration 5-1, we can calculate the gas-liquid mass-transfer coefficient for hydrogen. Assume that the gas phase consists mainly of hydrogen and the molecular diffusivity of hydrogen in the liquid phase is 10^{-8} m²/sec. Equation (5-100) can be used to evaluate $(k_\ell)_{H_2}$:

$$(k_\ell)_{H_2} = 0.31 \left(\frac{981 \times 0.89 \times 0.008}{(0.9)^2}\right)^{1/3} \left(\frac{10^{-8} \times 0.9}{0.008}\right)^{2/3} = 0.0005 \text{ m/sec} .$$

Using Equation (5-101), we can estimate $(k_\ell a_\ell)_{H_2}$. For the pilot-scale reactor,

$$(k_\ell a_\ell)_{H_2} = 0.6\,(10^{-4})^{0.5} \left(\frac{0.008}{0.9}\right)^{-0.12} \left(\frac{10}{0.9}\right)^{-0.62}$$

$$\times\,(80)^{0.17}\,(981)^{0.93}\,(0.11)^{1.1} = 0.27 \text{ sec}^{-1} .$$

The corresponding value for the large-scale reactor is 0.7 sec^{-1}.

If a_ℓ is of the order of 0.02 m^{-1}, the values of gas-liquid mass-transfer coefficient calculated using Equations (5-100) and (5-101) differ considerably. This may be due to the empirical diameter scale-up factor $D^{0.17}$.

5.5.5 *Other Design Parameters*

A. PRESSURE DROP

The most recent work on the measurement of the pressure drop in a three-phase fluidized-bed column is by Javdani et al. [31]. They measured the pressure drop in the flow of a nitrogen-Blandol oil-coal slurry. The data showed that the pressure drop increases with the gas velocity and the coal concentration in the slurry. However, no correlation has been reported so far in the literature on the pressure drop in a three-phase fluidized bed.

B. MINIMUM GAS VELOCITY REQUIRED FOR FLUIDIZATION

Normally in a three-phase fluidized bed, the particles form the fluidized phase with the liquid as fluidizing medium, while the energy required to keep the particles suspended is provided by the relatively fast-moving gas phase. The bed is fluidized when the superifical velocity past the particles is greater than their terminal settling velocity. Narayanan et al. [43] studied the minimum gas velocity required to suspend particles in a stagnant liquid medium.

The critical gas velocity, U_{cg}, is given by the following relations:

$$U_{cg} = 0.043 \left(\frac{D}{2}\right)^{n'} e^{-0.1 C_s'} U'_{og} \quad \text{for } C_s' < 10 \tag{5-105}$$

and

$$U_{cg} = 0.0125 \left(\frac{D}{2}\right)^{n'} e^{-0.03\, C_s'} U'_{og} \quad \text{for } C_s' > 10 , \tag{5-106}$$

where D is the reactor diameter (in inches), C_s, is the percentage solid concentration in the liquid [i.e. 100×(kg solid)/(kg liquid)],

$$n' = \begin{cases} 0.2 & \text{for } d_p < 200\ \mu\text{m}, \\ 0.5 & \text{for } d_p > 200\ \mu\text{m}, \end{cases} \tag{5-107}$$

and U_{cg} is the superficial critical gas velocity. U_{og} is obtained from the solutions of the equations

$$U'_{og} + 3.687\sqrt{U'_{og} H_o} = \phi \quad \text{for } U'_{og} < 0.067 \text{ m/sec} \tag{5-108}$$

and

$$U'_{og} + 5.378\sqrt{U'^{\,0.38}_{og} H_o} = \phi \quad \text{for } 0.067 < U'_{og} < 0.2134 \text{ m/sec}, \tag{5-109}$$

where

$$\varphi = \sqrt{2g(\rho_s - \rho_\ell)\left(\frac{2d_p}{3\rho_\ell} + \frac{H_s H_o}{\rho_s + H_s \rho_\ell}\right)} \tag{5-110}$$

where $H_s = C'_s/100$ and H_0 is the static slurry height in the column.

Kato [34] presented a graphical correlation for the critical gas velocity required for complete suspension of glass spheres and sand particles.

C. CRITICAL SOLID HOLDUP IN THREE-PHASE FLUIDIZED BED

Roy et al. [54] have given the following correlations for the critical solid holdup ϵ_{cs}, which is the maximum holdup of solids in a three-phase fluidized bed:

$$\epsilon_{cs} = 6.84 \times 10^{-4}\, \text{Re}_g N_b^{-0.23} (U_{to}/U_b)^{-0.18} \gamma'^{-3.0} C_{\mu\ell} \text{ for } \text{Re}_g < 600, \tag{5-111}$$

$$\epsilon_{cs} = 0.1072\, \text{Re}_g^{0.2} N_b^{-0.23} (U_{to}/U_b)^{-0.18} \gamma'^{-3.0} C_{\mu\ell} \text{ for } \text{Re}_g > 600. \tag{5-112}$$

ϵ_{cs} is the weight of solids divided by the weight of suspension, Re_g is the superficial gas Reynolds number based on an empty cross-section of column, U_{to} is the Stokes free-settling velocity, U_b is the bubble velocity, N_b is the bubble number defined as $\sigma/U_b\mu_\ell$, and γ is the relative wettability factor for solids with respect to quartz (i.e., $\gamma_s/\gamma_{\text{quartz}}$, where γ_s is the wettability factor for the solids, related to the contact angle θ' between the solid and the liquid by $\gamma_s = 1 + \cos\theta'$. $C_{\mu\ell}$ is the viscosity correction factor, defined by the relation

$$C_{\mu\ell} = 1 - 0.5892 \log_{10} \mu_\ell + 0.1026 \log_{10}^2 \mu_\ell, \tag{5-113}$$

where μ_ℓ is in centipoises.

D. IMPORTANT SCALE-UP FACTORS

In coal-liquefaction slurry reactors, some important scale-up factors are:

1. Flow regime.
2. Backmixing in the three phases, and the holdups of the gas and liquid phases.
3. Mass-transfer resistance to hydrogen absorption and for desorption of product gases.
4. Vaporization of the liquid phase.
5. Temperature rise across the reactor (adiabatic temperature rise); the quench locations, the number of quench streams, and the amount and the quality of quench.

Flow Regime.

For designing and operating the pilot-scale reactor, it is very important that the pilot-scale reactor is operated in the same flow regime in which the commercial reactor will be operated. The heat and mass-transfer effects, backmixing characteristics of the various phases, and holdups of the three phases are not likely to be the same in two different scales (in fact they are considerably different). Often, however, exact reproduction is not possible due to various design effects. Effects which cause channeling, bypassing, and recirculation depend on factors such as the length-to-diameter ratio of the column, the feed-distributor design, the quench-gas design, etc. If the commercial-reactor flow conditions are not reproduced in a pilot-scale reactor, then the prevailing heat, mass-transfer effects, backmixing, and particle-entrapment effects in the large-scale unit will be very difficult to estimate.

Backmixing in the Three Phases and Gas and Liquid Holdup.

In a particular flow regime, the backmixing depends on the velocities of the three phases, and properties of the three phases such as the density, viscosity, and surface tension; the shape, size, and configuration of catalyst and other solid particles; and the column diameter. The holdups of the respective phases also depend upon the same parameters.

In a bubble column, the backmixing in the liquid and solid phase is a moderately strong function of column diameter. It is often difficult to get the same degree of backmixing in the liquid and solid phases unless the gas flow rate is reduced to some extent. Because of the commercial reactor has a larger diameter than the pilot-scale reactor, a greater degree of backmixing for both solid and liquid phases will most likely prevail in the former.

The commercial reactors may exhibit considerable backmixing in the gas phase (see Illustration 5-1). At the normal gas and liquid velocities used in the pilot-scale reactors, the gas phase usually moves as plug flow. Thus the backmixing characteristics in the gas phase for the large-scale reactors and the small-scale reactors may also be significantly different.

The gas and liquid holdups are relatively mild functions of reactor diameter, and these can be reproduced in the small-scale reactors by adjusting the flow rates of the three phases. Since the gas velocity used in commercial reactors is considerably larger than in the pilot-scale reactor, the gas holdup may be expected to be higher in the commercial reactor. Reproduction of the gas holdup can be obtained by increasing the superficial gas velocity in pilot-scale reactors, the difference can be accounted for via the models used for simulation.

Gas-Liquid Mass-Transfer Resistance.

In coal-liquefaction processes, the availability of hydrogen in the slurry phase is a very important criterion in the complex reaction network. In order to increase the yield of the desired products, it is necessary to have a low mass-transfer resistance. The nature of the flow regime and the superficial gas velocity are two important parameters which affect the gas-liquid mass-transfer resistance.

The gas-liquid mass-transfer coefficient can be increased by increasing the superficial gas velocity U_{og}. The same conditions also increase the vaporization of the volatile liquid phase under certain conditions and concentrate the less volatile coal liquid within the slurry phase wherein the catalytic reactions occur.

Vaporization of Liquid Phase.

The extent of vaporization depends on the operating conditions such as the reactor temperature, operating pressure, and composition. It has also been found from experiments that as the gas velocity is increased, the vaporization also increases. An excessive amount of vaporization could result in coking. By operating at somewhat lower reactor temperatures and higher pressures, vaporization can be reduced.

Temperature Rise, Quench Location, and Quench Amount and Quality.

In some of our unpublished work, we found that the adiabatic temperature rise in the commercial-size reactor may be somewhat less than in a pilot-scale reactor. The gas holdup in the commercial reactor is usually larger than that in the pilot-scale reactor, due to higher gas velocities. This causes a reduction in consumption due to the slurry-phase reaction, which in turn reduces the heat generation. The temperature rise in a commercial reactor is therefore lower than that in the case of a small-scale reactor for the same space time and degree of backmixing.

In the commercial reactors, gaseous quenches may be added in order to control the outlet temperature of the reactor. The gas flow rate has a dominant effect on the nature of the flow regime, backmixing and holdup of the three phases, and heat- and mass-transfer resistances; thus the location of quench points could play a significant role in the performance of a commercial reactor. The best location of quench points to maintain a reasonable temperature profile has been studied by Parulekar and Shah [45, 46]. The amount of quench will depend upon the operating conditions. Usually, hydrogen will be preferred as a quench gas.

Finally, it should be noted that one of the purposes of the pilot-plant reactor is to test the viability of the desired design and operating conditions in the large-scale reactor. It is, therefore, important to identify the important scale-up variables and determine the design of a small-scale reactor which will meet the necessary conditions. The above discussion indicates that for given feed temperature, pressure, coal concentration in feed slurry, catalyst pellet shape and size; design of gas distributor, and type of coal, the important reactor design variables are: (1) superficial and interstitial (based on open cross-section) gas velocities, (2) gas-to-liquid volumetric ratio, (3) column diameter, (4) column length, and (5) feed-slurry composition.

5.6 Nomenclature

5.6.1 *Latin Letters*

A	Heavy distillate
A	A constant in Equation (5-1)
A	Dimensionless concentration of heavy distillate as defined in (5-17)
a_ℓ	Liquid-phase specific mass-transfer area at gas-liquid interface, m^{-1}
B	Light distillate
B	A constant in Equation (5-1)
B	Dimensionless concentration of light distillate as defined in (5-17)
C	Concentration of moisture- and ash-free coal in Equation (5-3)
C	Volatile products
C	Dimensionless concentration of C as defined in Equation (5-17)
$C_{A\ell}$	Concentration of A, kg/(m^3 liquid)
$C_{B\ell}$	Concentration of B, kg/(m^3 liquid)
C_C	Concentration of C, kg-mol/m^3
C_{H_2}	Concentration of hydrogen, kg-mol/(m^3 gas)
C_i	Concentration of moisture-free coal, (kg mf coal)/ (m^3 slurry)
C_p	Specific heat kJ/kg K

C_s	Concentration of solids, kg/(m^3 slurry)
C_s'	Percentage solid concentration in liquid, 100× (kg solid)/(kg liquid)
C_T	Total gas-phase molar concentration, kg-mol/(m^3 gas)
$C_{\mu \ell}$	Viscosity correction factor defined in Equation (5-113)
D	Diameter of column, m
D	Diffusivity, m^2/sec
D_H	Axial thermal-dispersion coefficient, kJ/m sec K
$D_{i\ell}$	Diffusivity of component i in liquid phase, m^2/sec
d_o	Orifice diameter, m
d_p	Particle diameter, m
$\bar{e}$	Dimensionless mass-dispersion coefficient
E_H	Eddy thermal diffusivity, m^2/sec
E_z	Axial mass-dispersion coefficient, m^2/sec
E_1–E_4	Activation energies for reactions (5-5) through (5-8) respectively, kJ/kg-mol
E_4'	Activation energy for hydrogen consumption reaction, kJ/kg-mol
f, f_1	Quantities defined by Equation (5-17)
Fr	Froude number
G	Gas flow
G	Superficial mass velocity, kg/m^2 sec
g	Gravitational acceleration, m/sec^2
g_c	Conversion factor
h	Slurry-wall heat-transfer coefficient (see Table 5-1)
H	Henry's-law constant, [kg-mol/(m^3 liq.)] / [kg-mol/(cm^3 gas)]
H'	Dimensionless concentration of hydrogen
H^*	Preexponential factor in the expression for Henry's-law constant, [kg-mol/(m^3 liq.)] / [kg-mol/(m^3 gas)]
H_s	$C_s'/100$
H_o	Static slurry height in column (m)
H_2	Hydrogen
K	Thermal conductivity of slurry in Table 5-1
$\mathcal{K}_o$	A constant in Equation (5-50), m$^{-0.17}$ sec^{-1}
k_ℓ	Liquid-side mass-transfer coefficient, cm/sec
$k_\ell a_\ell$	Volumetric liquid-side mass-transfer coefficient, sec^{-1}
k_1, k_1^*	Kinetic constant and frequency factor for reaction (5-5), (m^3 slurry)/kg cat.)(sec)

k_2, k_2^*	Kinetic constant and frequency factor for reaction (5-6), (m^3 gas) (m^3 slurry)/(sec)(kg cat.) (kg A)
k_3, k_3^*	Kinetic constant, and frequency factor for reaction (5-7), (m^3 gas) (m^3 slurry)/(sec)(kg cat.) (kg B)
k_4, k_4^*	Kinetic constant and frequency factor for reaction (5-8), (m^3 gas) (m^3 slurry)/(m^3 liq.) (kg cat.)(sec)
k_4'	Arrhenius factor for hydrogen-consumption reaction (units in Table 5-2)
L	Length of column (m)
m	Ratio of the amount of quench to the amount of gas in mainstream
M_{H_2}, M_C	Molecular weights of hydrogen and product gases (C) respectively, kg/kg-mol
n	Mass flux, kg/m^2 sec
n	Order of the reaction in Equation (5-3)
n'	Defined in (5-107)
N_b	Bubble number
N_{Bo}	Bond number defined in (5-88)
N_{Ga}	Galileo number defined in (5-89)
P	Pressure, MPa
p_3	Defined by Equation (5-7)
Pe	Peclet number
Pe_θ	Thermal Peclet number
Pr	Prandtl number in Table 5-1
R	Dimensionless dissolution rate constant in Equation (5-3)
r	Ratio of superficial liquid velocity to superficial gas velocity
R_g	Universal constant MPa m^3/kg-mol K or kJ/kg-mol K
R_ℓ, R_o	Defined in (5-41)
R_4'	Dimensionless reaction constant for hydrogen-consumption reaction
Re_p	Particle Reynolds number defined in (5-28)
r_1	Rates of reaction (5-5), kg/(m^3 liq.)(sec)
r_2, r_3, r_4	Rate of reaction, (5-6), (5-7), (5-8), kg-mol/(m^3 liq.)(sec)
R_1–R_4	Quantities defined by Equation (5-17)
S	Slurry flow
Sc	Schmidt number
Sh	Sherwood number
SRT	Slurry residence time, hr
T	Temperature, K

U	Net velocity of a phase, m/sec
$\overline{U}$	Dimensionless superficial velocity
U_b	Bubble velocity, m/sec
U_{cg}	Critical gas velocity for fluidization, m/sec
$U_{og}, U_{o\ell}, U_s'$,	Superficial velocity of gas, liquid, and solid phase respectively, m/sec
U_{slip}	Slip velocity between gas and liquid or gas and liquid-solid slurry, m/sec
U_t	Settling velocity with respect to liquid, m/sec
U_t'	Terminal settling velocity of particles, m/sec
U_{to}	Stokes free-settling velocity, m/sec
U_{og}^*	Superficial gas velocity at incipient fluidization, m/sec
U_{og}'	Defined in (5-108) and (5-109), m/sec
W_s	Weight fraction of solid in feed slurry, kg/(kg slurry)
We	Weber number
x	Distance from inlet of reactor, m
$Y_{\text{A}}, Y_{\text{H}_2}$	Consumption of heavy distillate and hydrogen respectively, kg/(kg feed slurry)
$Y_{\text{B}}, Y_{\text{C}}$	Yields of light distillate and volatile products respectively, kg/(kg feed slurry)
z	Dimensionless distance from the bottom of column
ΔH_R	Combined heat of reaction for reactions 1, 2, 3, and 4 expressed in terms of hydrogen consumption, kJ/(kg-mol H_2 consumed)
ΔH_s	Heat of solution, kJ/kg-mol

5.6.2 *Greek Letters*

Γ_R	Defined in (5-41) and (5-37)
Γ_s	Defined in (5-41)
γ'	Relative wettability factor
γ_{quartz}	Wettability factor for quartz
γ_s	Wettability factor for solids
δ_s, δ_R	Defined in (5-41)
ϵ_{cs}	Critical solid holdup
$\epsilon_g, \epsilon_\ell \epsilon_s$	Holdup of the gas, liquid, and solid phase respectively
ϵ_g^*	Gas-phase holdup at incipient fluidization
η	Dimensionless distance of quench location from the bottom of the reactor
θ	Dimensionless temperature

θ'	Contact angle between solid and liquid
μ	Viscosity, kg/m sec
υ	Kinematic viscosity, m^2/sec
ρ	Density, kg/m^3
σ	Surface tension of liquid, kg/sec^2 or N/m
τ	Space time
ϕ	Defined in (5-110)

5.6.3 *Subscripts*

b	Bulk slurry
C	Volatile products
g	Gas phase
H_2	Hydrogen
ℓ, L	Liquid phase
q	Quench
s	Solid phase
$s, s\ell$	Solid-liquid slurry
w	Wall
0	Feed condition
1	Reactor-outlet condition
1	Prequench-section conditions
2	Postquench-section conditions

REFERENCES

1. Akita, K. and Yoshida, F., *Ind. Eng. Chem. Proc. Des. Dev.,* **12** (1), 76–80 (1973).
2. Berkowitz, N., "The Distribution of Coking Properties in British and Commonwealth Coals," *Fuel,* **29**, 138–143 (1950).
3. Berkowitz, N., "A Physical Approach to the Theory of Coking", *Fuel,* **28**, 97–102 (1949).
4. Briggs, D. E., et al., "Behavior of Liquefied Coal in Filtration Coal Processing Technology," CEP Technical Manual, *AIChE,* **5**, 52–59 (1979).
5. Brown, G. G., *Unit Operations,* Wiley, New York, 1956, p. 78.
6. Calderbank, P. H., *Trans. Inst. Chem. Eng.,* **37**, 173 (1959).
7. Calderbank, P. H., and Moo-Young, M. B., *Chem. Eng. Sci.,* **16**, 39 (1961).
8. Castillo, C. and Williams, M. W., "Rheology of Very Concentrated Coal Suspensions," *Chem. Eng. Commun.* **3**, 529–547 (1979).
9. Catalytic, Inc., "SRC Process-Operation of Solvent Refined Coal Pilot Plant at Wilsonville, AL," 1976 Annual Report, FE-2270-15, June 1977.

10. Catalytic, Inc., "SRC Process-Operation of Solvent Refined Coal Pilot at Wilsonville, AL," 1977 Annual Report FE-2270-31, May 1978.
11. Chen, J.C., *I&EC Proc. Des. Dev.*, **5**, 322 (1966).
12. Collier, J. G., and Pulling, D. J., AERE Report 3809, 1962.
13. Collier, J. G., *Convective Boiling and Condensation*, McGraw-Hill, New York (1972).
14. Cova, D. R., *Ind. Eng. Chem. Proc. Des. Dev.*, **5** (1), 21 (1966).
15. Crewe, G. F. and Simons, H. P., "Heating and Cooling of Flocculated and Non-flocculated Slurries in Turbulent Heat Exchangers," *Can. J. Chem. Eng.*, **51**, 556–561 (1973).
16. Cronauer, D. C., Shah, Y. T., and Ruberto, R. G., *Ind. Eng. Chem. Proc. Des. Dev.*, **17**, 281 (1978).
17. Deckwer, W. D., Burckhart, R., and Zoll, G., *Chem. Eng. Sci.*, **29**, 2177 (1974).
18. Diboun, M. and Schugerl, K., *Chem. Eng. Sci.*, **22**, 147 (1967).
19. Droege, J. W. and Chouhan, S. P., "Viscous Properties of Coal-SRC Mixtures, Bench-Scale Coal Liquefaction Studies," EPRI-AF-612, 1978.
20. Exxon Research and Engineering Co., "EDS Coal Liquefaction Process Development Phases IIIB/IV," FE-2893-17, 1978.
21. Fair, J. R., Lambright, A. J., and Anderson, J. W., *Ind. Eng. Chem. Proc. Des. Dev.*, **1**, 33 (1962).
22. Govier, G. W. and Aziz, K., *The Flow of Complex Mixtures in Pipe*, Van Nostrand Reinhold Co., New York, 1972.
23. Guin, J. A., Pitts, W. S., Tarrer, A. R., and Parther, J. W., *ACS Div. Fuel Chem. Preprints*, **22** (2), 214 (1977).
24. Han, K. W., Dixit, V. B., and Wen, C. Y., *Ind. Eng. Proc. Des. Dev.*, **17**, 16 (1978).
25. Henley, J. P., Guin, J. A., Tarrer, A. R., Pitts, W. S., Styles, G. A., and Prather, J. W., *ACS Div. Duel Chem. Preprints*, **21** (5), 59 (1976).
26. Hikita, H., and Kikukawa, H., *Chem. Eng. J.*, **8**, 191–197 (1974).
27. Hikita, H., and Kikukawa, H., *J. Chem. Eng. Japan*, **8** (5), 412–413 (1975).
28. Hovmand, S., and Davidson, J. F., *Trans. Inst. Chem. Eng.*, **46**, T190 (1968).
29. Imafuku, K., Wang, T.-Y., Koide, K., and Kubota, H., *J. Chem. Eng. Japan*, **1** (2), 153 (1968).
30. Institute of Gas Technology, *Preparation of a Coal Conversion System*, Technology Data Book, FE-2286-32, February 1979.
31. Javdani, K., Schwalbe, S., and Fischer, J., in *Proceedings of the Second Pacific Chemical Engineering Congress, Colorado, August 28–31, 1977*, **1**, 103 (1977).
32. Johanson, E., "Solvent Refining of Illinois Number 6 and Pittsburgh Number 8 Coals," Report submitted to Electric Power Research Institute, reproduced and distributed by National Technical Information Service, U.S. Department of Commerce, Springfield, VA, Report No. EPRI 389 (1), (1975), pp. 30, 46.
33. Jones, G. L., "Second Order Models for SRC Yields from Kentucky and Powhatan coals," C&M Report No. 627 RK015, Gulf Res. & Dev. Co., Pittsburgh, September 1979.
34. Kato, Y., *Kagaku Kogaku*, **3**, 210 (1965).
35. Kato, Y., Nishiwaki, A., Fukuda, T., and Tanaka, S., *J. Chem. Eng. Japan*, **5** (2), 112 (1972).

36. Khong, T. D., Malhotra, S. L., and Blanchard, L. P., "Effects of Shear Rate, Penetration and Temperature on the Viscosity of Asphalts, "Report of Institute of Petroleum, IP-78-001, London (U.K.), 1978.
37. Kito, M., Shimada, M., Sakai, T., Sugiyama, S., and Wen, C. Y., in Fluidization Technology, D. L. Keairns, Ed., Hemisphere Publishing Corp., Washington, DC, **1**, 411 (1976).
38. Lee, M. H., Guin, J. A., and Tarrer, A. R., *Ind. Eng. Chem. Proc. Des. Dev.,* **17**, 127 (1978).
39. Matsuura, T., Ph.D. Thesis, Technische Universität, Berlin, 1965.
40. McCabe, W. L. and Smith, J. C., *Unit Operations of Chemical Engineering,* 3rd ed., McGraw-Hill, New York, 1976, pp. 154–157.
41. McGinnis, E. L., "SRC-II Dissolver Solids Study," Report No. 625RK117, Gulf Science and Technology Co., Pittsburgh, PA, September 1979.
42. Mhaskar, R. D., Shah, Y. T., and Paraskos, J. A., *Ind. Eng. Chem. Proc. Des. Dev.,* **17**, 27 (1978).
43. Narayanan, S., Bhatia, V. K., and Guha, D. K., *Can. J. Chem. Eng.,* **47**, 360 (1969).
44. Parulekar, S. J., M. S. Thesis, University of Pittsburgh, 1979; also Parulekar, S. J. and Shah, Y. T., "Steady State Behavior of Gas-Liquid-Solid Fluidized Bed Reactor," *Chem. Eng. J.,* **20**, (1), 21–33 (August, 1980).
45. Shah, Y. T. and Parulekar, S. J., "Modeling and Simulation of the Thermal Behavior of SRC-II Reactors," Chemicals and Mineral Division Report No. 627RK017, Gulf Res. & Dev. October 3, 1979.
46. Parulekar, S. J. and Shah, Y. T., "Steady State Thermal Behavior of an Adiabatic Three Phase Fluidized Bed Reactor–Coal Liquefaction under Slow Hydrogen Consumption Reaction Regime," *Chem. Eng. J.* (in press).
47. Pavlica, R. T. and Olson, J. H., *Ind. Eng. Chem.,* **62** (12), 45–58 (1979).
48. Pitts, W. S., M.S. Thesis, Auburn University, 1976.
49. Pittsburgh and Midway Coal Mining Co., "Solvent Refined Coal (SRC) Process," **3**, FE-496-113, June 1974–April 1977, June 1977.
50. Pittsburgh and Midway Coal Mining Co., "Solvent Refined Coal (SRC) Process," FE-496-141, 1977.
51. Prather, J. W., Tarrer, A. R., Johnson, D. R., and Guin, J. A., *ACS Div. Fuel Chem. Preprints,* **21** (5), 144 (1976).
52. Proceedings of the 1978 Coal Chemistry Workshop, SRI International, CONF-780372.
53. Reith, T., Renken, S., and Israel, B. A., *Chem. Eng. Sci.,* **23**, 619 (1968).
54. Roy, N. K., Guha, D. K., and Ras, M. N., *Chem. Eng. Sci.,* **19**, 215 (1964).
55. Shah, Y. T., Cronauer, D. C., McIlvried, H. G., and Paraskos, J. A., *Ind. Eng. Chem. Proc. Des. Dev.,* **17**, 288 (1978).
56. Shah, Y. T., Gas-Liquid-Solid Reactor Design, McGraw-Hill, New York, 1979, Chapters 3, 7, and 9.
57. Shah, Y. T., Mhaskar, R. D., and Paraskos, J. A., *Ind. Eng. Proc. Des. Dev.,* **15** (3), 400 (1976).
58. Sieder, E. N. and Tate, G. E., "Heat Transfer and Pressure Drop of Liquids in Tubes," *Ind. Eng. Chem.,* **28**, 1430–1435 (1936).
59. Thurgood, J. R. and Youngblood, E. L., "Coal Slurry Preheaters," IKNL-5550, 1979, pp. 18–34.

60. Traeger, R. K. and Curlee, R. M., "Preheater Studies in Coal Liquefaction," SAND-78-1124, 1978.

61. Traeger, R. K., Bickel, T. C., and Curlee, R. M., "Preheater Studies in Coal Liquefaction," Annual Report – October 1977–September 1978, Sandia Laboratories (DOE Contract AT (29-1)-789), Report No. SAND-79-0150, May 1979.

62. U.S. DOE, *Coal Liquefaction Preheater Studies,* Proceedings of the Department of Energy Project Review Meeting, Oak Ridge, TN, March 21, 1979.

63. Vasalos, I. A., Bild, E. M., Tatterson, D. F., and Wallin, C. C., "H-COAL Fluid Dynamics Topical Report," Part 1: Literature Search Report No. FE-2588-6 prepared by Amoco Oil under Contract No. EF-77-C-01-2588, U.S. Department of Energy, May 3, 1978.

64. Weber, W. H. and Basu, A., "Coal Dissolution Studies Utilizing the Slurry Preheater at the Wilsonville SRC Pilot Plant," *Coal Liquefaction Preheater Studies,* Oak Ridge Natl. Lab., 1979.

65. Whitehurst, D. D., Mitchell, T. O., Farcasiu, M., and Dickert, J. J., Jr., *The Nature and Origin of Asphaltenes in Processed Coal,* Vol. 1, EPRI AF-1298, 1979.

66. Whitehurst, D. D., Farcasiu, M., Mitchell, T. O., and Dickert, J. J., Jr., "The Nature and Origin of Asphaltenes in Processed Coals. The Chemistry and Mechanisms of Coal Conversion to Clean Fuel," EPRI Annual Report on Project 410-1, (EPRI AF-480), July 1977.

67. Wright, C. A., and Schmalzer, D. K., "Short Residence Time SRC-I: Background and Recent Work," Report 4, Coal Liquefaction Preheater Studies, Oak Ridge Natl. Lab., 1979.

68. Yoshida, F. and Akita, K., *AIChE J.,* **11**, 9 (1965).

6. Refining of Coal Liguids

G.J. Stiegel, Y.T. Shah,
S. Krishnamurthy,
and S.V. Panvelker

INTRODUCTION

The composition of liquids derived from coal has been shown in Chapter 2 to be substantially different than that of most petroleum fuels. Specifically, the former possess higher concentrations of polynuclear aromatics, trace metals, and oxygen- and nitrogen-containing heterocyclics [10, 133]. The extent to which the above components are present depends not only upon the rank of the coal but also upon the nature of the liquefaction process. In light of the above differences, currently available petroleum-refining technology may not be readily applicable to coal liquids, and modifications may be required [88].

Major efforts have been directed at various aspects of coal-liquid upgrading during the past few decades and have been documented in several reviews [35, 139, 162]. These review works report the technological developments in coal-liquid hydrogenation and are not currently applicable because of recent advances in coal-liquefaction technology.

Recently, Cusamano et al. [26, 27] summarized the development of several coal-liquefaction processes and discussed various aspects of product upgrading for each process. Potential problem areas in the refining of coal liquids by conventional processes and recommendations for further research were presented. Uehara and Fu [149] recently published a brief review of the upgrading of coal liquids from the more advanced liquefaction processes (e.g. SRC-II, EDS, H-coal).

Reaction engineering plays a very important role in the refining of coal liquids. This chapter discusses various aspects of the refining of coal liquids and

Yatish T. Shah, Reaction Engineering in Direct Coal Liquefaction, ISBN 0-201-07381-1

the influence of reaction engineering on the selection of the final process. The chapter first presents a brief discussion of the compatibility of coal liquids and petroleum fuels because of the possibility of blending these two feedstocks during the initial stages of refining. This is followed by discussions on the stability of coal liquids and a review of model compound studies which are relevant to refining of coal liquids. Finally, the applicability of various refining processes such as hydrotreating, hydrocracking, and reforming to the refining of coal liquids is discussed.

6.1 Compatibility of Coal-Derived Liquids and Petroleum Fuels

As coal liquefaction progresses towards commercialization during the next decade, coal liquids will initially supplement decreased petroleum imports. During the developmental stages of coal liquefaction, the quantities produced may not be sufficient to justify dedicated refineries. Furthermore, these initial supplies will most likely be produced via several liquefaction processes, as evidenced by the concurrent development of the Exxon donor-solvent (EDS), solvent-refined coal (both SRC-I and SRC-II), and H-coal liquefaction processes. This, together with the use of different coals, will result in products with different compositions and physical properties which will have to be processed to different extents. Until a constant source of supply is established to justify dedicated refineries, blends of coal liquids and petroleum feedstocks will most likely be processed commercially.

In blending petroleum fuels derived from crudes of different origins, compatibility of the blending stocks is important. It will be even more so in the blending of petroleum fuels with coal liquids. Materials are incompatible if they form residues upon mixing and/or in subsequent storage. The methods and criteria employed to determine the compatibility of fuels vary considerably. Techniques which are frequently used involve comparing the amount of residue produced from a given blend with that from a reference blend or determining the total amount of residue deposited upon blending [22].

Fuels containing similar components are more likely to be compatible than those containing dissimilar components. Table 6-1 presents a comparison of three coal liquids and three petroleum fuels. The petroleum fuels can be seen to possess higher concentrations of hydrogen and sulfur and lower concentrations of nitrogen and oxygen. The petroleum fuels also contain more saturated compounds and fewer polynuclear aromatics than coal liquids. Figure 6-1 presents a plot of the aromaticity of various fuels versus their carbon-to-hydrogen atomic ratio. It is clear from this figure that coal liquids can contain up to twice the C/H atomic ratio of petroleum fuels, due to their higher aromaticity.

6.1.1 *Recent Investigations*

Investigations of the compatibility of coal liquids with petroleum fuels are scarce. The most recent and currently applicable investigations have been conducted by Bendoraitis et al. [10], Cabal et al. [22] and Stein et al. [133–135]. In these studies, coal liquids produced by the SYNTHOIL, H-coal, and SRC-I processes

Table 6-1

Comparison of Coal Liquids and Petroleum Fuels [10] *

Cut No.	Description	Coal liquids: SRC-I, wt%	Coal liquids: H-coal, wt%	Coal liquids: SYNTHOIL, wt%	Petroleum liquids: FCC clarified slurry oil, wt%	Petroleum liquids: No. 6 fuel oil, wt%	Petroleum liquids: 811-K+ (1000°F+) light Arabian residuum, wt%
1	Saturates	0.04	9.31	6.60	10.37	30.50	11.0
2	MNA† + DNA‡ Oil	0.18	22.81	17.70	6.81	20.30	18.6
3	PNA§ Oil	0.90	17.15	12.10	32.04	13.20	19.7
4	PNA soft resin	10.83	13.15	17.00	29.89	16.10	28.0
5	Hard resin	2.40	1.67	3.30	2.95	2.30	4.5
6	Polar resin	4.47	1.92	3.70	4.35	4.20	6.1
7	Eluted asphaltenes	23.94	11.34	24.50	8.78	7.90	8.9
8	Polar asphaltenes	14.64	5.20	7.00	1.98	3.80	2.3
9	Polar asphaltenes	6.14	2.14	1.30	0.54	0.30	0.2
10	Polar asphaltenes	5.94	2.17	1.90	0.14	0.08	0.04
11	Polar asphaltenes	13.90	5.80	2.80	0.06	0.06	0.06
12	Polar asphaltenes	6.67	1.34	0.20	0.04	0.02	0.02
13	Noneluted asphaltenes	9.95	6.00	1.90	2.05	1.20	0.6
	Ultimate analyses, wt%:						
	Carbon	87.93	89.00	87.62	90.52	86.4	83.88
	Hydrogen	5.72	7.94	7.97	7.64	11.2	9.97
	Oxygen	3.50	2.12	2.08	0.76	0.30	0.48
	Nitrogen	1.71	0.77	0.97	0.44	0.41	0.40
	Sulfur	0.57	0.42	0.43	1.16	1.96	4.19

*Gradient elution chromatographic analyses.

†Mononuclear aromatic.
‡Dinuclear aromatic.
§Polynuclear aromatic.

were evaluated for compatibility with No. 2 distillate fuel, No. 6 fuel oils, heavy coker gas oil, and fluid-catalytic-cracker (FCC) clarified slurry oil. Each petroleum fuel and coal liquid was characterized by gradient-elution chromatography (GEC) in order to determine its composition; some of these results are presented in Table 6-1. For compatibility, the criteria employed required that the blends yield less than 0.3 vol% residue.

A summary of some of the results obtained in the compatibility tests are presented in Table 6-2. As shown, the full-range products of SYNTHOIL and H-coal liquids are incompatible with the petroleum fuels, whereas the −783 K fractions

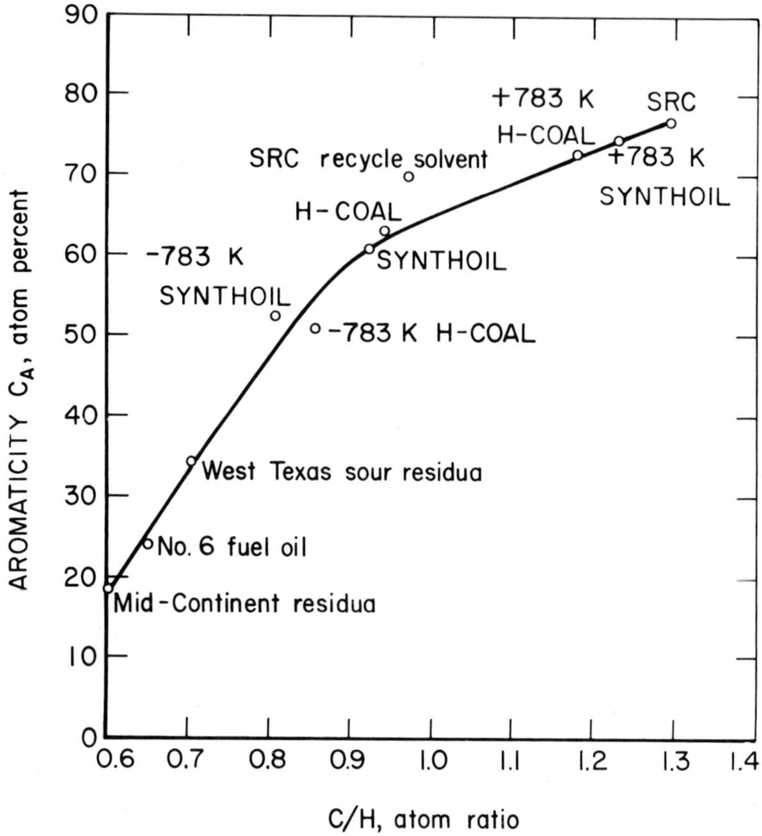

Figure 6-1. Plot of aromaticity versus C/H atom ratio for various hydrocarbon feedstocks [10].

are completely compatible. A comparison of the GEC analyses of the −783 K fraction of the coal liquids with those of the petroleum fuels reveals a similarity in their compositions. In particular, the C/H ratio and the aromatic content of this fraction of the coal liquids more closely approximate the petroleum fuels than do those of the whole product. This is also evident from the data in Figure 6-1. The full-range coal liquids contained more asphaltenic material than the petroleum fuels, and distillation was effective in removing these components from the raw coal liquid and rendering them compatible with petroleum fuels.

The concentration of the coal liquid constituting the blend has been found to have significant effects on compatibility. Using SYNTHOIL and heavy coker gas oil, it has been shown that at high concentrations of SYNTHOIL, the coker gas oil and SYNTHOIL were compatible; however, at low concentrations the two were

Table 6-2

Compatibility of Coal Liquids and Petroleum Fuels

	Residue after centrifuging, vol%			
	No. 2 distillate fuel	No. 6 fuel oil	Heavy coker gas oil	FCC-clarified slurry oil
SYNTHOIL (Pittsburgh Seam, Ireland Mine):*				
Full-range product	18	50	15	0.4
– 783 K fraction	0.05	0.05	0.05	–
H-coal (IL No. 6, Burning Star Mine):*				
Full-range product	26	50	30	20
– 783 K fraction	0.05	0.05	0.05	–
H-coal (+478 K):				
Raw	–	100	–	–
Mild hydrotreatment	–	6	–	–
Moderate hydrotreatment	–	0.05	–	–
SRC-I: IL No. 6, Burning Star Mine:*	Insoluble	Insoluble	Insoluble	20
IL No. 6, Monterey Mine*	–	–	Insoluble	–
SRC-I recycle solvent:				
Raw	Trace	–	–	–
Hydroprocessed	Nil	–	0.05	–
Recycle Solvent/SRC-I (Wyodak):†				
Raw	12	14	–	–
Hydrotreated (203 sm^3/m^3)	5	2	–	–
Hydrotreated (364 sm^3/m^3)	3	0.3	–	–
Hydrotreated (494 sm^3/m^3)	0.3	1.0	–	–
Hydrotreated (548 sm^3/m^3)	13	11	–	–
Recycle Solvent/SRC-I (IL No. 6 Burning Star Mine):†				
Raw	50	–	50	–
Hydrotreated (311 sm^3/m^3)	–	–	15	–
Hydrotreated (317 sm^3/m^3)	15	–	4	–
Hydrotreated (503 sm^3/m^3)	–	–	9	–

*1:1 blend.
†2:1 blend.

incompatible [10, 22]. These results are graphically shown in Figure 6-2. As indicated, small amounts of petroleum fuels in coal liquids may not adversely affect processing, but the processing of small amounts of coal liquids in petroleum may not be possible.

The compatibility of SRC-I derived from Illinois #6 Burning Star, Monterey, and Wyodak coals with petroleum fuels has also been investigated [10, 22, 133–135]. The SRC-I products were found to be incompatible with all petroleum fuels except the FCC-clarified slurry oil, where partial solubility was observed. This was

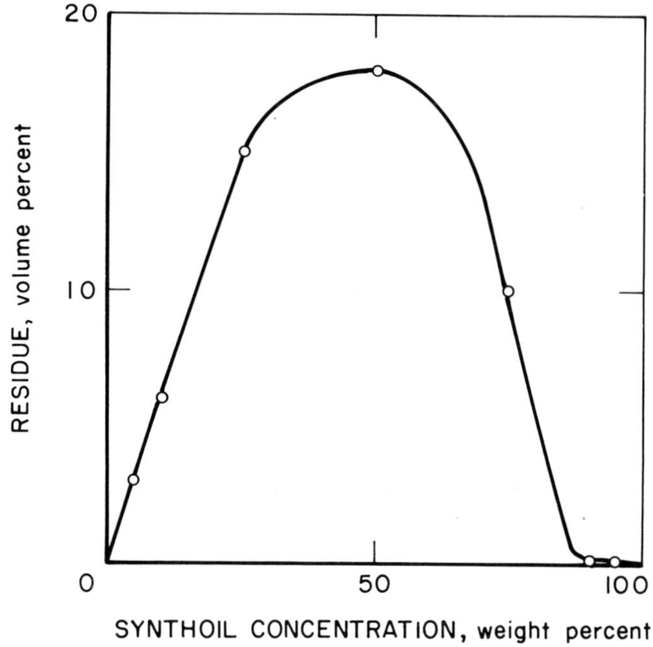

Figure 6-2. Compatibility of SYNTHOIL with heavy coker gas oil [10].

attributed to the fact that this petroleum fuel contains more aromatic components, which are concentrated in the higher GEC fractions, than the other fuels.

Hydrotreating of raw coal liquids prior to blending with petroleum fuels has been shown to improve compatibility. In a study using a +478 K fraction of H-coal, SRC-I recycle solvent, and SRC-I recycle solvent mixture [133–135], all blends containing raw coal liquids were incompatible with the same petroleum fuels. However, as shown in Table 6-2, hydrotreating of the coal liquid prior to blending significantly reduced the formation of residue. Of the hydroprocessing severities employed, blends containing moderately hydrotreated coal liquids resulted in the lowest yield of residue. These results suggest that an optimum set of hydroprocessing conditions exist which minimizes the incompatibility of coal liquids with petroleum fuels. Figure 6-1 can also be used to provide a rough estimate of the amount of hydrogen required to convert an incompatible fuel into a compatible one.

Coal liquids (480–810 K fraction) derived from Illinois #6 coal via the Exxon donor solvent process have also been found to be incompatible with several commercial petroleum heavy fuel oils [39]. As with the other coal liquids discussed above, hydrotreatment was shown to result in a compatible blend over the range from 1 to 50 wt%.

6.1.2 *Summary*

The results presented above provide a preliminary insight into the compatibility of coal liquids with petroleum fuels. The available information, however, is insufficient to assess the viability of processing blends of coal liquids and fuels. As the results presented in Figure 6-2 indicate, this type of processing may not be technically feasible because of the formation of large quantities of residue. More research is needed in this area to define the ranges of compatibility for the products from the more promising coal liquefaction processes with various petroleum fuels.

The blending of coal liquids with petroleum fuels can also have considerable impact upon subsequent refining operations. Besides changing the composition and physical properties of the process feedstocks, the blending of two components which are incompatible and produce a residue can result in material buildup in a reactor which can decrease the normal life of catalysts and seriously affect the conversion and selectivity of the process. Plugging due to this buildup can also cause unwarranted and costly downtime. Several preliminary studies of subsequent refining of compatible blends have been conducted; however, more will be required to define operating parameters before such processing can commence on a commercial scale. Some of the reported studies are discussed in the following sections.

6.2 Stability of Coal Liquids

The stability of coal liquids upon storage and/or heating is an important factor which can seriously affect further processing and final product specifications. Several investigations [21, 40, 47, 61, 64, 75, 93, 109] have focused on these problems to some degree using coal liquids produced by the SYNTHOIL, H-coal, COED, and SRC-I processes or coal carbonization tars. These raw coal liquids were invariably found to be unstable upon storage. Significant increases in viscosity and higher-molecular-weight components have been observed along with the formation of gum and other deposits.

The effects of storage time, temperature and atmosphere, agitation, and light on the aging characteristics of SYNTHOIL have been evaluated in preliminary studies by Finseth. et al [40] and Karn et al. [61]. In both studies, the viscosity of the coal liquid increased substantially with time when stored under an oxygen atmosphere, whereas it increased only slightly under nitrogen. The results presented in Table 6-3 show that the oil content of the original liquid decreases with time while the asphaltene content increases. It is interesting to note that the benzene-insoluble fraction and the oxygen content of the coal liquids increase substantially on storage under oxygen. This indicates that oxygen is definitely involved in the aging process. The following reaction mechanisms were proposed by Finseth et al. [40] to account for the observations:

Coupling of small neutral molecules by oxygen.

$$A + B \xrightarrow{O_2} A{-}O{-}B + H_2O. \tag{6-1}$$

Table 6-3

Analyses of an Aged* Coal Liquid [40]

	Oil, wt%	Asphaltenes, wt%	Benzene-insolubles, wt%	O_2 Concentration, wt%
Original sample	70	23	7	3.11
N_2 storage	67	26	7	3.19
O_2 storage	55	25	20	3.69

*The coal liquid was aged for 30 days at 363 K.

Coupling of radicals with radicals or neutrals.

$$A\cdot + B\cdot \rightarrow AB, \tag{6-2}$$

$$A\cdot + B \rightarrow AB\cdot, \tag{6-3}$$

$$A\cdot + B \rightarrow AB' + C\cdot\ . \tag{6-4}$$

The mechanism (6-1) accounts for the observed increase in the oxygen concentration, whereas the mechanisms (6-2) to (6-4) account for the observed increase in viscosity of the coal liquid stored under nitrogen. ESR examination of the samples suggest that the radical-radical recombination reaction is not the preferred route. The validity of the other reactions could not be confirmed. It was concluded that the aging mechanism involved the condensation of small heteroatom-rich molecules to form higher-molecular-weight insolubles and asphaltenes in the presence of oxygen.

Lower storage temperatures and lower initial viscosities of the stored liquids have been shown to reduce the rate of viscosity increase [61]. The effect of agitation on the viscosity of samples stored under nitrogen was negligible, whereas under an oxygen atmosphere, agitation caused significant increases in viscosity. No effect of laboratory light on the liquid viscosity was observed.

Similar changes in the viscosity and toluene-insoluble content of coal liquids stored under air and inert atmospheres were observed for a coal-carbonization tar [64]. Results showed a substantial decrease in the concentration of hydroxyl groups and an increase in the number of carbonyl groups. A slight decrease in the concentration of carboxyl groups was also observed. The conversion of phenolic "OH" to a quinonoid-type structure due to oxidation was offered as a possible explanation.

The reaction kinetics of coal-tar autooxidation were investigated by Lin et al. [75]. The ease and extent of the reactions were found to be dependent upon the composition of the liquid, the reaction temperature, and the oxygen partial pressure. The reaction of neutral oils was found to involve long-chain radicals, and the rate was proportional to the concentration of hydrocarbon reactant. The oxidation of tar acids and bases was found to be first order in both reactant concentration and oxygen partial pressure and was diffusion controlled.

Additional studies [93, 109] have revealed that hydrotreatment of coal liquids prior to storage results in stable products. This is probably due to the saturation and removal of trace amounts of unstable species which react to form high-molecular-weight components.

The stability of coal liquids, especially those from the more advanced processes, needs to be investigated further to ascertain the chemical nature of these materials under typical processing conditions. An evaluation of the stability of compatible blends of coal liquids and petroleum fuels is also necessary. The effects of reaction conditions and product specifications for processing unstable feedstocks which have undergone chemical changes during storage must be evaluated. The use of stabilizing agents to inhibit the formation of gum and higher-molecular-weight components in the coal liquids and their effects on downstream processing should be investigated.

6.3 Thermal Cracking

Thermal cracking uses heat to decompose large hydrocarbon molecules without the aid of a catalyst. Although catalytic cracking processes have replaced some of the thermal processes, thermal cracking still plays a vital role in the petroleum industry. Thermal cracking processes have been divided into two classes, namely, viscosity breaking and coking, which differ in their operating conditions.

6.3.1 *Chemistry*

During thermal cracking, large, high-boiling molecules thermally decompose into smaller, lower-boiling components. However, under certain conditions some of the reactive intermediates formed during this process may combine with each other to yield molecules which are larger than those originally present in the feed. The reactions which occur are classified into two types: (1) primary reactions in which large molecules decompose into smaller ones, and (2) secondary reactions which involve the recombination or polymerization of reactive components to form high-molecular-weight products which may gain prominence in the absence of free-radical terminators such as hydrogen. The products from the secondary reactions may, if sufficient reaction time is available, decompose again into smaller molecules.

Paraffins are the most susceptible to thermal cracking, followed by olefins, diolefins, naphthenes, and aromatics. The yield of light products increases at first with reaction time; however, after a certain period of time it decreases as a result of secondary reactions. Pressure is a variable that is employed to retain heavy molecules in the cracking zone and to control the molecular density of the products. Temperature is the most important variable in the process, since it affects not only the decomposition reactions, but also the extent to which secondary reactions occur.

Thermal cracking reactions are generally modeled as first-order reactions if the decomposition is limited to low conversion per pass (20–25%), as in conventional cracking operations. The model is given by

$$k_1 = \frac{1}{t} \ln \frac{100}{100-x} \qquad (6\text{-}5)$$

where k_1, the reaction rate constant, is related to temperature by the Arrhenius relationship. This equation has been used to correlate data on a wide variety of feedstocks.

6.3.2 *Process Description*

Viscosity breaking, or visbreaking, is a mild form of thermal cracking and is used primarily to reduce the viscosity and lower the boiling range and pour point of residual stocks. Typical operating conditions for this class of thermal cracking are 700–797 K and 0.34–0.69 MPa. Residence times are relatively short.

Coking, in contrast to viscosity breaking, is a severe form of thermal cracking. Its purpose is to produce low-carbon gas oil for catalytic cracking with gas, gasoline, and coke as secondary products. The primary feedstocks for coking are residuals which resist cracking by other means. Typical operating conditions are 755–838 K and approximately atmospheric pressure to 0.14 MPa; however, they are also dependent upon the type of process employed. Reaction times are extremely long compared to the visbreaking process, especially if coke is the desired product.

Thermal cracking processes can be operated in either a vapor-phase or a mixed-phase mode. The process is classified as vapor phase if the operating pressure is low enough so that pure vapor is produced at the operating temperature of the reactor. Vapor-phase processes offer advantages of producing higher-antiknock motor fuels, relatively low capital costs, enhanced safety due to low pressures, and adaptability to low-boiling-range distillates. Such processes are generally conducted at pressures less than 0.35 MPa and at temperatures of 810–867 K. Residence times are generally of the order of 1 second; however, recently developed processes use even shorter times. Under these conditions, dehydrogenation reactions increase, thus yielding more olefins and aromatics.

At higher operating pressures, vaporization is suppressed, resulting in a mixed-phase operation. These processes offer the advantage of lower gas production, easier treatment of product gasoline, and lower fuel consumption. Operating pressures of 2.4 MPa or greater are employed to maintain a homogeneous phase, and the temperature usually ranges from 672 to 755 K depending on the feedstock.

6.3.3 *Thermal Cracking of Coal Liquids*

Thermal cracking of coal liquids has been studied by several investigators, but not to the extent of catalytic upgrading techniques. Liquids derived from the COED, H-coal, SYNTHOIL, and SRC processes as well as liquids derived from basic liquefaction studies have been examined. A variety of reactors and operating conditions have also been employed. A summary of these investigations is presented in Table 6-4.

Korosi et al. [63] investigated the steam cracking of a coal liquid derived from a Kentucky hvAb coal by the SYNTHOIL process. The SYNTHOIL product was first fractionated to yield a material with a boiling range of 523–653 K, which was subsequently hydrogenated to produce a boiling range of 453–673 K. From the pyrolysis run, methane and ethylene were found to be the predominant gas products, whereas BTX products were formed in the liquid phase. The authors concluded that in order to obtain desirable products, the polynuclear aromatic components must first be hydrogenated to their alicyclic analogs. Partially hydrogenated species were found to dehydrogenate and revert back to their fully aromatic state.

The use of hydrogen-donor compounds during the fluid coking of a raw, unfiltered coal liquefaction product was investigated by Wilson and Angelo [153]. By providing sufficient hydrogen-donor compounds with the feedstock, polymerization of free radicals was inhibited, resulting in a reduction of coke and an increase in liquid yield. Analytical results have shown that five-sixths of the hydrogen transferred in the coker from the donor solvent was donated to the liquid and the remainder was converted to molecular hydrogen. The results also indicate that the solvent undergoes only slight degradation in the coker.

The thermal cracking of hydrotreated COED light and heavy oil and SRC-II recycle solvent, all derived from Western Kentucky coal, has been investigated by Krishnamurthy et al. [65–67]. Under the conditions investigated, ethylene, propylene, and methane were the dominant gas products, and their yield increased with temperature and residence time. For these feedstocks, the primary reactions were the cracking of saturates, the dehydrogenation of partially saturated aromatics, and the removal of peripheral groups on the aromatic ring structures (dealkylation). At higher temperatures and longer residence times, secondary reactions which lead to the formation of larger molecules were found to occur.

The results of the above investigation also indicate that hydrogenation of polyaromatics to their alicyclic analogs is required for the production of desirable products. From their study, it may be noted that the boiling-point distribution for the COED heavy oil is similar to that of the SRC-II recycle solvent, although the H/C ratios are quite different (1.63 and 1.24 respectively). Under similar reaction conditions, the yield of desirable products (i.e. ethylene, propylene, BTX) decreases in the following order: COED light oil, COED heavy oil, SRC-II recycle solvent.

The data for the pyrolysis of hydrotreated COED light- and heavy-oil liquids [65–67] were modeled using *n*th-order kinetics and the following reaction mechanism:

Table 6-4
Investigation of Thermal Cracking of Coal-Derived Liquids

Investigators	Reactor	Temp., K	Pressure, MPa	Residence time	Feedstocks
Korosi et al. [63]	Single-pass tubular	1148	0.185	0.3 sec	Hydrogenated fraction of SYNTHOIL
Wilson and Angelo [153]	Fluid coker	755–977	0.207	0.1–1 WHSV	Coal liquefaction product
Krishnamurthy et al. [65, 66]	Transfer line	773–1053	0.1	0.03–0.13 sec	COED light and heavy oil, SRC-II recycle solvent
Wiser [159]	Flow reactor	723–687	1.72–10.34	—	SYNTHOIL

$$\text{oil} \xrightarrow{k} \text{gas} + \text{coke}, \tag{6-6}$$

where

$$\frac{dC}{dt} = -\,k\;C^n\,. \tag{6-7}$$

In both cases, the reaction order for the lumped model was found to be 1, which is in agreement with Equation (6-5). The reaction rates for the COED light and heavy oil are given by

$$r_L = 5.8 \times 10^4 \exp\,[-76.06/RT]\, C_L \,, \quad \text{(moles/liter-sec)} \tag{6-8}$$

$$r_H = 3.1 \times 10^4 \exp\,[-70.3/RT]\, C_H \,, \quad \text{(moles/liter-sec)} \tag{6-9}$$

Although pyrolysis reactions are generally first order, the model presented does not adequately describe the experimental observations. In particular, the model does not account for the recombination of free radicals that result in the formation of higher molecular-weight aromatic compounds.

The use of steam-coil cracking to process COED and H-coal liquids has recently been employed to evaluate the potential for using coal liquids for direct olefin production [98–100]. The COED oil [96] employed was derived from Western Kentucky coal and hydrotreated subsequent to liquefaction. The hydrotreated oil was fractionated into straight-run naphtha (IBP–450 K), straight-run mid-distillate (450–616 K), and straight-run gas oil (616–727 K). The naphtha and mid-distillate were either subjected directly to or hydrocracked prior to steam-coil cracking. The straight-run and hydrocracked naphtha fractions yielded similar product distributions; however, the hydrocracked naphtha showed a slight increase in benzene production with a corresponding decrease in fuel-oil yield. The combined production of ethylene, propylene, and butylenes was approximately 29 g per 100 g of feed. From a comparison of the analyses, the straight-run naphtha used by Peters [98] and the hydrotreated COED light oil used by Krishnamurthy et al. [65]. are very similar in their boiling-point distribution and elemental analysis. Although Peters [98] did not disclose the operating conditions of his process, a qualitative comparison can be made of the results of the two studies. For the two highest temperatures (i.e. 1013 and 1053 K) and over the range of residence times from 0.042 to 0.086 sec employed by Krishnamurthy et al. [65], the production of ethylene, propylene, and butylene ranged from 20 to 36 g per 100 g of feed. As can be seen, the results of the two investigations are comparable.

Steam cracking of the hydrocracked mid-distillate resulted in slightly higher ethylene yields than that for the straight-run mid-distillate [98]. The hydrocracked feedstock also showed a slight increase in benzene with a corresponding decrease in fuel oil. Peters concluded that neither the COED naphtha nor the mid-distillate was very attractive as an ethylene cracker feed because of the low ethylene yield and excessive coking due to high concentrations of aromatics in the feed.

Straight-run naphtha and mid-distillate from H-coal syncrude derived from Illinois No. 6 coal have also been investigated by Peters [100]. The H-coal straight-run naphtha had carbon and hydrogen concentrations similar to the COED naphtha of Peters [100] and the COED light oil of Krishnamurthy et al. [65]. For the H-coal naphtha, the olefin yield was approximately 35 g/(100 g feed), which was about a 20% increase over the yield from the COED naphtha. The yield of olefins from the H-coal mid-distillate was comparable to that of the COED mid-distillate [i.e. 21 g/(100 g feed)] ; however, the H-coal feed yielded approximately twice the amount of BTX per 100 g of feed.

The thermal cracking of SYNTHOIL in the presence of hydrogen (i.e. hydropyrolysis) has been investigated using a flow reactor [159]. Preliminary results using SYNTHOIL, which contained 45% material boiling above 773 K, showed yields of gaseous and light liquid products (333–653 K) to be 26% and 74%, respectively, at the highest-severity conditions employed.

6.3.4 *Summary*

As can be deduced from the above, there is a dearth of information on the thermal cracking of coal liquids. In particular, data are needed on the products from the more advanced coal liquefaction processes (i.e. H-coal, EDS, SRC-I, SRC-II). Because of differences in their compositions, whole product as well as various fractions will have to be investigated over a wide range of process conditions in order to ascertain the optimum product distribution.

From preliminary results, thermal cracking does not appear to be attractive for production of chemicals at the present time. The high concentrations of aromatics in the feedstocks minimize the extent of cracking and cause serious coking problems. The use of donor solvent may offer a solution to these problems; however, the economics of this idea must be critically evaluated. Also, the compatibility of the donor solvent with the feedstock would play a major role in the design of cracking reactors. If chemical feedstocks are desired from the distillate fractions of coal liquids, then high-temperature, short-contact-time reactors will probably yield the best results with a minimum amount of coke production.

Residual fractions of coal liquids (i.e. +780 K) are the most likely candidates for thermal cracking processes. These refractory materials are probably best suited for conventional fluid coking processes for the production of gas oil for catalytic cracking. The coke and undesirable products from the cracking units can then be gasified to produce the hydrogen required for other refining processes.

6.4 Catalytic Hydrotreating

Catalytic hydrotreating has been applied to a wide variety of feedstocks for the purpose of improving product stability and/or preparing feedstocks for subsequent processing by removing undesirable elements such as sulfur, nitrogen, oxygen, halogens, and metals and by increasing their hydrogen content. A substantial amount of research on catalytic hydrotreatment of petroleum feedstocks has been done, and much of this technology is now being applied to the refining of coal liquids. Because of the high concentrations of sulfur, oxygen, nitrogen, and metals and the low H/C ratios of coal liquids, hydrotreating of coal liquids will definitely be necessary before they can be refined to useful fuel and chemicals.

6.4.1 *Hydrotreating Catalysts*

Hydrotreating catalysts currently used commercially are composed of various combinations of metals such as cobalt, molybdenum, nickel, tungsten, and vanadium impregnated on alumina or silica-alumina supports. The metals employed are usually in the form of oxides and must be activated prior to use by presulfiding with hydrogen sulfide, carbon disulfide, mercaptans, etc. in order to convert the metal from oxidic to sulfidic form.

Cobalt-molybdenum catalysts supported on alumina are the most widely used hydrotreating catalysts in the petroleum industry because of their high selectivity towards sulfur removal, resistance to poisoning, and ease of regeneration. These catalysts are very effective in removing sulfur atoms; however, their efficiency in removing nitrogen is poor. The need to remove nitrogen has led to the use of nickel-molybdenum and nickel-cobalt-molybdenum catalysts, which are more efficient. Because the coal liquids differ considerably in composition from petroleum liquids and the former contain much higher concentrations of nitrogen, conventional hydrotreating catalysts may not be optimal for the refining of coal liquids.

6.4.2 *Reaction Mechanisms and Kinetics*

In the catalytic hydrotreating process, the primary reactions that occur are the following:

$$\text{(nitrogen-containing heterocyclic compounds)} + H_2 \rightarrow \text{(nitrogen-free hydrocarbons)} + NH_3\ , \tag{6-10}$$

$$\text{(sulfur-containing heterocyclic compounds)} + H_2 \rightarrow \text{(sulfur-free hydrocarbons)} + H_2S, \tag{6-11}$$

$$\text{(oxygen-containing heterocyclic compounds)} + H_2 \rightarrow \text{(oxygen-free hydrocarbons)} + H_2O. \tag{6-12}$$

In addition to the above reactions, a certain degree of saturation of aromatics occurs concurrently.

The removal of nitrogen, sulfur, and oxygen occurs through a sequence of hydrogenation and hydrogenolysis steps. Numerous studies have been reported on the mechanisms of heteroatom removal from compounds that are present to significant extents in coal liquids. A discussion of some pertinent investigations in each category are presented below to help in better understanding the chemistry of heteroatom removal.

A. HYDRODENITROGENATION

Nitrogen is generally present in coal liquids as five- and six-membered heterocyclic compounds. The process of hydrodenitrogenation (HDN) involves contacting the compound with a catalyst containing various combinations of Ni, Co, Mo, or W on SiO_2 and/or Al_2O_3 supports in the presence of hydrogen under high pressure. Through a sequence of hydrogenation and hydrogenolysis reactions, the nitrogen is ultimately removed as ammonia. Fewer studies have been made on nitrogen removal from five-membered than from six-membered structures. This is perhaps because it is easier to remove nitrogen from five-membered compounds than from their six-membered counterparts [41].

Cox and Berg [25] studied the hydrodenitrogenation of several five-membered nitrogen compounds. Denitrogenation of these compounds was found to follow second-order irreversible kinetics. Saturated heterocyclics were easier to "destructively hydrogenate" than unsaturated heterocyclics, which implies that hydrogenation of the heterocyclic ring is a necessary prerequisite for nitrogen removal.

The relative rates of nitrogen removal from several heterocyclics were studied by Flinn et al. [41]. Details of this and other investigations are presented in Table 6-5. These authors found nitrogen removal from indole to be easier than from quinoline. The removal of nitrogen from aniline and *n*-butylamine was the easiest step in both networks. The rates of nitrogen removal from indole, aniline, and *n*-butylamine are comparable to that from quinoline at low space velocities and can be attributed to the difficulty in hydrogenolysis of refractory intermediates such as *o*-ethyl aniline. Based on first-order irreversible kinetics, a comparison of the rate constants for nitrogen removal from the above compounds is made below:

Compound	Rate constant, sec^{-1}
Quinoline	1.9×10^{-4}
Indole	8.6×10^{-4}
Aniline and *n*-butylamine	$(2.5-3.3) \times 10^{-3}$

The following mechanism for the denitrogenation of indole has been proposed by Stern [138]:

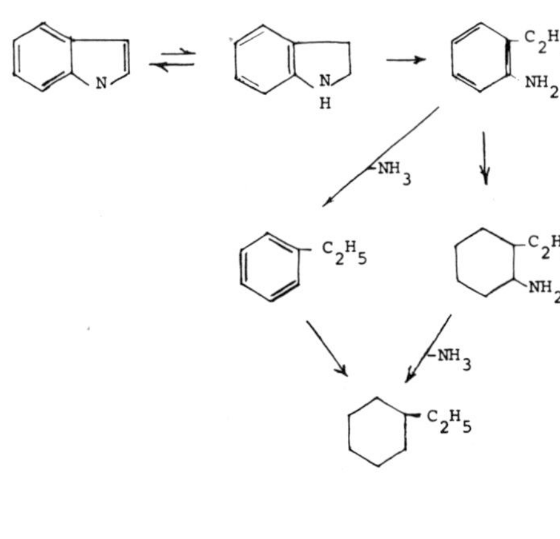

Table 6-5

Summary of Denitrogenation Studies

Feedstock	Pressure, MPa	Temp., K	Catalyst	Equipment	Reference
Quinoline, isoquinoline	8.1	423–673	$CoMo/Al_2O_3$	Fixed-bed reactor with recycle of product gas	Doelman & Vlughter [34]
Aniline, indole *n*-butylamine, & quinoline blended with a middle distillate	6.89	589	NiW/Al_2O_3		Flinn et al. [41]
Quinoline	8.1	473–723	$CoMo/Al_2O_3$	Batch autoclave	Madkour et al. [78]
Pyridine, piperidine, & *n* hexylamine in mixed xylenes	5.17–6.89	589	$CoNiMo/Al_2O_3$ (presulfided)	Fixed-bed reactor	McIlvried [79]
Pyridine, quinoline, aniline, pyrrole, & indole in high-purity paraffin oil	20.26	623–673	$CoMo/Al_2O_3$	Batch autoclave	Aboul Gheit & Abdou [1]
Pyridine in *p*-xylene	8.10	523–673	$CoMo/Al_2O_3$ (presulfided H_2S)	Fixed-bed reactor	Goudriaan et al. [50]
Quinoline & acridine in a highly paraffinic white oil	3.44–13.78	615–640	NiMo, CoMo NiW/Al_2O_3 (unsulfided & presulfided)	Batch autoclave	Shih et al. [122, 123]
Quinoline	0.34–6.89	503–693	$NiMo/Al_2O_3$	Flow microreactor	Satterfield et al. [120]
Thiophene, pyridine	0.44–1.12	373–773	CoMo, NiMo, NiW/Al_2O_3 & NiW/SiO_2-Al_2O_3	Flow microreactor	Satterfield et al. [118]
Pyridine, piperidine	1.12	473–698	NiMo, $CoMo/Al_2O_3$		Satterfield & Cocchetto [119]
Pyridine	15.75–75.31	523–648	Mo, $CoMo/Al_2O_3$	Tubular reactor	Sonnemans et al. [127–130]
Quinoline in a high-purity paraffin oil	20.26	623–673	$CoMo/Al_2O_3$	Batch autoclave	Aboul Gheit [2]
Mixture of fused-ring thiophenes, furans, quinolines, indole, and alkylphenols	2.07–10.34	573–723	CoMo (presulfided)	Fixed-bed reactor	Rollmann [113]
Acridine & carbazole in xylene	10.133	473–633	Mo/Al_2O_3	Fixed-bed reactor	Nai et al. [86]
Indole in petroleum	2.128	588	NiW/Al_2O_3 (sulfided)	Batch reactor	Hartung et al. [52]
Indole, pyrrole, & carbazole in dodecane	6.895	623	CoMo, NiMo, Re, $CoRe/Al_2O_3$ (sulfided)	Batch reactor	Stern [138]
Pyridine & piperidine *n*-pentylamine in gasoline	5.171	493–613	(sulfided & unsulfided)	Flow reactor	Stengler et al. [137]
Pyridine	10.342	553	MoS_2	Batch autoclave	Landa et al. [70]

Table 6-5 (Continued)

Summary of Denitrogenation Studies

Feedstock	Pressure, MPa	Temp., K	Catalyst	Equipment	Reference
Indole, quinoline, substituted quinolines, dibenzothiophene, & naphthalene in paraffin oil (also hexadecane)	3.447	623	$NiMo/\gamma\text{-}Al_2O_3$	Batch autoclave	Bhinde [18]
	16.212–20.265	623–615	$NiMo/\gamma\text{-}Al_2O_3$	Flow microreactor	
Mixture of NSO heterocyclics in heptane	6.89	675	$NiMo/\gamma\text{-}Al_2O_3$	Flow reactor	Badilla-Ohlbaum et al. [8]

The rate constants for some of the steps in the reaction scheme are given in Table 6-6. Whereas NiMo and CoMo catalysts are affected by the formation of *o*-ethyl aniline, the Re and CoRe catalysts appear to facilitate the hydrogenolysis of *o*-ethyl aniline. The rate constants, assuming first-order kinetics, for the conversion of a mixture of pyrrole, indole, and carbazole are given in Table 6-7. For all the catalysts examined, the rate constants decrease with increasing complexity of the nitrogen-containing molecule.

The denitrogenation of pyrrole and indole using a $CoMo/Al_2O_3$ catalyst has been investigated by Aboul Gheit and Abdou [1]. The rate expressions for the decomposition of these compounds at 20.68 MPa are

$$r_{\text{pyrrole}} = 4.5 \times 10^8 \exp\left[-73.67/RT\right], \tag{6-13}$$

$$r_{\text{indole}} = 4.2 \times 10^5 \exp\left[-77.44/RT\right]. \tag{6-14}$$

Table 6-6

Rate Constants for the Conversion of Indole, 2,3-Dihydroindole and *o*-Ethyl Aniline [138] at 623 K and 10.34 MPa

	$k_1 \times 10^2$, $(\text{min})^{-1}$		
Catalyst	Indole	2,3-Dihydroindole	*o*-Ethyl aniline
CoMo	3.46	3.01	0.838
NiMo	3.65	3.3	1.06
Re	2.39	1.26	7.82
CoRe	1.75	1.86	1.51

Table 6-7

Rate Constants for the Conversion of Pyrrole, Indole, and Carbazole at 623 K and 10.34 MPa [138]

	Rate constant $k_1 \times 10^2$, $(min)^{-1}$		
Catalyst	Pyrrole	Indole	Carbazole
CoMo	5.54	1.07	0.26
NiMo	3.3	1.16	0.12
Re	7.7	0.89	0.23
CoRe	6.93	0.74	0.10

A comparison of the rate constants reported by Stern [138] and Aboul Gheit and Abdou [1] using $CoMo/Al_2O_3$ (comparable metals loading) at 623 K for the conversion of pyrrole and indole is shown in Table 6-8. The differences in these values may be explained by the fact that the catalyst employed by the former was presulfided, whereas that of the latter was not. Also, the former maintained 5% H_2S in the gas phase during the entire reaction, which probably helped to maintain the catalyst in its active, sulfided state.

Bhinde [18] has also investigated the hydrodenitrogenation of indole and mixtures of indole and quinoline. Indole was found to exist in equilibrium with 2,3-dihydroindole with an equilibrium constant of 0.083 at 623 K and 3.45 MPa. This appears to compare favorably with the value of 0.167 reported by Stern [138] at 623 K and 6.89 MPa. The mechanism proposed and the corresponding rate constants in (g oil)/(g cat.)(min), assuming first-order kinetics, are shown below:

Table 6-8

Comparison of Rate Constants for Pyrrole and Indole

	Stern [138]	Aboul Gheit & Abdou [1]
Pressure, MPa	6.89	20.68
$k_{pyrrole}$, hr^{-1}	3.324	0.299
k_{indole}, hr^{-1}	0.642	0.131

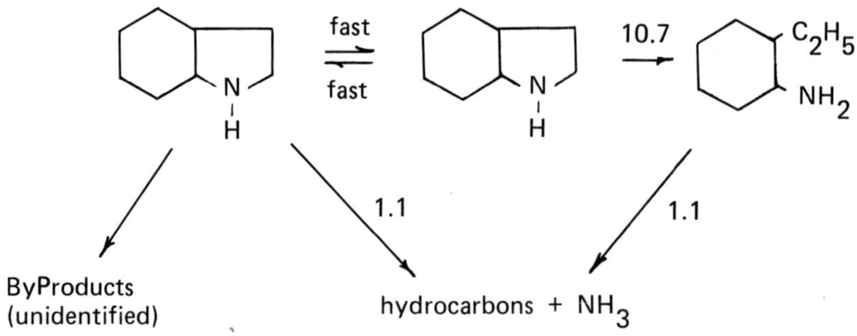

In this mechanism, the formation of ethyl cyclohexane and indolines is not ruled out. The order of the reactions with respect to hydrogen were not ascertained.

In studies made on mixtures of indole and quinoline the rate constants in the indole reaction network were reported to decrease in the presence of quinoline; however, the reverse was not true. This was attributed to indole being a relatively weak base when compared with quinoline, which due to its strong basicity is strongly adsorbed on the acidic catalyst sites. Thus, a reevaluation of the definition of "ease of removal" of nitrogen from mixtures of nitrogen-containing heterocyclics is necessary.

For the denitrogenation of carbazole, Flinn et al. [41] proposed the following mechanism:

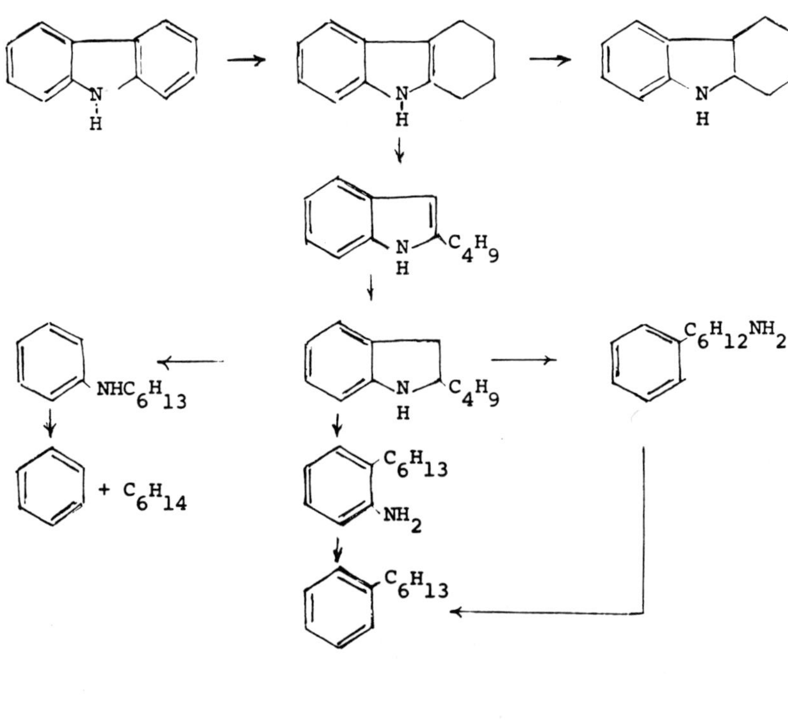

The results are primarily qualitative and do not include an evaluation of the selectivity or rate-limiting steps. The mechanism also does not consider the hydrogenation of hexahydrocarbazole to decahydrocarbazole to be important. This route appears to be important in the presence of a MoS_2 catalyst at 10.34 MPa and in the temperature range of 533–633 K, as shown in the mechanism proposed by Nai et al. [86]:

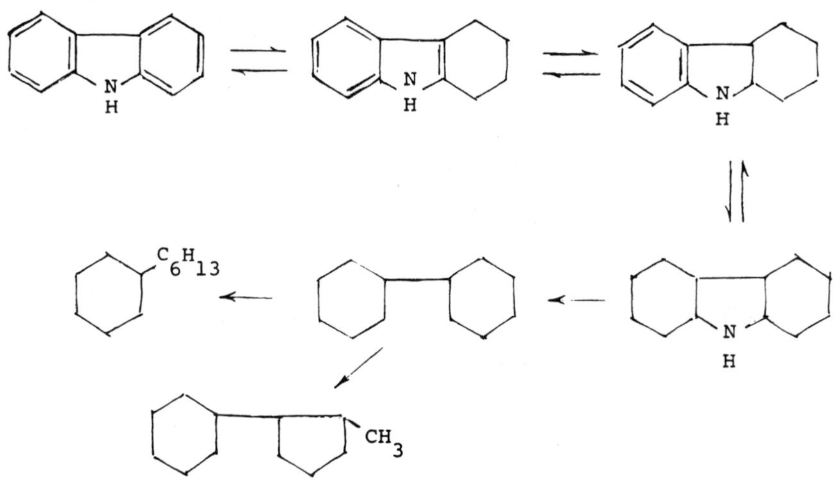

Although the above two mechanisms propose different steps leading to the removal of nitrogen, they are similar in that they agree that hydrogenation of at least one ring should precede the removal of nitrogen as ammonia, a fact that has been noted by earlier workers [25–41].

In an investigation of the hydrogenation of pyridine using a MoS_2 catalyst [68], the reactions appeared to proceed selectively through the formation of piperidine; however, a considerable amount of alkyl pyridines was detected in the products. This can be attributed to reactions between products of piperidine ring opening and piperidine. In another study, which examined the HDN of pyridine, piperidine, and *n*-pentylamine using both sulfided and unsulfided catalysts [137], it was found that hydrogenolysis of the piperidine ring in the presence of an unsulfided catalyst was the rate-controlling step up to 623 K. At higher temperatures, hydrogenation of pyridine appeared to be rate controlling. The hydrogenation of pyridine and the hydrogenolysis of *n*-pentyl amine were found to be first order and irreversible, while the hydrogenolysis of piperidine was found to follow zero-order kinetics.

McIlvried [79] proposed the following mechanism for nitrogen removal from pyridine:

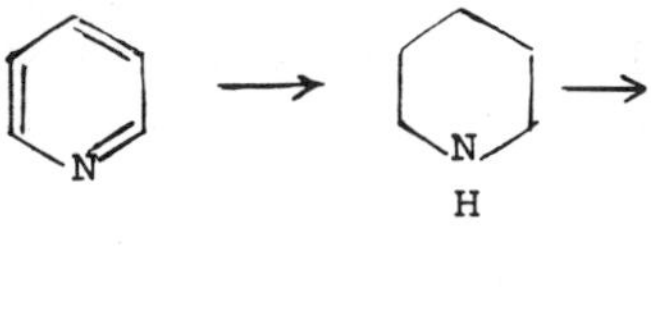

$$\rightarrow C_5H_{11}NH_2 \rightarrow C_5H_{12} + NH_3. \quad (6\text{-}15)$$

The hydrogenation of pyridine and the denitrogenation of *n*-pentyl amine were found to be rapid, while the hydrogenolysis of piperidine was slow. Due to the strong adsorption of nitrogen-containing compounds on the catalyst surface, the rate of denitrogenation decreases as the nitrogen content of the feed increases. On this basis, McIlvried [79] proposed the following Langmuir-Hinshelwood expression for the denitrogenation of piperidine in xylene:

$$r = -\frac{k_N P_N P_{H_2}}{1+K_N P_N + K_A P_A} . \tag{6-16}$$

Assuming equally strong adsorption of piperidine and ammonia, the above equation can be reduced to

$$r = c\,\frac{k_N P_N}{1+K'_S P_S + K'_N P_{No}} . \tag{6-17}$$

The rate of hydrogenation was found to fit the following model

$$r = -\frac{k_{H_2} P_P P_{H_2}}{1+K'_A P_A} \tag{6-18}$$

In the above models it is apparent that while ammonia alone is strongly adsorbed on the hydrogenation sites, both ammonia and piperidine are adsorbed on the hydrogenolysis sites, a fact which appears to be justified by experimental findings. Sonnemans et al. [126] found, using an unsulfided catalyst, that piperidine could undergo disproportionation reactions which might affect the overall reaction scheme. The following rate expression was proposed for the hydrogenation of pyridine over unsulfided MoO_3 and CoMo on alumina catalysts:

$$r = \frac{k_{H_2} P_P P^n_{H_2}}{P_{Po}} . \tag{6-19}$$

The dependence of the rate was found to be first order in hydrogen at 523 K and 1.5 at 573–648 K, which is in agreement with previous results [79]. The presence of the initial partial pressure of pyridine (P_{P0}) in the denominator indicates strong adsorption of pyridine and nitrogen-containing products on the catalyst. Independent adsorption experiments showed that nitrogen bases were strongly adsorbed; however, they failed to show equally strong adsorption of all other bases.

The effects of presulfiding and the presence of H_2S on the denitrogenation of pyridine have been brought out by Goudriaan et al. [50] and are shown in Figures 6-3 and 6-4. It is apparent from Figure 6-4 that the presence of H_2S enhances the hydrogenolysis of piperidine over a presulfided catalyst. The effect of H_2S on the

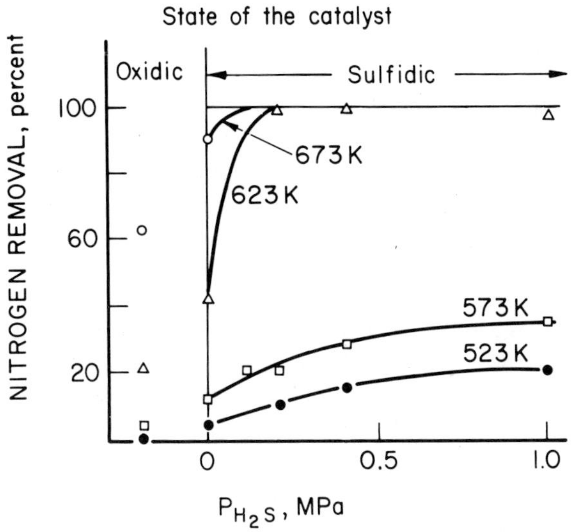

Figure 6-3. Hydrodenitrogenation of pyridine as a function of H_2S partial pressure [50].

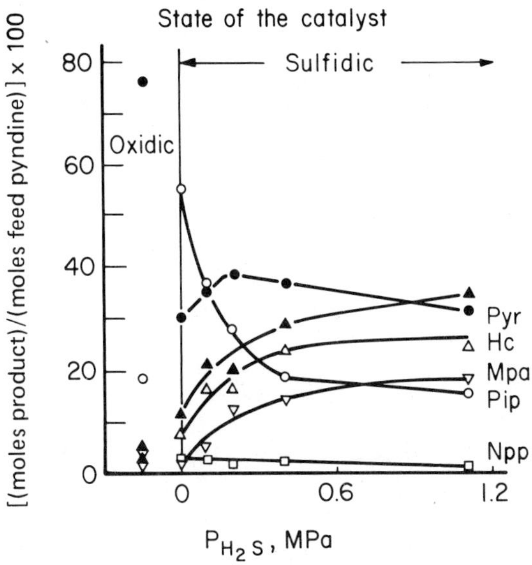

Figure 6-4. Reaction intermediates in the hydrodenitrogenation of pyridine as a function of H_2S partial pressure [50]. Pyr = pyridine; Pip = piperidine; Mpa = monopentylamine; Npp = N-pentylpiperidine; Hc = pentene, pentane.

hydrogenation of pyridine using the same catalyst is marginal. Thus, the rate-limiting step postulated by McIlvried [79] can be accelerated in the presence of H_2S.

Under certain experimental conditions, the equilibrium between pyridine and piperidine can affect the overall reaction rate [119]. For the denitrogenation of pyridine, the hydrogenolysis of piperidine (assuming the hydrogenation of pyridine to piperidine is rate controlling) is rapid, and the rate of denitrogenation is unaffected by reaction equilibrium. However, when the hydrogenolysis step is rate limiting, an increase in temperature enhances the rate of this reaction by shifting the equilibrium between pyridine and piperidine to the left, thereby decreasing the concentration of piperidine, which in turn decreases the rate of its hydrogenolysis. Thus, the rate may go through a maximum with temperature. The above argument reveals the existence of thermodynamic limitations in reaction networks where the hydrogenolysis of the –C–N– bond is rate limiting and the availability of the reactant for this reaction is limited.

Aboul Gheit [2] has reported the following mechanism for the denitrogenation of quinoline:

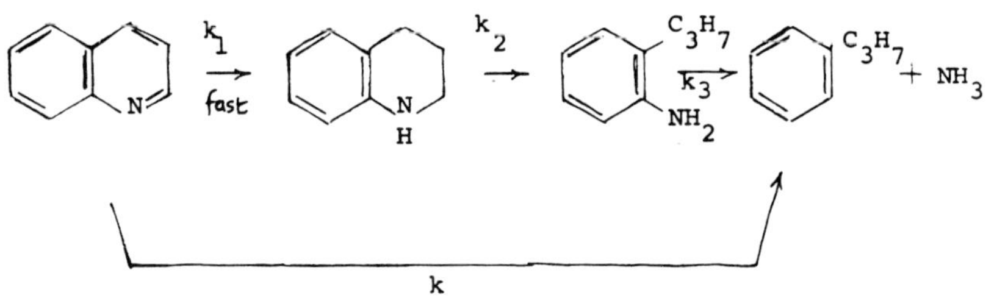

Based on first-order irreversible kinetics, the rate constants in sec^{-1} for the various steps in the reactions are

$$k_2 = 4.42 \times 10^3\, e^{-103.4/RT}\,, \tag{6-20}$$

$$k_3 = 5.73\, e^{-59.4/RT}\,, \tag{6-21}$$

$$k = 1.48 \times 10^5\, e^{-119.7/RT}\,. \tag{6-22}$$

Using an unsulfided $CoMo/Al_2O_3$ catalyst, the rate of hydrogenation of quinoline to 1,2,3,4-tetrahydroquinoline was found to be very rapid, 96% conversion being attained in the first 30 minutes of reaction. As reported by Goudriaan et al. [50], this type of catalyst does not possess good hydrogenolysis capability, as can be seen by a low value of the frequency factor for this step.

Satterfield and Cocchetto [119] have found quinoline to exist in equilibrium with 1,2,3,4-tetrahydroquinoline in the temperature range of 503–693 K over a sulfided NiMo/Al_2O_3 catalyst. At low temperatures the formation of 1,2,3,4-tetrahydroquinoline and, subsequently, *o*-propyl aniline was observed, whereas at high temperatures the formation of 5,6,7,8-tetrahydroquinoline and, subsequently, decahydroquinoline was favored. At extremely high temperatures, a significant fraction of quinoline reacted to form high-molecular-weight polymeric materials, some of which were identified as refractory nitrogen-containing heterocyclics.

On comparing the relative rates of hydrodenitrogenation of pyridine and quinoline, it is found that although the reactants are in equilibrium with their first (heterocyclic) hydrogenated intermediate, this equilibrium is more rapidly attained for quinoline. On the other hand, the hydrogenolysis of *n*-pentyl amine is more rapid than that of *o*-propyl aniline, indicating that the latter is a refractory intermediate. For this reason nitrogen removal from isoquinoline can be expected to be much more rapid than from quinoline.

The reaction mechanism for hydrodenitrogenation of quinoline at 3.45 MPa and 615 K over a sulfided NiMo/Al_2O_3 catalyst has been quantified by Shih et al. [122] and is presented below:

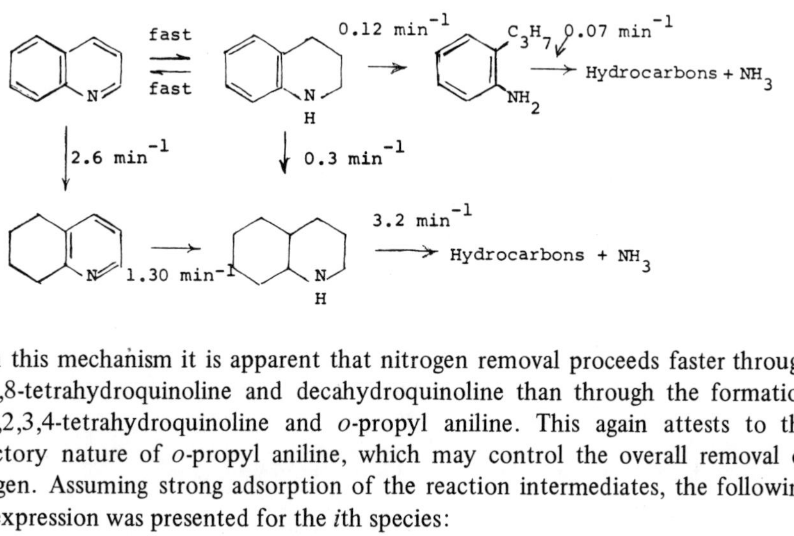

From this mechanism it is apparent that nitrogen removal proceeds faster through 5,6,7,8-tetrahydroquinoline and decahydroquinoline than through the formation of 1,2,3,4-tetrahydroquinoline and *o*-propyl aniline. This again attests to the refractory nature of *o*-propyl aniline, which may control the overall removal of nitrogen. Assuming strong adsorption of the reaction intermediates, the following rate expression was presented for the *i*th species:

$$r_{ij} = \frac{k_{ij} P^n_{H_2} C_i}{(1+K_{H_2} P_{H_2} + \Sigma K_i C_i)^2} \,. \tag{6-23}$$

For equal adsorption of all reaction intermediates the above equation simplifies to

$$r_{ij} = \frac{k_{ij} P^n_{H_2} C_i}{(1+K_{H_2} P_{H_2} + K_N C_{No})^2} \,. \tag{6-24}$$

Table 6-9

Constants for the Hydrogenation Reactions during the Hydrodenitrogenation of Quinoline [122] *

Reaction:	k_{ij}, (g oil)/(min) (psi^2) (g cat.)	K_{H_2}, psi^{-1}	K_N, (g oil)/g-mol
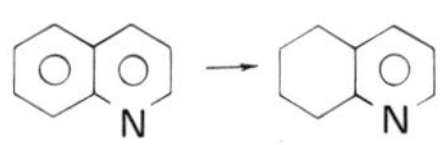	6.97	0.8	640X10^6
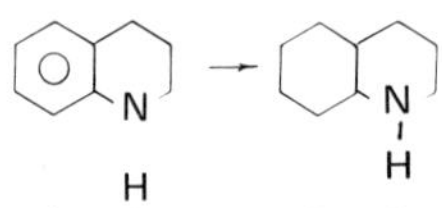	1.41	0.8	640X10^6
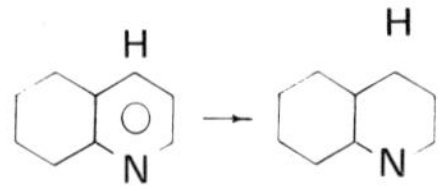	5.02	0.8	640X10^6
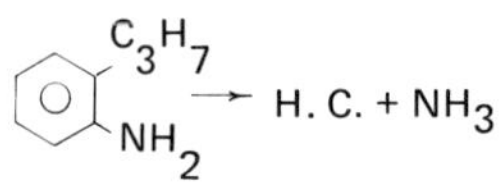	0.21	0.8	640X10^6

*Rate expression (6-24), n=2, C_{No} in g-mol/(g oil), P_{H_2} in psi. Reaction conditions: NiMo/Al_2O_3, 615 K, 0.05 wt% of CS_2 in white oil, 0.5 wt% of catalyst in white oil.

The exponent on the partial pressure of hydrogen was found to be as follows:

n = 2 for hydrogenation reactions and nitrogen removal from o-propyl aniline,

n = 1 for hydrogenolysis of 1,2,3,4-tetrahydroquinoline,

n = 0 for hydrogenolysis of decahydroquinoline.

The adsorption coefficients and the activation energies for the various steps in the reaction are given in Tables 6-9 to 6-11. Shih et al. concluded that the hydrodenitrogenation of quinoline occurs through interactions of the reaction intermediates with the metal and acid sites on the catalyst, the former promoting hydrogenation and the latter hydrogenolysis.

The denitrogenation of dimethyl quinolines has been found to proceed in a manner very similar to that of quinoline [18]. As reported for other nitrogen-containing heterocyclics, the reaction involving the hydrogenation of the heterocyclic ring was found to be reversible.

Table 6-10

Constants for the Hydrogenolysis Reactions in the Hydrodenitrogenation of Quinoline [122]*

Reaction	k_{ij}	K_{H_2}	K_{NO}
1,2,3,4,-tetrahydroquinoline	$175 \dfrac{\text{g-mol}}{\text{(g cat.) (min) (psi)}}$	$0.8\ \text{psi}^{-1}$	$6.4 \times 10^6 \dfrac{\text{g oil}}{\text{g-mol}}$
decahydroquinoline	$9.0 \times 10^5 \dfrac{\text{g-mol}}{\text{(g cat.) (min)}}$	$0.08\ \text{psi}^{-1}$	$6.4 \times 10^6 \dfrac{\text{g oil}}{\text{g-mol}}$

*Rate expression (6-24), $n = 1$ for hydrogenolysis of 1,2,3,4-tetrahydroquinoline, $n = 0$ for hydrogenolysis of decahydroquinoline. C_1 and C_{NO} in g-mol/(g oil), P_{H_2} in psi. Reaction conditions: NiMo/Al_2O_3, 615 K, 0.05 wt% CS_2 in white oil, 0.5 wt% catalyst in white oil.

In a companion study, Shih et al. [123] examined the denitrogenation of quinoline and acridine using various combinations of Ni, Mo, and W on Al_2O_3 supports. The mechanism for acridine hydrodenitrogenation [163] which was used in this work for catalyst evaluation is shown below:

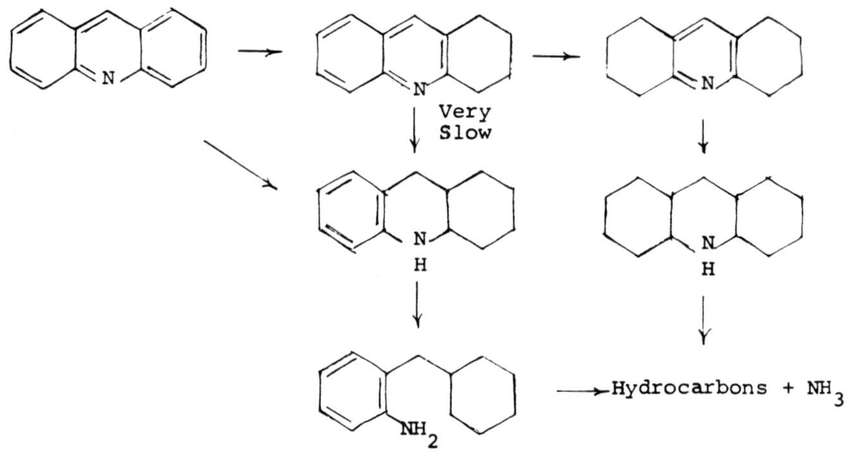

In extensive studies under various conditions, presulfiding and the addition of CS_2 to the system enhanced the nitrogen-removal reactions. $NiMo/Al_2O_3$ was, in general, found to be superior to $CoMo/Al_2O_3$ or NiW/Al_2O_3 catalysts for both hydrogenation and hydrogenolysis. However, in light of earlier findings [122] where hydrogenolysis was presumably occurring on the support, more justification is needed to establish the superiority of $NiMo/Al_2O_3$ in promoting such reactions.

Table 6-11

Estimated Activation Energies for the Various Steps in the Hydrodenitrogenation of Quinoline [122]

	Energy, kJ/mol
Total nitrogen removal	104.65
Quinoline → 5,6,7,8-tetrahydroquinoline	75.35
1,2,3,4-Tetrahydroquinoline → decahydroquinoline	66.97
5,6,7,8-Tetrahydroquinoline → decahydroquinoline	71.16
1,2,3,4-Tetrahydroquinoline → *o*-propyl aniline	146.5
o-Propyl aniline → hydrocarbons + NH_3	100.46
Decahydroquinoline → hydrocarbons + NH_3	129.76

With increasing complexity of the model compound, the size of the network and the number of possible alternate pathways increases. This is borne out by an alternative mechanism for acridine which is shown below [86]:

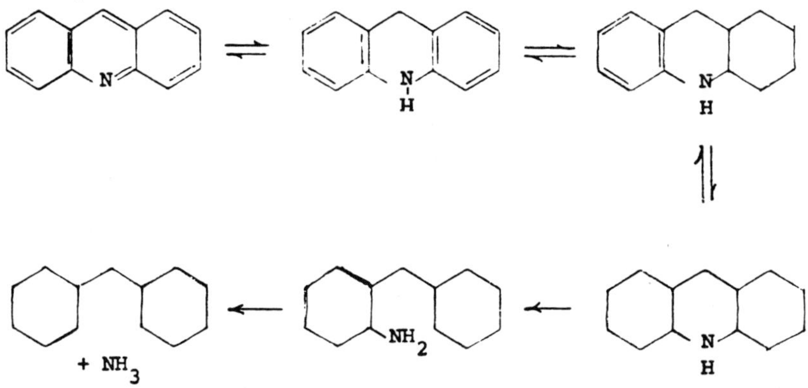

This network assumes sequential hydrogenation commencing with the heterocyclic ring with equilibrium between steps.

B. HYDRODESULFURIZATION

Sulfur is present in coal liquids mostly as single and multiring thiophenic compounds. The removal of sulfur occurs under conditions that are favorable for nitrogen removal. However, sulfur is removed as H_2S with minimum hydrogenation of adjacent aromatic rings. Table 6-12 summarizes the relevant details of several major investigations.

Amberg and coworkers [32, 33, 62, 94, 96] presented a comprehensive study of the desulfurization of thiophene in a pulse microreactor and a low-conversion steady-state flow reactor. Chromia as well as sulfided commercial $CoMo/Al_2O_3$ and molybdenum sulfides were used as catalysts. The product distribution obtained for different reactants is shown in Table 6-13. The conversion of butenes to butane is slow in comparison with that of butadiene to butenes. No butadiene was observed when commercial cobalt molybdate was used. The proposed reaction network is shown in Figure 6-5. The absence of hydrogenated thiophenes in Figure 6-5 seems to indicate that the C–S bond cleavage is the first step in the network.

Desikan and Amberg [33] studied the reaction of tetrahydrothiophene and found *n*-butane thiol to exist under certain conditions. This compound was not found among the products from the desulfurization of thiophene. Several investigators have reported the presence of hydrogenated products [74, 82, 83] and products containing sulfur, i.e., mercaptans and disulfides [24]. This suggests that the desulfurization of thiophene proceeds via two routes, the main route being the one shown in Figure 6-5.

Table 6-12

Summary of Hydrodesulfurization Studies

Compound	Temp., K	Pressure, MPa	Catalyst	Remarks	Reference
Thiophene	473–673	0.103	Oxides & sulfides of Co and Mo	Flow reactor	Pease and Keighton [97]
Thiophene	520–690	0.103	Commercial CoMo/Al_2O_3, chromia, molybdenum sulfides	Pulse microreactor, differential reactor	Owens and Amberg [94–96] Desikan and Amberg [32, 33] Kolboe and Amberg [62]
Thiophene	508–538	~ 0.103	Commercial CoMo/Al_2O_3	Recycle flow reactor	Satterfield and Roberts [117]
Thiophene	523–586	0.00267–0.02133*	CoMo/Al_2O_3	Recycle flow reactor	Lee and Butt [74]
Thiophene	523–623	0.103	Commercial CoMo/Al_2O_3	Differential reactor	Morooka and Hamrin [84]
Benzothiophene & substituted benzothiophenes	573–673	0.103	Commercial CoMo/Al_2O_3	Flow reactor	Givens and Venuto [48]
Benzothiophene & dibenzothiophene	573–673	0.103	CoMo/Al_2O_3	Flow reactor	Bartsch and Tanelian [9]
Dibenzothiophene	573	10.5	Commercial CoMo, NiMo, & NiW on γ-Al_2O_3	Flow reactor, batch autoclave	Houalla et al. [53]
Ethyl mercaptan, butyl mercaptans, 1-hexane thiol, 1-octane thiol, 1-dodecane thiol, debutyl sulfides, etc.	453–613	10.5	MoS_2		Landa and Mrnkova [68]

*Partial pressure of thiophene.

Table 6-13

Product Distribution over Chromia Catalyst in a Pulse Microreactor at 638 K [94]

	Product (% of C_4 hydrocarbons)*				
Reactant	Isobutane	*n*-Butane	1-Butene* + isobutene	*trans*-2-Butene	*cis*-2-Butene + butadiene
Thiophene	0	8.6	30.1	27.4	33.9†
1, 3-Butadiene	0	2.0	23.1	36.7	38.2†
1-Butene	0	8.9‡	42.1	26.6	22.4
trans-2-butene	0	1.9	22.3	52.4	23.5
cis-2-butene	0	2.0	21.2	37.6	39.2
Isobutene	2.0	0	98.0	0	0
n-Butane	–	100	0	0	0
n-Butane§	–	–	20.0	48.0	32.0

*Isobutene was assumed to be absent when isobutane was absent.
†Butadiene was present.
‡Value may be high.
§Equilibrium in gas phase.

H_2S has been found to have a strong inhibiting effect on thiophene conversion and butene hydrogenation, but only a very small effect on double-bond shifts and cis-trans isomerization [95]. The effect of H_2S was very large on butene hydrogenation. From studies made on methylthiophenes [32] as well as selective poisoning of catalysts [33] it can be concluded that the hydrogenation of olefins on sulfided cobalt molybdate catalysts takes place on strongly acidic sites on which H_2S and thiophene can be strongly adsorbed. These strong sites can catalyze desulfurization to a limited extent when the compound to be desulfurized is sufficiently basic. The bulk of the desulfurization takes place on weakly acidic sites either by the addition of H_2 across a C–S or S–S bond or by the removal of –SH from C–SH, thereby forming a carbonium ion.

It has also been found that the rate of conversion of 3-methylthiophene is higher than that of 2-methylthiophene [32], which means that the rate-determining step in the sequence is not the chemisorption of hydrogen. Moreover, these two compounds have almost the same heats of chemisorption, and so the chemisorption of thiophenes cannot be the rate-determining step. The rate-limiting step is believed to be a surface reaction such as C–S bond scission on catalyst sites. Assuming zero-order kinetics, apparent activation energies for the various reactions in Figure 6-5 are presented in Table 6-14.

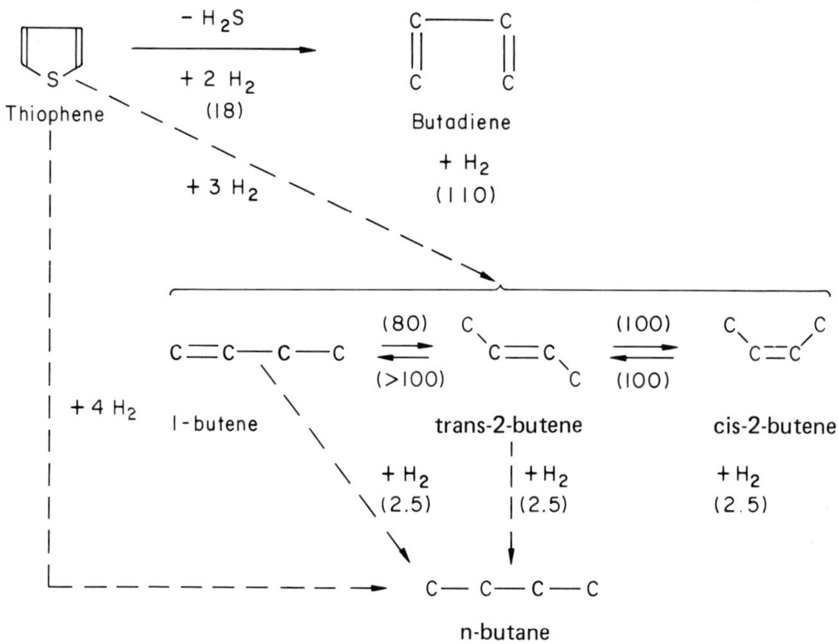

Figure 6-5. Reaction network for hydrodesulfurization of thiophene [94]. Figures in parenthesis are the approximate rate constants, μmol/g sec catalyst, of the individual steps using chromia at 688 K.

Satterfield and Roberts [117] correlated their results for thiophene desulfurization by the equation

$$r_T = \frac{kP_T P_{H_2}^n}{(1+K_T P_T + K_{HS} P_{HS})^2}. \tag{6-25}$$

Their data were found to fit equally well for $n = 0$ and $n = 1$. The corresponding values of the activation energy were 37.67 and 15.49 kJ/mol, respectively. The above equation can be derived by assuming the presence of two types of sites, with thiophene and hydrogen sulfide competing for one type, and hydrogen adsorbing on the second type. Thiophene undergoes a two-point adsorption, whereas H_2S is adsorbed on a single site. The surface reaction of adsorbed H_2 with adsorbed thiophene is hypothesized to be the slow step in the reaction sequence. The rate of butane formation was correlated by

$$r_B = \frac{k' P_B P_{H_2}}{1+K'_B P_B + K'_{HS} P_{HS}}. \tag{6-26}$$

Table 6-14

Apparent Activation Energies for the Various Reactions [94]

Reaction	Temp. range, K	E, kJ/mol
Thiophene → C_4	533–603	100.46
Butadiene → butene	443–503	167.44
1-butene → butane	463–543	64.88
cis → butane	443–523	64.88
trans → butane	583–633	64.88
	473–523	~ 10.46
1-butene → 2-butene	423–498	60.7
trans → 1-butene	453–553	~ 71.16
cis → 1-butene	408–493	~ 46.05
trans → *cis*	453–513	27.21
cis → *trans*	523–673	35.58
	463–523	27.21
	393–553	

Givens and Venuto [48] have examined the chemistry of the hydrodesulfurization of benzothiophene and substituted benzothiophenes using commercial CoMo/Al_2O_3 catalyst. Hydrogenation of the thiophene ring was found to be rapid:

The mechanism postulated for the desulfurization of 2-methyl benzothiophene is shown in Figure 6-6. The formation of mercaptans could not be ascertained because of the high reactivity of these species. It was also observed that alkylation of the thiophene ring and alkyl migration on the thiophene ring could occur, whereas migration or dealkylation of methyl groups on the benzene ring did not occur to an appreciable extent. Table 6-15 presents the conversion and selectivity for the desulfurization of substituted benzothiophenes examined by the above investigators.

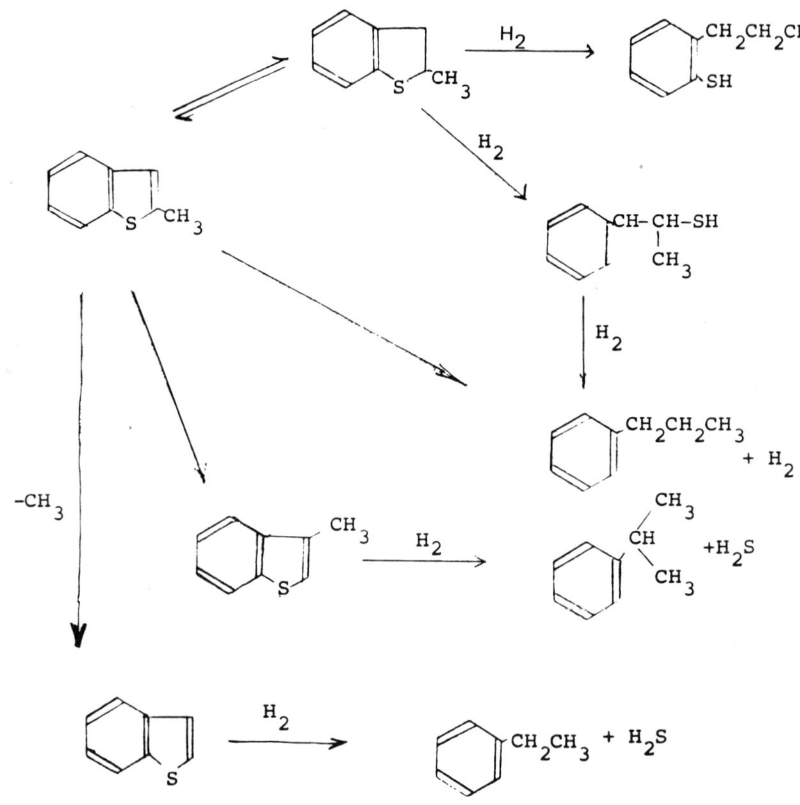

Figure 6-6. Desulfurization of 2-methyl benzothiophene [48].

Dibenzothiophene (DBT) is relatively difficult to hydrodesulfurize, and the reaction network involved has not been accurately identified. Several investigators [9, 89, 90] have found biphenyl to be the only product obtained, but the presence of hydrogenated products like cyclohexylbenzene or bicyclohexyl has also been reported [54]. Houalla et al. [53] have shown that the product distribution depends upon the type of catalyst, concentration of H_2S in the reactants, and hydrogen concentration. Using a CoMo/Al_2O_3 catalyst, they found a high yield of biphenyl, cyclohexylbenzene, and bicyclohexyl present in the products. Batch reactor studies revealed the presence of 1,2,3,4-tetrahydrodibenzothiophene and 1,2,3,4,10,11-hexahydrodibenzothiophene, indicating that some hydrogenation precedes hydrodesulfurization. The selectivity of hydrodesulfurization versus hydrogenation decreased when NiMo and NiW were used as a catalyst instead of CoMo. An increased concentration of H_2S in the feed favored hydrogenation, as shown in Table 6-16. The network for hydrodesulfurization of DBT shown in

Table 6-15

Conversion and Selectivity in Desulfurization of Substituted Benzothiophenes [48]*

Position (methyl group)	None	7	2	2,7	3	3,7	2,3	2,3,7
Fraction converted	9.91–0.99	0.6	0.74	0.54	0.43	0.47	0.39	0.47
Fraction desulfurized	0.91–0.99	0.57	0.66	0.43	0.32	0.24	0.15	0.16
Selectivity, hydrodesulfurization/conversion	1.00	0.95	0.89	0.80	0.75	0.51	0.38	0.34

*Reaction at 823 K, atmospheric pressure over cobalt-molybdena catalyst pretreated in hydrogen.

Figure 6-7 has been reported by the above authors. The figures in parenthesis are the pseudo first-order rate constants in m^3/(kg cat.)(sec).

Houalla et al. [53] also examined the desulfurization chemistry of methyl-substituted DBT. They found the reactivity to decrease considerably when methyl groups were incorporated in the positions nearest to the sulfur atom (i.e., the 4 and 6 positions). The yield of hydrogenated products from such compounds was found to be more than that from DBT substituted with methyl groups at the 2 and 8 or the 3 and 7 positions.

Table 6-16

Effect of H_2S on Selectivity for Hydrogenation versus Hydrodesulfurization of Dibenzothiophene [53]*

$(C_{H_2S}/C_{H_2})_{feed}$	Space velocity, $\frac{m^3 \text{ feed}}{(\text{kg}_{\text{cat.}})(\text{sec})}$	Yield of cyclohexyl benzene, mol% of hydrocarbon product
0.00	3.52×10^{-5}	7.4
0.00	2.02×10^{-4}	1.9
0.01	3.34×10^{-5}	9.0
0.01	1.92×10^{-4}	2.0
0.13	3.47×10^{-5}	13.0
0.13	1.99×10^{-4}	3.8

*Reaction at 573 K, 10.33 MPa over sulfided $CoO\text{-}MoO_3/\gamma\text{-}Al_2O_3$.

C. HYDRODEOXYGENATION

Coal liquids are known to contain oxygen-containing compounds. The type of bonds that have been detected are hydroxylic (–OH), carboxylic (–COOH), carbonylic ($-\overset{O}{\overset{\|}{C}}-$), and etheric (–O–). Carboxylic, carbonylic, and etheric bonds are easily broken under thermal conditions; however, hydroxylic groups retain their integrity and are thus found in coal liquids produced by noncatalytic liquefaction processes [119].

During the upgrading step wherein nitrogen and sulfur are removed, oxygen removal also occurs concurrently, resulting in the formation of water. A summary of relevant studies in this area is presented in Table 6-17.

The hydrogenation and subsequent hydrogenolysis of oxygen compounds on sulfided hydrotreating catalysts are very poorly understood at the present time. The hydrodeoxygenation of phenol on a sulfided catalyst can be postulated to occur in the following manner, the ultimate products being cyclohexane and water:

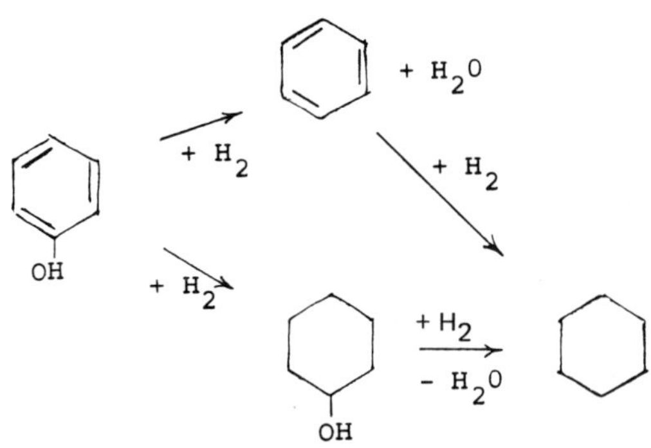

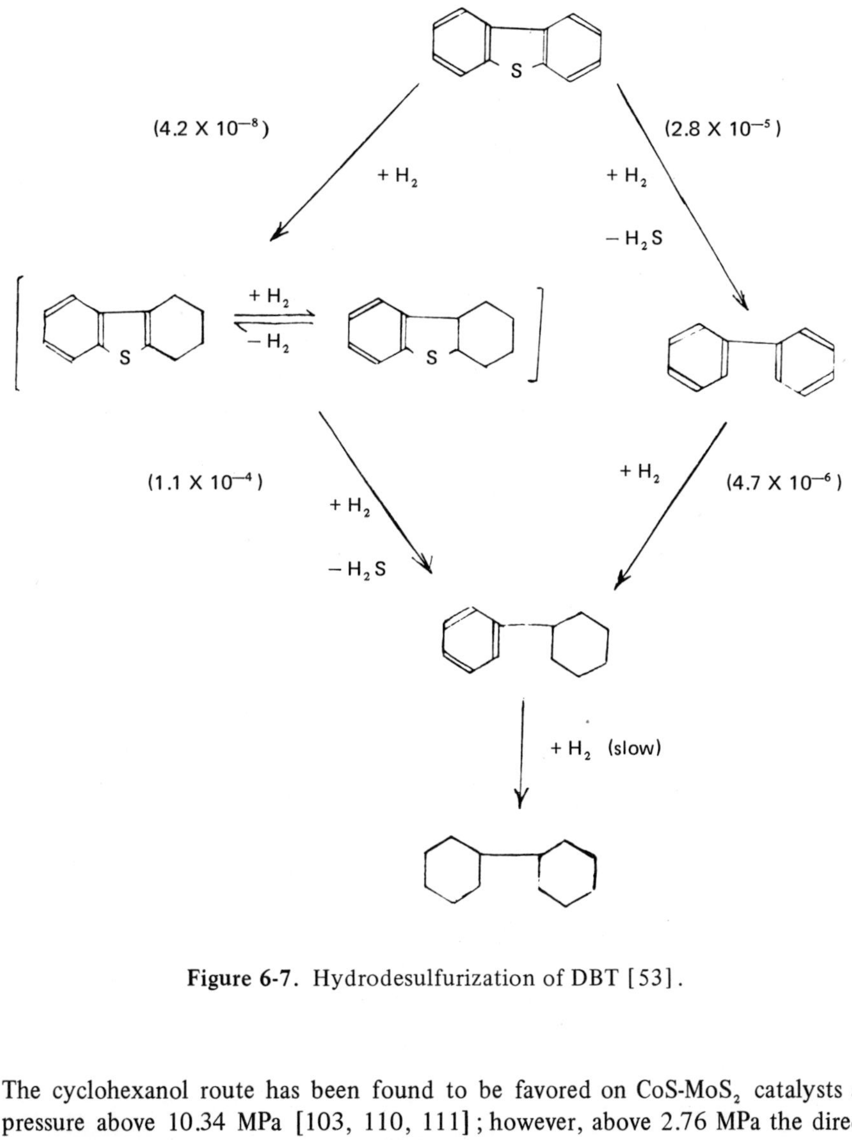

Figure 6-7. Hydrodesulfurization of DBT [53].

The cyclohexanol route has been found to be favored on CoS-MoS_2 catalysts at pressure above 10.34 MPa [103, 110, 111]; however, above 2.76 MPa the direct dehydroxygenation route is favored [82]. Thus, two mechanisms are apparent for the removal of the –OH group from phenol depending upon the operating conditions. A similar conclusion was reached regarding the dehydroxylation of cresols [5]. The relative rates of hydrodeoxygenation of phenols are given in Table 6-18. It is evident from this table that the rates of removal of oxygen from cresols, with the exception of *o*-cresol, are more rapid than that for phenol. Similarly, the rates of deoxygenation of xylenols are lower, and those of the naphthols are greater, than that for phenol. Further studies are needed to better understand these differences. A noteworthy point shared by all of the above studies is the absence of

Table 6-17

Summary of Hydrodeoxygenation Studies

Compound type	Temp., K	Pressure, MPa	Catalyst	Reference
Phenols	623	3.13	MoS_2	Bobyshev et al. [19]
Dibenzofuran	573–773	10.34	MoO, MoS, MoS_2	Hall & Cawley [51]
Benzofuran Dibenzofuran	473–613	10.13	MoS_2	Landa et al. [69]

cyclic alcohols in the reaction products. This is due to the rapid hydrogenolysis at high pressures of such compounds to the corresponding naphthene and water.

Oxygen-containing heterocyclics have also been detected in coal liquids. Typical examples are furan, benzofuran, and dibenzofuran. The relative rates of oxygen removal from these compounds decrease as the number of aromatic rings attached to the heterocyclic ring increases. Here again, totally hydrogenated derivatives of the above compounds are readily hydrogenolyzed. Landa et al. [69] postulated the hydrodeoxygenation of benzofuran to occur by the following mechanism:

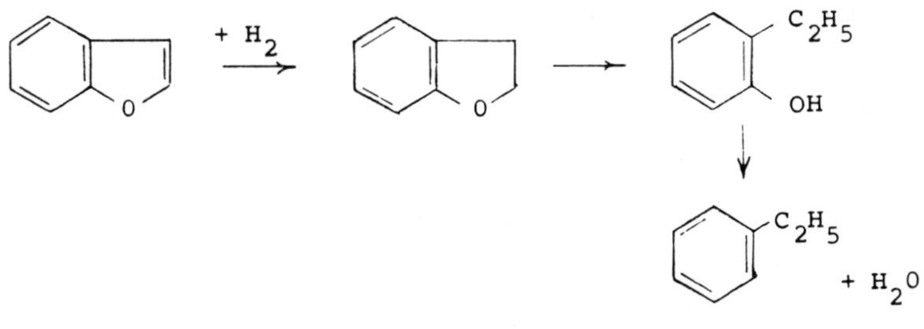

For the hydrodeoxygenation of dibenzofuran over a MoS_2 catalysts, two mechanisms, shown in Figure 6.8 have been postulated [51]. Mechanism A was found to be dominant in the 573–623 K temperature range, the formation of *o*-cyclohexyl phenol being favored to a greater extent than *o*-phenyl cyclohexanol. The selective removal of oxygen with the formation of diphenyl, mechanism B, was favored at higher temperature (>623 K). Partially hydrogenated derivatives, such as 1,2-dihydrodibenzofuran, 1,2,3,4-tetrahydrodibenzofuran, and 1,2,3,4,10,11-hexahydrodibenzofuran, were not detected in the reaction products. This confirms the earlier belief that such intermediates are rapidly converted at the experimental conditions used in the above study. Also, the route phenol → cyclohexanol → cyclohexane would be favored at these conditions, resulting in high yields of cyclohexane.

D. HETEROATOM REMOVAL FROM MIXTURES OF MODEL COMPOUNDS

Although the kinetics and mechanism of heteroatom removal from various compounds that occur in coal liquids have been examined individually, few investigations of heteroatom removal from mixtures of such compounds are reported in the literature. Certain basic nitrogen-containing heterocyclics are strongly adsorbed on catalysts. This is also true of sulfur and oxygen compounds. In addition, the products resulting from the removal of heteroatoms, namely NH_3, H_2S, and H_2O,

Table 6-18

Relative Rate of Hydrogenation of Phenols to Aromatic Hydrocarbons [19]

Phenol hydrogenated	Relative rate	Phenol hydrogenated	Relative rate
Phenol	100	2,4-Xylenol	70.2
o-Cresol	60.8	Thymol	65.8
m-Cresol	108	Carvacrol	44.9
p-Cresol	126	α-Naphthol	160
3,4-Xylenol	62.5	β-Naphthol	208
3,5-Xylenol	65.5	Thiophenol	2845

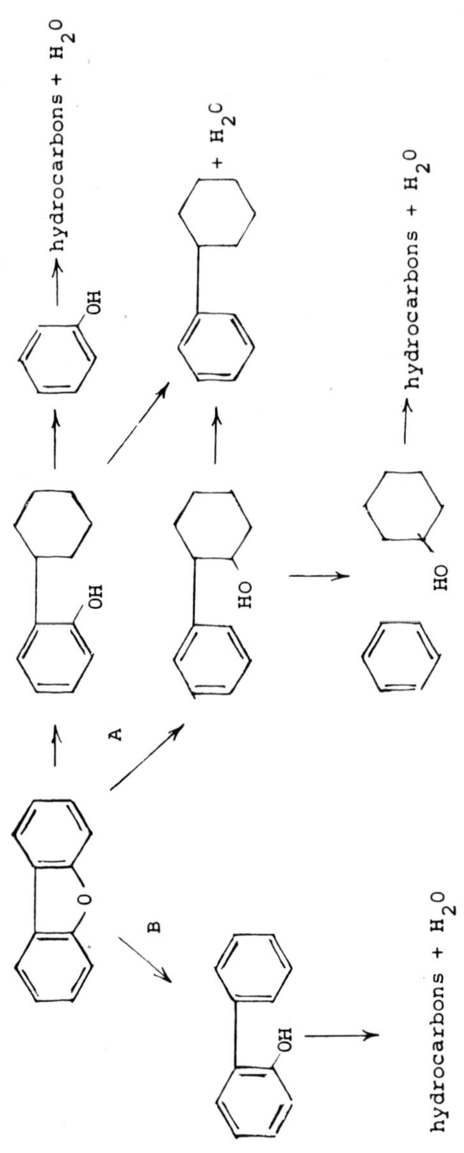

Figure 6-8. Two mechanisms for Hydrodeoxygenation of dibenzofuran [51].

can also affect catalyst performance. In an early investigation of such interactive effects, pyridine was found to inhibit the desulfurization of thiophene [118]. The effect of sulfur on nitrogen-containing compounds was temperature dependent. At low temperatures, thiophene competitively adsorbed on the catalytic sites, thereby retarding the rate of pyridine denitrogenation. At higher temperatures, H_2S, which is a product of thiophene desulfurization, improved the activity of the catalyst towards hydrogenolysis of piperidine.

The relative rates of heteroatom removal from a mixture of nitrogen-, sulfur-, and oxygen-containing heterocyclics blended to represent coal liquids have been examined by Rollmann [113]. The composition of the mixture used in the above study is given in Table 6-19.

When nitrogen or oxygen was bonded to an aromatic ring, saturation of that ring was necessary before –C–N– or –C–O– bond scission could occur. This was not true in the case of sulfur-containing heterocyclics. Assuming that the rates of conversion of the various components in the feed follow irreversible first-order kinetics, the following ordering of reactivity has been ascertained: sulfides $\gg$ *p*-alkylphenols > benzothiophenes > quinolines ∿ *o*-alkyl phenols ∿ indoles ∿ benzofurans ∿ naphthalenes > dibenzofurans $\gg$ mononuclear aromatics.

Badilla-Ohlbaum et al. [8] also examined the hydroprocessing of mixtures of nitrogen-, oxygen-, and sulfur-containing compounds. The results obtained were in agreement with those reported by Rollmann [112]. Attempts to establish the mechanisms for the removal of the various heteroatoms were hindered by the complexity of the system as well as the inability to detect several reaction intermediates.

For mixtures of quinolines and naphthalene, the hydrodenitrogenation of quinoline was found to be unaffected by the presence of naphthalene; however, the

Table 6-19

Feed Compositions Examined by Rollmann [113]

Standard feed	Moles	wt%
Dibenzothiophene	1.0	10.4
Dibutylsulfide	1.0	8.2
Quinoline	1.0	7.3
Dibenzofuran	1.0	9.5
Naphthalene	1.0	7.2
2-Methylnaphthalene	2.0	16.0
2,3-Dimethylnaphthalene	1.0	8.8
Hexadecane	1.5	19.1
1,2,4-Trimethylbenzene	2.0	13.5

Substitute components

Benzothiophene	*p*-Cresol
Dioctylsulfide	4-Propylphenol
Thiophenol	2-Ethylphenol
	2,3-Benzofuran
Indole	2-Phenylphenol
o-Ethylaniline	*cis*-2-Phenyl-l-cyclohexanol
	1-Naphthol

hydrogenation of naphthalene is significantly reduced by the presence of quinoline [18]. This inhibiting effect was also found to be time dependent – initial severe inhibition followed by a subsequent moderate inhibition. The initial inhibiting effect is due to the presence of strong bases, such as quinoline and other reaction intermediates, which are strongly adsorbed on the catalyst sites. As these compounds react, only the ultimate products, namely *o*-propyl aniline and NH_3, are adsorbed on the catalyst, thereby exposing more of the catalyst sites to naphthalene hydrogenation. This explains the increase in the first-order rate constant with time.

It is not surprising that several mechanisms have been proposed for heteroatom removal from a given compound. A mechanism postulated for a given compound depends on several factors, namely, the pressure and temperature, the type and state of the catalyst, the addition of CS_2 or H_2S, the concentration of the reactant, the presence of other reactants, and the use of proper analytical techniques for separation and identification. In view of the fact that model compound studies are used to simulate complex processes, which are otherwise difficult to understand, it is necessary to select the operating conditions for these studies in such a manner as to make them truly representative of the actual process.

E. SUMMARY

From the above studies, it may be summarized that: (1) six-membered heterocyclics hydrogenate and hydrogenolyze with more difficulty than their five-membered counterparts, and the difficulty increases for both with increasing number of rings (the increase being due to steric hindrance); (2) hydrogenation is a prerequisite for denitrogenation, but not for sulfur and oxygen removal; (3) H_2S has a beneficial effect in the removal of nitrogen, whereas water formed by deoxygenation hinders nitrogen and sulfur removal through catalyst deactivation. The effects of NH_3 are not currently known.

At present, very little is understood about the complex interactions between nitrogen-, sulfur-, and oxygen-containing heterocyclics, aromatics, and catalyst surfaces. Single compound studies are helpful in evaluating mechanisms, but they do not represent reality. Future studies in these areas should include the effect of catalyst and support type on the various hydrotreating reactions, the interactions of model compounds with the catalyst surface, and the development of catalysts that resist poisoning and hydrotreat coal liquids through selective hydrogenation of the heterocyclic ring.

6.4.3 *Hydrotreatment of Coal Liquids*

Coal liquids contain high concentrations of heteroatoms and metals, and possess low H/C ratios compared to petroleum crudes. In order to improve the quality of these liquids for further processing to useful fuels and chemicals, these undesirable properties must be minimized by hydrotreatment. Because of the high nitrogen content of coal liquids, nitrogen removal has attracted much attention. In a brief review of the hydrodenitrogenation of petroleum and coal liquids, nickel-molybdenum on alumina catalysts were rated the best for denitrogenation [72]. The evaluation of proprietary catalysts and process systems developed for petroleum refining is essential to the development of catalysts for denitrogenation of coal liquids. Proprietary information on the structure and properties of catalyst supports, other minor components in the catalyst, and special operational techniques needs to be analyzed and reported. A partial list of some of commercial catalysts used for hydrotreating coal liquids and their properties is given in Table 6-20.

In a preliminary investigation of the hydroprocessing of several coal liquids using presulfided $NiMo/Al_2O_3$ catalysts, liquids from lower-rank coals were found to require less reaction time to yield a specified product [73]. The nitrogen concentrations of the feedstocks were reduced substantially, with hydrogen consumptions ranging from 0.9 to 3.1 wt% of feed. A correlation of nitrogen content versus asphaltene content was suggested as a way to monitor the upgrading process. Also, under identical processing conditions, liquids derived from low-rank coals were easier to upgrade in terms of nitrogen removal and asphaltene conversion than those from higher-rank coals.

Several investigators have studied the catalytic hydrogenation of low-temperature carbonization tar oils (Table 6-21). Although these oils are not likely to be a major feedstock in any coal-liquid refinery, such investigations will contribute to the processing of other coal liquids. The data obtained by Anderson et al. [6] indicate that the hydrogenolysis of these tars produce middle oil, gasoline, and gas by a combination of simultaneous and consecutive cracking and hydrogenation reactions. The overall kinetics for gasoline production were found to be first order in tar concentration.

The hydrogenation of anthracene oil has also been the subject of several investigations, but like the low-temperature carbonization tars, this oil is not expected to be a major feedstock for coal-liquid refineries. This oil, however, is quite often used as a startup vehicle solvent for coal liquefaction and in many instances is hydrogenated prior to use. A summary of several pertinent investigations is presented in Table 6-22. Using the partial-wetting model developed for trickle-bed reactors, Sivasubramanian [124, 125] developed the following correlation for the hydrodenitrogenation of anthracene oil:

$$\ln \frac{C_{N0}}{C_N} = k\,(\mathrm{LHSV})^{-x} \quad . \tag{6-27}$$

x was found to range from 0.306 to 0.932, depending upon the operating conditions used, but no correlation with operating conditions was established. The rate constants and activation energies were found to vary from 0.223 to 1.585 hr^{-x} and 50.9 to 65.1 kJ/mol, respectively.

Table 6-20

Summary of Catalysts and Supports Investigated for Hydrotreatment of Coal Liquids

Mfr & type	Co	MoO	wt% NiO	W	Al_2O_3	SiO_2	Pore volume, cm^3/g	Density, g/cm^3 Real	Density, g/cm^3 Particle	Surface area, m^2/g	Ave. pore diam, Å	References
Amer. Cyanamid												
HDS-3A		15	3				0.54	3.41	1.20	173	125	7,10,37,38,43,46, 47,55,56,58,121, 132,135,152
HDS-9A		12.1	2.6				0.43	3.23	1.35	145	119	10,14,37,38,101,136
HDS-20A	5.9	17.0			99.8	0.2	0.516	3.78	1.28	207	100	15,17,30,100,133
HDS-86			7.1	17.5	66	34	0.28	3.20	1.70	198	56	10,133,136
HDS-1441A	2.8	13.8			96.5	3.5	0.57	3.67	1.19	323	71	10,133,134,136
HDS-2A	3.2	15.1			99.9	0.1	0.67			270		60
Harshaw												
Ni-4301			6	19			0.37			228		7,11,15,76,77
Ni-4303			6	19			0.54			152		7,11,15,16,77
Ni-4401			6	19	50	25						7,11,15
Ni-3250			50		propr.		0.34			150		11,16
Ni-1600	3.4		3.4		—	—				1		11
Ni-1601*	3.4		3.4		active	—				78		11
Ni-1800†			10			—				3		11
Cr-0304‡					—	—	0.26			120		11,16
Fe-0301§					100		0.31			41		11,160
W-0101				10	100		0.37			75		11,16
W-0801				10	100		0.36			145		11,16
HT-100E		16.8	3.8		100		0.54			190		11,15,16
CoMo-0401T	3	9								160		3,6,15,31,77,98, 100,133,156,157
CoMo-0402T	3	15			100		0.40			200		13,46,47,76,77
NiMo-618X		14.8	2.7				0.6	3.60	1.14	140	171	10,136
HT-400E	3	15			100		0.50			220		13,14,15,16,60 100,101
HT-500			3.17				0.49			210	93	7,13,15,16

Table 6-20 (Continued)

Summary of Catalysts and Supports Investigated for Hydrotreatment of Coal Liquids

Mfr & type	Co	MoO	wt% NiO	W	Al_2O_3	SiO_2	Pore volume, cm^3/g	Density, g/cm^3 Real	Density, g/cm^3 Particle	Surface area, m^2/g	Ave. pore diam, Å	References
Houdry												
HZ-1					45	53				100–150		76,77
Others												
Ketjen 1535		15.1	3.1		95	5	0.52	3.55	1.24	439	50	10,17,133,134,136
Nalcomo 474	3.5	12.5			82.4	1.5	0.463		131	240–270	66	3,115,116,131, 132,151
Nalco 72-4710A	3.5	12.5			82.4	1.5	0.441			298	50	115,116,131,132
Nalco 72-4710B	3.5	12.5			82.4	1.5	0.556			303	50	115,116,131,132
Nalcomo 477	3.5	12.5			100		0.55			250	88	17,121
Nalco NM 502		14	4				0.53			240	88	
Shell 344	2.4	9.9					0.6			195	123	14,15,16
Sphericat 550			5.1	20.2	100							150
Sphericat 474	3.5	12.5			100		0.981			270	60,1700	3, 151
Supports												
Harshaw 1404T					97	3						76,77
Catapal HP-20												124,125
Ketjen 007-1.5E					100		0.647			244	80,2200	124,125
Ketjen 000-3P					100		0.0647			314	6.6	124,125
Ketjen 003-1.5					100		0.834			254	68,300	124,125
Ketjen 006-1.5					98.66	1.34	0.7			240	117	13,14
Ketjen 005-2E					99.53	0.37	0.73			200	146	13
Ketjen 000-3E					99.39	0.55	0.64			150	171	13
Ketjen LA-3P					98.37	1.5	0.64			230	111	13
Norton 6176					99.85	0.12	0.8–1.1			250	152	13

*3.4% FeO_2. †1% Cu. ‡33% Cr_3O_3. §20% Fe_2O_3.

Table 6-21

Summary of Hydrogenation Studies on Low-Temperature Carbonization Tar

	Reactor	Temperature, K	Pressure, MPa	Residence time	Results
Brahmachari et al. [20]	Packed reactor, MoS_3 pellets	473–673	11.03	0.5–1 LHSV	Aromatics reduced from 56 to 22%. Olefins reduced from 22 to 2–4%. Low LHSV and high temp. favor aromatic reduction.
Morparia and Sarkar [85]	Packed bed, $CoMo/Al_2O_3$, H_2/oil ratio 400–500	623–723	8.1–12.2	0.5–1.5 hr^{-1}	Optimum condition found to be 673 K and 10.1 MPa. No effect of H_2/oil ratio. Low pressure yielded more mid-distillate. Excessive cracking at high temp.; less hydrogenation at low temp.
Mirza et al. [80]	Packed bed, downflow, single- and two-stage system, $CoMo/Al_2O_3$, NiW/Al_2O_3	613–653	5.06–10.1	0.12–2 VHSV	Single-stage CoMo: S and N_2 removal were 55% and 40%. Single-stage NiW: S and N removal were 76–89% and 56–62%. Two-stage system (CoMo in first stage): S and N_2 removal and aromatic saturation lower for CoMo in 2nd stage than for NiW. Two-stage processing allows higher space velocities to be maintained with same product quality.
Anderson et al. [6]	Batch autoclave	623–748	3.45–20.7		Gasoline and gas yields increased linearly with temp., while middle oil and residue decreased. Highest-quality product, containing 60% aromatics, obtained at 723 K and 17.2 MPa (47% yield). Highest yield (77%) obtained at 748 K and 20.7 MPa.

Considering the hydrodesulfurization of anthracene oil as many simultaneous first-order reactions and combining them according to whether they are fast or slow, the following model was developed by Sooter and Crynes [131, 132]:

$$C_S/C_{S_o} = \alpha \exp[-k_1 t] + (1-\alpha)\exp[-k_2 t] \,. \tag{6-28}$$

Fitting the data over the temperature range employed, α increased from 0.63 to 0.81, and the activation energies for k_1 and k_2 were found to be 186.9 and 22.6 kJ/mol respectively. The fast reactions have an activation energy which is chemical-reaction controlled; however, the activation energy for the slow reactions is indicative of diffusion control.

An nth-order model for the hydrogenation of anthracene oil has been proposed by Satchell [115, 116]. This model assumes different reactivities for differ-

Table 6-22

Summary of Investigations on the Hydrogenation of Anthracene Oil

	Reactor	Catalyst	Pressure, MPa	Temp., K	Residence time	Results
Satchell [115, 116]	Trickle bed 1.25 cm o.d., 83.8 cm long; 25.4–50.8-cm catalyst bed	Nalcomo 474 Nalco 72–4710A Nalco 72–4710B	3.45–10.39	589–700	1350–5400 sec	Catalyst deactivation significant at higher temperature, pore diam. (50–66 Å) and H_2 flow had negligible effect on HDN.
Sivasubramanian [124, 125]	Trickle bed; 50.8-cm catalyst bed	Co & Mo impregnated on Catapal HP-20, Ketjen 007-1.5E, Ketjen 000-3P supports	10.3	616–672	0.46–1.86 VHSV	HDN decreased with decreased catalyst surface area, HDN not affected by pore diam. (50–80 Å). Bimodal catalysts offer no advantages over unimodal catalysts.
Sooter [131, 132]	Trickle bed 1–27 cm i.d.; 30–32 g catalyst	$CoMo/Al_2O_3$	6.89	589–673	1850–5400 sec	HDS increased with large pore diam. No effect of H_2 flow rate on HDS.
Wan and Crynes [151]	Trickle bed	Nalcomo 474, Sphericat 474 (bimodal)	345–138	589–700	900–7200 sec	HDN greater for Nalcomo 474. Pore diffusion considered most likely resistance. Effectiveness factors 0.3, 0.6. Nalcomo 474 maintained better activity.

ent boiling fractions where the rate constants for each can be calculated from the following equation:

$$k_i = \frac{1}{n-1}\left[\left(\frac{1}{N_{pi}}\right)^{n-1} - \left(\frac{1}{N_{fi}}\right)^{n-1}\right]\theta_v . \tag{6-29}$$

Rearranging the above equation and summing over all fractions yields the concentration of nitrogen in the total product:

$$N_p = \sum_i \frac{W_i}{[k_k r_i (n-1)\theta_v + (1/N_{fi})^{n-1}]^{1/(n-1)}} \tag{6-30}$$

For temperatures above 644 K, a second-order reaction resulted in the best statistical fit of the data. Over the temperature range investigated, the activation energy, 72.1 kJ/mol, was independent of temperature. With limited data, the rate constant was found to be dependent on pressure to the 0.76 power.

Numerous investigations have concentrated on products from more advanced coal-liquefaction processes such as COED, H-coal, EDS, SRC-I, and SRC-II. A summary of relevant studies is given in Table 6-23. For the hydrotreatment of COED oil derived from Utah A-seam coal using HDS-3A (NiMo) catalyst, first-order kinetics were found to describe heteroatom removal reasonably well [152]. For deoxygenation, the rate constants ranged from 2.78×10^{-5} to 2.78×10^{-3} sec^{-1}. As shown in Figure 6-9, a sharp break in the Arrhenius plot occurs around 663 K, giving rise to different activation energies. For the high temperature range ($T >$ 663 K) the activation energy was 20.4 kJ/mol, whereas for low temperatures ($T <$ 663 K), the activation energy was 160 kJ/mol, which is representative of thermal reactions. The same break in the Arrhenius plot also occurred for nitrogen removal, as shown in Figure 6-8, with the activation energy in the low and high ranges equal to 112.8 and 38.2kJ/mol, respectively. No explanation was offered for the observed breaks in the Arrhenius plots; however, a change in the controlling mechanism from reaction to diffusion control is likely. No correlation for sulfur removal could be obtained because conversion always exceeded 90%. For a constant space velocity and pressure, the hydrogen uptake for the C_4+ material was found to increase with increasing temperature to a maximum around 672 K and then to decrease at higher temperatures. This trend was attributed to the equilibrium reactions of polynuclear aromatics; however, the same occurrence has also been observed by other investigators. A model predicting this behavior will be discussed later in this section. Lower pressures resulted in reduced hydrogenolysis of the naphthenes, and the naphtha fraction of the product, although still somewhat high in heteroatoms, would be a good reformer feedstock if the heteroatom content could be reduced.

Unimodal and bimodal catalysts have been found to maintain their activity for sulfur removal from COED oil at temperatures below 700 K [3]. For tem-

Table 6-23

Summary of Studies on the Hydrogenation of Coal Liquids

Investigators	Reactor data	Pressure, MPa	Temp., K	Residence time	Feedstock
Bendoraitis et al. [10]	Batch shaker bomb	13.78	644–700	< 7200 sec	H-Coal (IL #6 Burning Star)
Callen et al. [23]					SYNTHOIL (WV Pittsburgh seam, Ireland mine)
Stein et al. [136]					SRC-I/recycle solvent SRC-I/clarified slurry oil
Berg & McCandless [11, 12, 16, 17]	Batch autoclave	13.78 (cold)	673–723	3600 sec	SYNTHOIL
		13.78 (cold)	703–713	3600 sec	SRC/recycle solvent
	Packed bed	5.5	688–723	1.05–4 LHSV	COED oil
		5.5–6.89	653–723	1–2 LHSV	SYNTHOIL
		6.89	673–748	1–2 LHSV	SRC/recycle solvent
deRosset et al. [29]	Trickle-bed reactor	13.78–20.68	648–723	0.1–1 WHSV	SYNTHOIL, H-coal, SRC-I, SRC-I filter feed
Gallagher et al. [43]	Packed bed	13.78–17.23	616–644	0.33–1 WHSV	H-coal, COED
Givens et al. [46, 47]	Fixed-bed reactor	10.34–13.78	616–727	0.3–1.6 VHSV	SRC-I (KY coal)
Jacobs et al. [55, 56]	Trickle-bed reactor	20.68	616–727	0.3–3 WHSV	COED oil
Lanning et al. [73]	Batch autoclave	16.88–18.27	616–672	$\leqslant 648 \times 10^4$ sec	Coal liquids derived by authors (Wyodak, Pittsburgh, IL #6, W. KY, Colstrip, lignite)
Low et al. [76, 77]	Batch autoclave	10.34–31.02	648–723	7200 sec	Solvent-refined lignite (ND lignite)

peratures above 700 K, excessive cracking resulted in coke deposition on the catalyst. As found in previous studies [151], the use of bimodal catalysts offered no advantage over unimodal catalysts.

As mentioned earlier [73], the coal from which the oil is derived has a major effect on the severity required to further upgrade the product. Using HDS-3A (NiMo) catalyst, it has been shown that hydrotreating COED oil derived from Pittsburgh coal required 2.5 times the severity required by oils derived from Utah and Illinois #6 coals [55, 56]. This is consistent with the fact that Pittsburgh coal is more difficult to liquefy initially. The oil derived from the Utah coal consumed

Table 6-23 (Continued)
Summary of Studies on the Hydrogenation of Coal Liquids

Investigators	Reactor data	Pressure, MPa	Temp., K	Residence time	Feedstock
Peters [98–101]	Trickle bed	6.38 (1st pass)	700 (1st pass)	3 LHSV	COED straight-run naphtha (IBP–850 K)
		6.2 (2nd pass)	671.89 (2nd pass)	1.2 LHSV	Hydrocracked mid-distillate naphtha
		5.5–10.34 (1st pass)	589–689	1.95–267 LHSV	H-coal straight run
		5.5 (2nd pass)	636 (2nd pass)	3.4 LHSV (2nd pass)	Hydrocracked mid-distillate naphtha
		6.89	633–689	2 LHSV	SYNTHOIL hydrocracked mid-distillate naphtha
		13.44–13.51	573–633	1.03–1.24 LHSV	SRC-II naphtha
Potts et al. [104]	LC-fining expanded bed	Not specified	689–705	0.40–0.94 WHSV	SRC-I (W. KY 9 & 14 coal)
Stein et al. [133]	Fixed-bed reactor	3.45–17.23	561–700	0.9–4 LHSV	H-coal distillate (IL #6 Burning Star)
				0.32–4.09 LHSV	SRC-I recycle solvent, light organic liquid (Wyodak, Monterey, (IL #6 AMax)
Stein et al. [134]	Fixed-bed reactor	3.45–17.23	627–694	0.25–3.0 LHSV	SRC-I wash solvent, SRC-I recycle solvent, H-coal distillate
Sullivan [140–143]	Fixed-bed reactor	5.5–17.23	644–672	0.5–2 LHSV	SRC-I/creosote oil, SRC-II whole oil, SRC-II naphtha, gas oil
Veluswamy et al. [150]	Fixed-bed reactor	17.23–20	573–673	0.5 LHSV	SYNTHOIL (WV)
Wiser [154–156]	Fixed-bed reactor	10.34–14.48	577–666	0.5 LHSV	SYNTHOIL
White et al. [152]		13.7–20.6	616–727	1.2 VHSV	Hydrotreated COED oil from Utah A seam coal

less hydrogen than the other feedstocks and yielded a superior product. The naphtha fractions were determined to be satisfactory reformer feedstocks, and it was anticipated that only mild severity would be required to obtain high-octane products. The mid-distillate would also be suitable as fuel oil or, after further pretreating, could be charged to a hydrocracker.

Johns et al. [58] reported on the hydrogenation of COED oil derived from high-volatile B bituminous coal (Colorado Bear Mine) using a 0.316-cm extruded HDS-3A NiMo catalyst. At low space velocities, heteroatom removal was complete, but with increasing space velocity, nitrogen removal decreased, whereas sulfur and

Table 6-23 (Continued)

Summary of Studies on the Hydrogenation of Coal Liquids

Investigators	Reactor data	Pressure, MPa	Temp., K	Residence time	Feedstock
Ahmed [3]	Trickle-bed reactor	10.34	644–727	2.99 LHSV	COED oil
Johns et al. [58]	Fixed-bed reactor	17.23	616–671	0.24–0.64 WHSV	COED oil from a high-volatile B bituminous coal
Eisen [37, 38], Nowack et al. [87]	Fixed-bed reactor	(1) 10.34 (2) 20.68	638–655 544–552	0.15–2 VHSV 0.75 VHSV	COED oils from W. KY & UT coals
Kang and Gendler [60]	Fixed-bed reactor	13.78–19.3	689–705	0.5–4 VHSV	H-coal oil from IL #6 coal
Kalfadelis [59], Shaw et al. [121]	Fixed-bed reactor	5.5–15.16	644	0.49–0.99 LHSV	H-coal oil and (< 568 K) fraction of SYNTHOIL
Tanner & Bell [148]	Batch autoclave	27.57	523–673		SRC-I from W. KY coal

oxygen removal were affected to a lesser extent. These investigators also concluded that hydrotreated COED oil is a satisfactory feed for commercial refineries.

Several evaluations have been conducted to assess the feasibility of producing JP-5 jet fuel from COED oil [37, 38, 87]. Using COED oil derived from Western Kentucky and Utah coals and a two-stage reactor with either HDS-3A or HDS-9A in the first stage and PHF-4 (Pt/Al_2O_3) in the second, it was concluded that it would be difficult to meet all specifications for JP-5 fuel. The performance of the fuels produced when used in test engines was, however, comparable to that of fuels derived from petroleum. No significant differences in the extent of upgrading were observed for the two feedstocks at similar conditions.

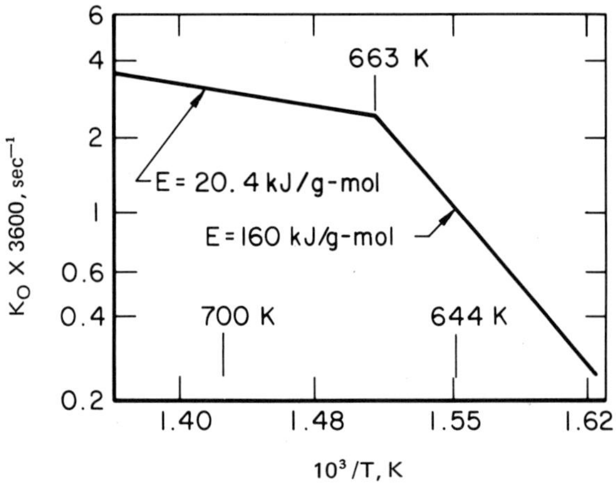

Figure 6-9(a). Arrhenius plot for the hydrodeoxygenation of a COED oil [152].

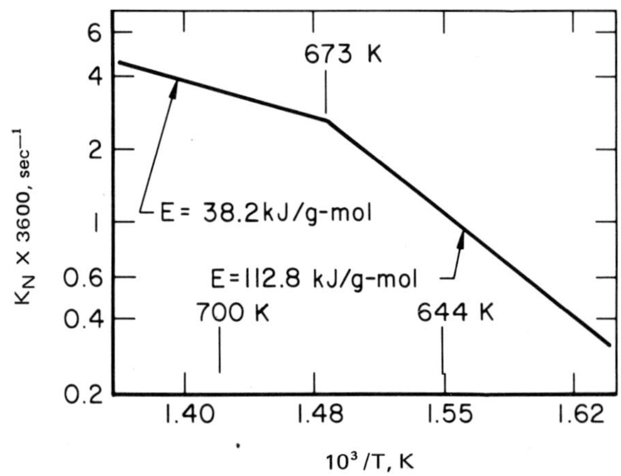

Figure 6-9(b). Arrhenius plot for the hydrodenitrogenation of a COED oil [152].

As shown in previous sections, presulfiding of catalysts has beneficial effects on the hydrotreating process. The effects of catalyst presulfiding have been evaluated using COED oil and three commercial catalysts [4]. Three presulfiding techniques were also assessed. The total amount of hydrogen sulfide supplied had a favorable effect on catalyst activity, and anything less than the stoichiometric quantity required to sulfide the metal components resulted in lower activities. Subjecting a sulfided catalyst to pure hydrogen resulted in a decrease in activity. The catalyst was also found to be sulfided during operation, but the maximum activity attained was dependent on the initial activity.

An extensive investigation of the upgrading of several different coal liquids in both batch and continuous-flow reactors [10, 22, 23, 133–135], using several commercially available catalysts, has recently been conducted. Raw SRC-I, H-coal, and SYNTHOIL were upgraded in a batch reactor using HDS-3A at 13.8 MPa, 700 K for one- and two-hour periods. For the H-coal feedstock, GEC analyses indicated that sulfur was uniformly removed from all fractions, whereas nitrogen and oxygen were preferentially removed from the higher GEC fractions. Data for the upgraded H-coal sample showed a 70% reduction in asphaltenes, and also that removal of asphaltenic sulfur is no more difficult than nonasphaltenic sulfur. For SYNTHOIL, heteroatom removal, and in particular sulfur removal, was more difficult than from H-coal feedstock. This was attributed to the fact that sulfur compounds in SYNTHOIL were more evenly distributed throughout the GEC fractions. Differences may also have been due to different starting coals, but this could not be substantiated from the data. Sulfur removal from the SRC-I feed was comparable to that for H-coal and SYNTHOIL; however, denitrogenation and deoxygenation were somewhat more difficult. Although differences were evident, it was concluded that sulfur and oxygen were equally easy to remove, whereas nitrogen was equally difficult to remove, from the three liquids. The concentrations of the polar and noneluted asphaltene fractions were higher in the SRC-I feed than in the H-coal feed, but hydroprocessing uniformly decreases these fractions for both feedstocks. The degree of sulfur removal was less from the lower GEC fractions, reflecting the fact that polyaromatic molecules were broken into smaller fractions without the removal of sulfur.

Because of their compatibility, blends of SRC-I with fluid-catalytic-cracker (FCC) clarified slurry oil were also processed in a batch reactor over HDS-3A catalyst. The resulting products had acceptable sulfur and oxygen concentrations which were below the maximum level for turbine fuels; however, practically no nitrogen was removed. From noncatalytic runs, it was found that clarified slurry oil acted as a hydrogen-transfer medium. In all cases, higher severities, i.e. higher temperatures, increased the sulfur and oxygen removal with a simultaneous increase in hydrogen consumption and gas production. GEC analyses of the various materials revealed that sulfur and oxygen are preferentially removed from the higher GEC fractions.

Initial studies in a trickle-flow reactor concentrated primarily on distillate fuels from the SRC-I process and H-coal products from both the fuel oil and syncrude modes of operation. SRC-I light organic liquid and wash solvent were processed using HDS-9A (NiMo) and HDS-1441A (CoMo). Heteroatom removal was found to be high in all cases except the low- severity runs. The SRC-I process solvent was hydrotreated using HDS-9A and HDS-1441A catalyst over a somewhat narrower range of process conditions. For the HDS-9A catalysts, all tests showed an increase in API gravity from 5.5 to 12–28° API, a reduction in aromatic carbon content from 71 to 24%, an increase in hydrogen content, and a substantial reduction in heteroatom concentration. The results obtained for HDS-1441A were quite similar to those of HDS-9A under the same operating conditions. The hydrogen consumption for this catalyst was slightly higher, and the higher gas yield which was obtained was attributed to excessive cracking due to the presence of a small amount of SiO_2 in the support.

The H-coal liquid from the fuel-oil mode of operation was processed using HDS-1441A, Ketjen 153S (NiMo), and HDS-20A (NiMo) catalysts. The HDS-20A catalyst has a lower surface area and a larger pore diameter than the HDS-1441A catalyst. Hydrogen consumption using HDS-20A catalyst was only 9 to 23 sm^3/m^3 higher than that for the HDS-1441A catalyst, with only a slight improvement in heteroatom removal. The Ketjen 153S catalyst, which had a similar pore diameter to the HDS-1441A catalyst, was used to evaluate the effects of nickel versus cobalt. The Ketjen 153S catalyst resulted in higher hydrogen consumption and consistently lower nitrogen levels in the products; however, the concentrations of oxygen and sulfur from the two were comparable. The hydrogen content of the liquid products was also slightly higher for the Ketjen 153S catalyst.

From the data obtained using distillate feedstocks, a kinetic model was developed for heteroatom removal and aromatic saturation [133]. Second-order kinetics was found to provide the best fit of the data obtained over the range of process conditions investigated and was attributed to the lumping of compounds with different reactivities. From this model, heteroatom removal is given by

$$1 - \frac{C_i}{C_{io}} = \frac{C_{io}k_i \exp\left[-E_i/RT\right] P^N}{\text{LHSV} + C_{io}k_i \exp\left[-E_i/RT\right] P^N} \tag{6-31}$$

Table 6-24 presents the kinetic constants obtained using the above model. As shown, the activation energy for sulfur removal is lower than that for nitrogen and oxygen removal for the three distillate feedstocks. The value of N was found to be in the range 0.7 to 1.5.

For H-coal liquid derived using the syncrude mode of operation, the data for nitrogen and oxygen removal using HDS-1441A and Ketjen 153S were found to fit a first-order kinetics model better than a second-order model. However, the second-order model provided a good fit for the data from the hydroprocessing of H-coal liquid (fuel-oil mode) by using HDS-1441A. In both cases, sulfur removal was found to follow second-order kinetics. The relevant results are presented in Table 6-24. These preliminary results indicate a difference in the type of nitrogen- and oxygen-containing compounds constituting the products from the H-coal syncrude and fuel-oil modes of operation.

In hydroprocessing operations, much of the hydrogen consumed is used to saturate aromatic components. The higher hydrogen content is reflected in a lower concentration of aromatics in the products. Figure 6-10 presents this relationship for the SRC recycle solvent and H-coal distillate. The following model was developed to correlate aromatic saturation data [133]:

Table 6-24

Kinetic Data for Heteroatom Removal

Feedstocks	Catalyst	Sulfur: k, sec^{-1} $wt\%^{-1}$ MPa^{-N}	Sulfur: E, kJ/mol	Sulfur: N	Nitrogen: k, sec^{-1} $wt\%^{-1}$ MPa^{-N}	Nitrogen: E, kJ/Mol	Nitrogen: N	Oxygen: k, sec^{-1} $wt\%^{-1}$ MPa^{-N}	Oxygen: E, kJ/mol	Oxygen: N
Second-order model										
H-coal distillate (fuel-oil mode)	HDS-1441A	1.55×10^{6}	106.13	0.84	3.41×10^{6}	132	1.3	2.77×10^{7}	148.9	1.49
SRC-I recycle solvent	HDS-1441A	4.13×10^{7}	129.85	0.96	6.05×10^{7}	148.81	1.04	2.50×10^{8}	159.4	0.71
	HDS-9A	9.09×10^{1}	66.01	1.36	1.18×10^{8}	149.2	0.96	4.70×10^{7}	153.1	0.85
H-coal distillate (syncrude mode)	HDS-1441A	2.77×10^{4}	85.22	0.9	2.08×10^{5}	118.45	1.33	9.86×10^{5}	127.41	1.15
	Ketjen 153S	—	—	—	2.37×10^{9}	169.4	1.5	5.89×10^{4}	107.05	0.82
33% SRC-I in recycle solvent	Harshaw 618X	4.83×10^{4}	95.02	1	1.08×10^{2}	73.69	0	1.62×10^{5}	119.6	0
90% SRC-I in recycle solvent	Harshaw 618X	2.78×10^{1}	60.77	1	3.78×10^{1}	73.69	0	6.78×10^{5}	132.0	0
33% SCT in recycle solvent	Harshaw 618X	3.75×10^{5}	110.26	1	3.28×10^{3}	93.52	0	1.91×10^{7}	147.6	0
First-order model										
					sec^{-1} MPa^{-N}			sec^{-1} MPa^{-N}		
H-coal distillate (syncrude mode)	HDS-1441A	—	—	—	9.36×10^{2}	93.7	1.08	2.07×10^{3}	92.8	0.93
	Ketjen 153S	—	—	—	2.44×10^{8}	155.78	0.92	1.64×10^{2}	77.45	0.82

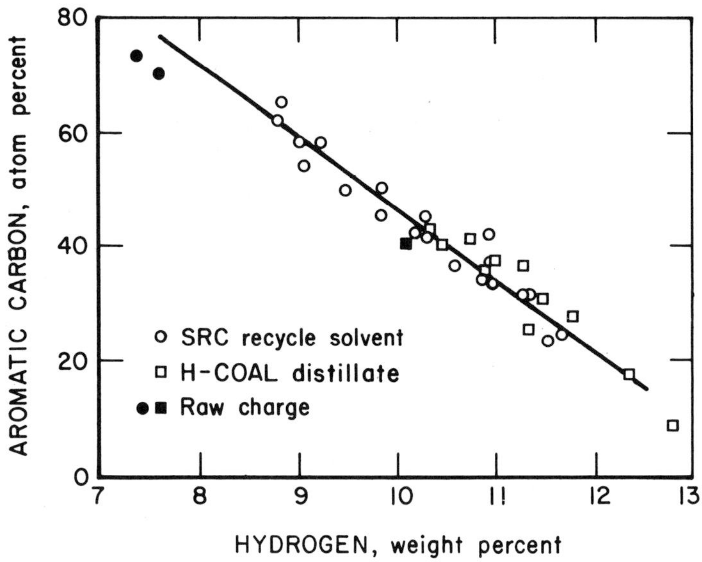

Figure 6-10. Aromatic carbon versus hydrogen content for coal-derived liquids [135].

$$f_A = \frac{k_r}{k_r + k_f P^N} + \left[f_o - \frac{k_r}{k_r + k_f P^N} \right] \exp\left[-\frac{k_f P^N + k_r}{\text{LHSV}} \right]. \tag{6-32}$$

The rate constants in the above model were assumed to follow the Arrhenius relationship for temperature dependence. The difference between the activation energies for the forward and reverse reactions is given by

$$E_r = E_f + \Delta H_{rx} , \tag{6-33}$$

where ΔH_{rx} is the heat of reaction and was assumed equal to 64.94 kJ/mol [57]. The values of the constants in Equation (6-34) were determined by a statistical fit of the data and are given in Table 6-25.

Using the above model, it is possible to predict the effect of pressure, temperature, and space velocity on the concentration of aromatic carbon in the products. Figure 6-11 illustrates the effect of pressure on the saturation of H-coal distillate. As shown, there exists an optimum temperature at a specified pressure for which the saturation of aromatics is a maximum. Lower pressures therefore result in higher equilibrium concentrations of aromatic carbon. Figures 6-12 and 6-13 illustrate the effects of the space velocity on aromatic saturation of SRC recycle solvent. These figures show that for LHSVs below approximately 1, there is a minimun in the curve which is indicative of maximum saturation. For space veloc-

Table 6-25

Parameters for Aromatic Saturation Model of Stein et al. [134]

Parameters	SRC-I recycle solvent/ H-coal distillate	SRC-I wash solvent
N	0.98	0.87
k_f, sec^{-1} MPa^{-N}	8.42×10^{-6}	9.43×10^{-5}
k_r, sec^{-1}	2.57×10^{-5}	1.01×10^{-4}
E_f, kJ/mol	114.7	93.6
E_r, kJ/mol	179.65	158.53
σ, %	2.9	3.12

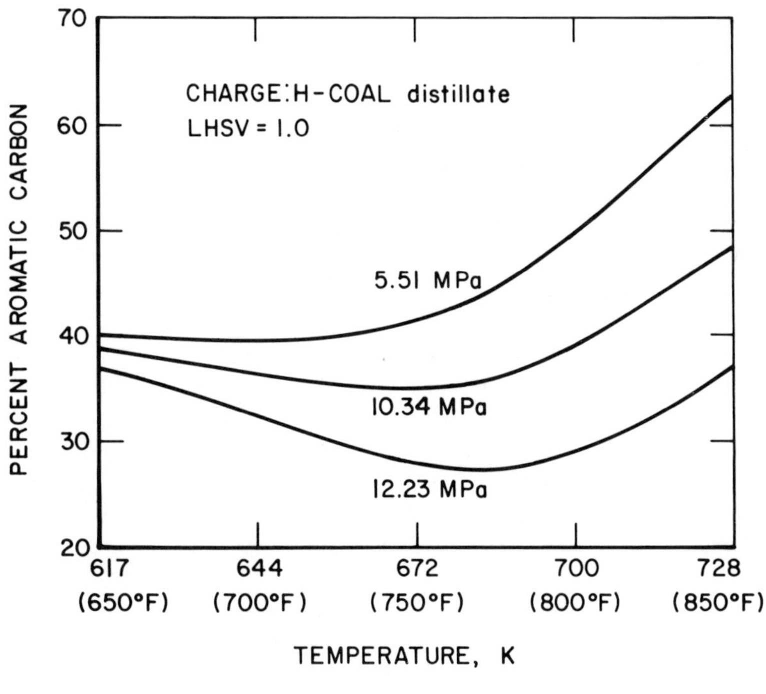

Figure 6-11. Predicted effect of pressure on aromatic saturation for H-coal distillate [133].

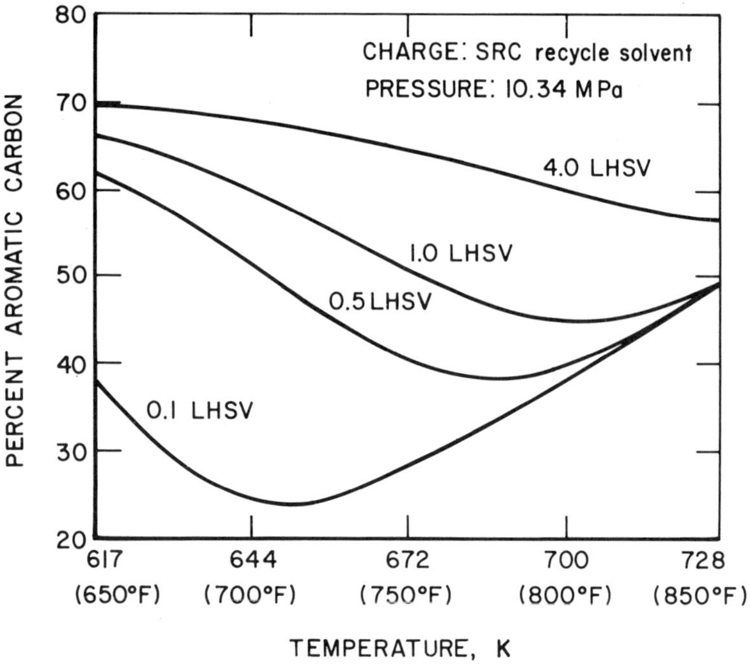

Figure 6-12. Predicted effect of LHSV on aromatic saturation for SRC recycle solvent [133].

ities below about 0.01, equilibrium would be attained for temperatures greater than 630 K; however, data at these low space velocities are needed to confirm these findings. This model may explain the observations of earlier investigations [152] which were discussed previously.

The hydrotreating of several blends of residual SRC-I (+810 K) with various oils has also been investigated. Because of their compatibility, SRC-I/FCC clarified slurry-oil blends were processed over HDS-1441A catalyst. Sulfur and nitrogen removal ranged from 68 to 93% and from 27 to 75%, respectively, with hydrogen consumptions of 142 to 410 sm^3/m^3. At a constant space velocity, higher temperatures resulted in less hydrogen consumption and a lower hydrogen concentration in the liquid products. This result is consistent with the aromatic-saturation model.

Various blends of SRC-I derived from various coals (i.e. Illinois No. 6 Burning Star, Monterey, and Wyodak) with recycle solvent were processed using HDS-1441A (CoMo) and Harshaw 618X (NiMo) catalysts. The SRC-I derived from the Monterey coal was found to be more difficult to upgrade and would require space velocities on the order of one-half of that for SRC-I derived from Burning Star coal to produce comparable products. The difference observed between these two

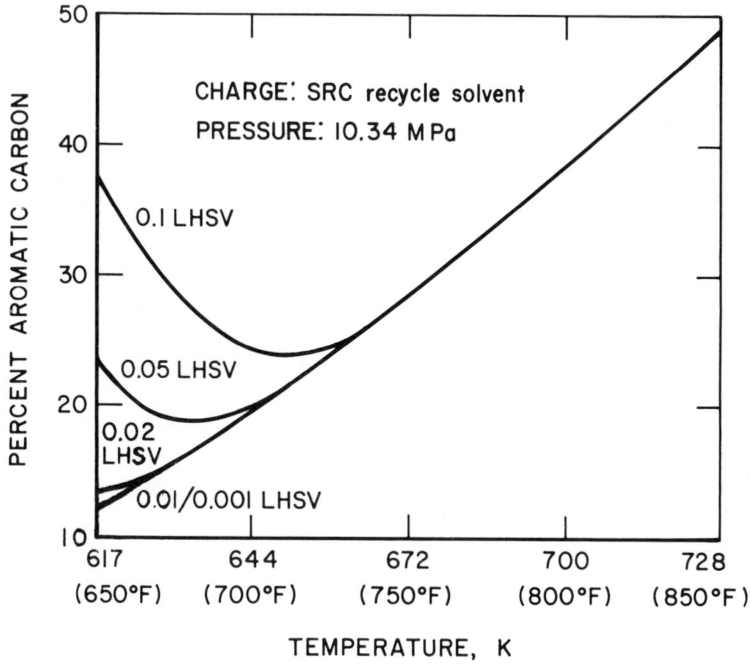

Figure 6-13. Predicted equilibrium aromatic saturation for SRC recycle solvent [133].

Illinois coals is probably due to the fact that the Monterey coal is higher in both inorganic and organic sulfur than the Burning Star coal. The Wyodak coal liquid was also found to be more difficult to upgrade than the Illinois No. 6 Burning Star product, which was attributed to the fact that the former feedstock possessed higher molecular weight components. Data for the conversion of asphaltene in SRC-I suggest the following mechanism:

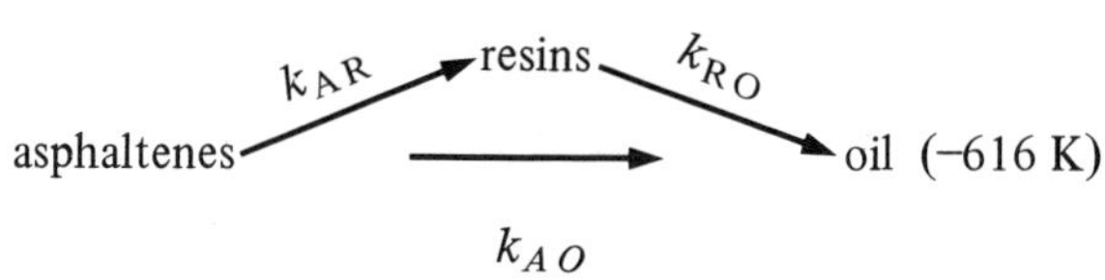

Using first-order kinetics with an Arrhenius temperature dependence for the rate constants and an aging parameter, the following expression was developed for the three rate constants:

$$k_{ij} = k_{ij}^{o} \exp\,[-at]\,\exp\,[-E_{ij}/RT]. \tag{6-34}$$

Interpretation of the data in terms of a time-dependent rate constant yielded $k_{AO} \sim 0$. Integrating the first-order rate expressions for the reactions in the above mechanism yields

$$C_{\mathrm{A}} = C_{\mathrm{AO}} \exp\left[-k_{\mathrm{AR}}/\mathrm{LHSV}\right], \tag{6-35}$$

$$C_{\mathrm{R}} = \frac{k_{\mathrm{AR}} C_{\mathrm{AO}}}{k_{\mathrm{RO}} - k_{\mathrm{AR}}} \exp\left[\frac{-k_{\mathrm{AR}}}{\mathrm{LHSV}}\right] + C_{\mathrm{RO}} - \frac{k_{\mathrm{AR}} C_{\mathrm{AO}}}{k_{\mathrm{RO}} - k_{\mathrm{AR}}} \exp\left[\frac{-k_{\mathrm{RO}}}{\mathrm{LHSV}}\right]. \tag{6-36}$$

The constants obtained by fitting the data to Equations (6-35) and (6-36) are given in Table 6-26. This model predicts that after 20 days, the catalyst possesses almost no activity for the conversion of asphaltenes.

Various blends containing 33–90% SRC-I in recycle solvent have also been processed using Harshaw 618X and HDS-1441A catalysts. The former catalyst had a large average pore diameter, whereas the average pore diameter for the latter was approximately 60% smaller. For blends having low concentrations of SRC-I, the larger-pore catalyst gave products with a hydrogen content similar to the smaller-pore catalyst but with a lower heteroatom content. At high concentrations of SRC-I, the HDS-1441A catalyst yielded a two-phase product. From the data, it was shown that the small-pore catalyst upgraded the distillate fraction, making it incompatible with the residual material. This confirms that pore-size distribution is a critical parameter for hydroprocessing residual feedstocks. With the exception of the data obtained from the runs yielding two-phase products, all the data were modeled using second-order kinetics; the results are presented in Table 6-24. The pressure dependence was assumed to be first order in this case. The results indicate that blends of SRC-I and recycle solvent could be significantly upgraded but hydrogen consumptions as high as 765 $\mathrm{sm^3/m^3}$ can be expected.

SRC products from the short-contact-time liquefaction process (SCT) were investigated using Harshaw 618 catalyst. This feedstock is higher in sulfur, asphaltenes, and resins than the regular residual SRC-I. Analytical results indicate that GEC fractions from hydroprocessed SCT feedstocks have higher-molecular-weight components with higher hydrogen contents than corresponding GEC fractions of regular SRC-I. The analyses also indicated that the nitrogen distribution was not

Table 6-26

Values for the Kinetic Constants in Equations (6-35) and (6-36)

	k_{ij}^{o}, $\mathrm{hr^{-1}}$	E_{ij}, kJ/mol	a, $\mathrm{day^{-1}}$
A → R	7.55×10^6	84.95	0.1633
R → O	9.69×10^5	73.25	

dependent upon the SRC process, whereas the oxygen content for SCT is higher in the asphaltene fractions. The products after hydrotreating have the same viscosity, even though they differ significantly at lower hydrogen contents. The data were modeled assuming second-order kinetics, and the results are presented in Table 6-24. The values of the rate constants at 670 K for the regular SRC-I are consistently higher than those of the SCT.

Veluswamy et al. [150] investigated the hydrogenation of SYNTHOIL, derived from a West Virginia coal, using a fixed-bed reactor. Sphericat 550 (NiMo) and HT-100E (NiMo) catalysts were employed because of their high hydrogenation and low cracking activity. An examination of the data at 20 MPa and 0.5 LHSV indicates that a decrease in aromaticity and an increase in hydrogen content of the products occur initially with increasing temperature. These trends were reversed with further increases in temperature, the maximum occurring at approximately 634 K. This occurrence was attributed to the shift in equilibrium between the hydrogenation and dehydrogenation reactions with increasing temperature and the extent of cracking. The former observation is supported by the model presented in Equation (6-36). Heteroatom removal increased with increasing temperature, with nitrogen and sulfur removal ranging from 18 to 70% and from 45 to 86%, respectively. The incomplete removal of the heteroatoms was attributed to stereochemical characteristics of the more complex heterocyclics and partially hydrogenated intermediates. Aromatic saturation and nitrogen removal increased with increasing pressure over the range 17.23–20.0 MPa; however, there was only a marginal increase in sulfur removal. The rate and extent of nitrogen and sulfur removal were dependent on the steric characteristics of the condensed aromatic components.

Several additional commercial catalysts have been evaluated for the denitrogenation of H-coal gas oil and fuel oil and blends of SRC-I and recycle solvent. The H-coal products, derived from Illinois No. 6 coal (Burning Star Mine), were hydrotreated using HDS-2A and HT-400E cobalt-molybdenum catalysts [60]. In processing the gas oil with HT-400E catalyst, no deactivation was observed after an age of 0.00024 (m^3 oil)/(g catalyst), whereas for the hydrotreatment of the fuel oil using HDS-2A catalyst, severe deactivation was observed and was attributed to carbon deposition caused by the heavier feedstock. Several developmental NiMo and CoMo catalysts prepared by commercial methods were also evaluated using a 60:40 recycle-solvent:SRC-I blend. The NiMo catalyst, as would be expected, had the highest denitrogenation activity. A NiMo catalyst with a unimodal pore distribution, containing a 30% higher metal loading than a bimodal NiMo catalyst, had the same activity as the bimodal catalyst. This result indicates that the macropores improve denitrogenation activity per unit weight of catalyst and may also reduce diffusional resistances. Additional investigation indicated that further increasing the macropore concentration at the expense of micropores did not increase the denitrogenation activity.

As part of a major research effort directed at upgrading coal liquids, deRosset et al. [29, 31] investigated the hydrotreatment of SYNTHOIL, H-coal, SRC-I filter feed, SRC-I filtrate, and SRC-I residual. A commercial UOP catalyst was employed; however, no data on the catalyst itself were presented. For the SYNTHOIL liquid, heptane-insolubles and nitrogen removal were found to increase significantly with increasing pressure. The heptane-insoluble materials were quite refractory, and their reaction did not follow first-order kinetics. In processing the SRC-I filter feed, catalyst deactivation was small, with heptane-insoluble conversions of up to 89% being obtained. The data were found to fit first-order kinetics and were relatively insensitive to pressure. For the SRC-I filtrate, the best results were obtained at the highest severity with 93% conversion to heptane insolubles. Some decline in catalyst activity was observed. Heptane-insoluble and nitrogen conversions for the SRC-I residual feed were 92% and 80%; however, catalyst deactivation was significant.

An evaluation of the upgrading of H-coal and EDS naphthas and H-coal gas oil has been conducted in a cocurrent downflow fixed-bed reactor [30, 144, 145, 147]. Commercial catalysts containing Group VI and VIII metals on a high-surface-area support were used; however, details were not disclosed. Both the H-coal and EDS naphthas were processed to yield products which are acceptable as reformer feeds. The H-coal gas oil was hydrotreated to remove nitrogen and improve its cracking characteristics in a fluid catalytic cracker. Three commercial UOP catalysts (UOP-DCB, UOP-DCA, and UOP-DSA) were evaluated on the basis of aromatic saturation rather than nitrogen removal. The UOP-DCB catalyst was best for increasing the hydrogen content of the liquids. After 482 hours of processing, only a small decline in activity was observed. It was also noted that the catalyst stability was improved at higher pressures.

Sullivan et al. [140, 141, 143] recently investigated the hydroprocessing of SRC-I and SRC-II products. Because of the nature of the SRC-I product, a 50:50 blend with creosote oil was processed. Catalyst deactivation was rapid and hydrogen consumption was high. For the SRC-II feedstock, the whole oil, the naphtha, and the −477 K gas oil were processed using ICR-106 (Ni, W, Si, Al) and ICR-113 (Ni, Mo, Si, A1) catalysts. The products from processing the whole oil over ICR-106 contained 95% naphthenes and only 1–2% aromatics, and were excellent for reformer feeds. Catalyst deactivation was also minimal. Oxygen appeared to be the most difficult heteroatom to remove, whereas the final nitrogen and sulfur concentrations were less than 10 ppm. The ICR-113 catalyst resulted in less aromatic saturation and nitrogen removal than the ICR-106 catalyst, and the products required further hydrotreatment before being acceptable as reformer feedstocks. The SRC-II naphtha processed over the ICR-106 catalyst yielded products that were comparable to that of the whole oil processed at the same conditions but with a higher space velocity. Because of the nature of the two fractions, it was concluded that the naphtha and gas-oil fractions could be processed simultaneously without any decline in catalyst activity. The hydrotreated naphtha was a suitable feedstock for a catalytic reformer or a fluid catalytic cracker. Although the ICR-106 catalyst was more active for denitrogenation and hydrogenation and deactivated at a slower rate than the ICR-113 catalyst, it is more expensive.

In another investigation, Givens et al. [46, 47] found that SRC-I in creosote oil using 0402T catalyst produced up to 46% distillate material with a hydrogen consumption of 178 sm^3/m^3. The data also indicate, as shown in Figure 6-14, that

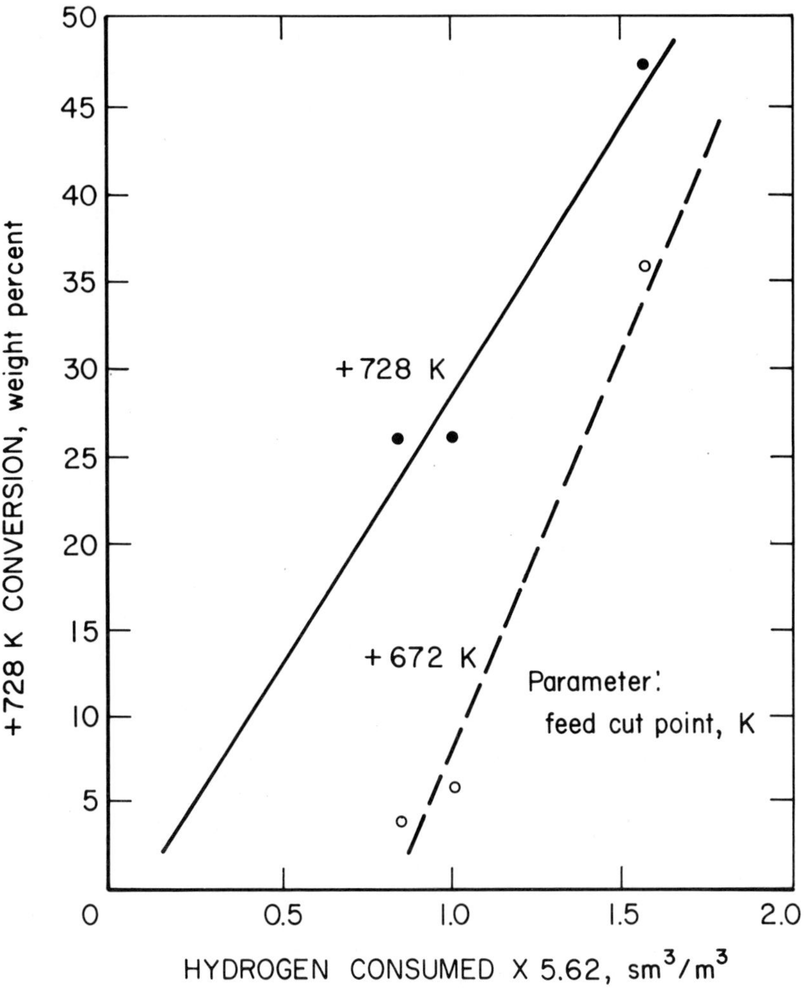

Figure 6-14. Hydrogen consumption for the conversion of +672 K and +728 K solvent-refined coal feedstocks [47].

an addition of 0.8 wt% H_2 is required before any conversion of the +672 K residual occurs. This figure also shows that the lighter (+672 K) residual is converted faster than the +728 K fraction. In a comparison of NiMo and NiW catalyst, it was found that HDS-3A catalyst had a higher activity and a selectivity for converting +728 K material to −728 K than Ketjenfine HC-5 catalyst; however, no significant difference in the conversion versus hydrogen consumption was observed. The metals distribution in the catalyst bed was also more uniform in an upflow than in a downflow mode of operation.

Results of processing various blends containing SRC-I in SRC-I process solvent, Koppers creosote oil, and Neville aromatic oil (petroleum-derived) in the LC-fining process (expanded bed) using HDS-1442B catalyst have been reported by Potts et al. [104]. Sulfur removal was excellent at all operating conditions; however, oxygen and nitrogen removal decreased with increasing SRC-I level in the blend and increasing space velocities. In a 30-day aging test, sulfur and nitrogen removal remained constant, but oxygen removal showed a declining rate. An evaluation of the effectiveness of certain proprietary NiMo, CoMo, and NiW catalysts (manufacturers and properties were not disclosed) showed that the NiMo catalyst resulted in greater conversion of +727 K material, better selectivity for production of 533–616 K products, and better hydrogenation and heteroatom removal.

Wiser [154–156] hydrotreated SYNTHOIL in both batch and fixed-bed reactors using NiMo, NiW, or CoMo catalysts. The batch-reactor studies were initially conducted for the purpose of catalyst screening. From the data, NiMo/Al_2O_3 catalyst was found to be most effective in reducing the nitrogen level, with a conversion of approximately 73%. In the fixed-bed reactor, nitrogen conversion increased further to about 76%. The NiMo catalyst gave higher conversions than the NiW catalyst in a fixed-bed reactor, which is in agreement with other investigators. Increasing the system pressure resulted in increased hydrogen uptake and decreased aromatic content. For temperatures up to 644 K, the percentage decrease in aromatics increased, but further increases in temperature resulted in the reverse trend. These investigators attempted to explain the observed results by cracking reactions; however, the aromatic saturation model discussed earlier can also be used to explain these observations. The data obtained were modeled by first-order kinetics according to the equation

$$\ln\left[\ln\left(\frac{1}{1-x}\right)\right] = \ln\left(\frac{k}{\mathrm{LHSV}}\right) - k_d t. \tag{6-37}$$

The expression for the apparent rate constant is given by

$$k = 2.72\times10^{-6} \exp[-E/RT] \ \mathrm{sec}^{-1}, \tag{6-38}$$

where E = 55.3 kJ/mol. The value of k_d was in the range $(1.33\text{–}2.5)\times10^{-5}$ sec^{-1}.

The catalytic hydrogenation of solvent-refined lignite (SRL) produced by the Pittsburgh and Midway Coal Company was investigated using seven commercially available catalysts [76, 77]. Initially, the SRL was reacted in THF, naphthalene, or tetralin; however, because of higher conversions in the absence of catalyst, tetralin was deemed the better solvent. When no solvent was used, hydrogenation was impeded, whereas with excess solvent, the solvent competed slightly with the SRL for hydrogen. A 1 : 1 ratio was found to be optimum. Of the catalysts tested with and without presulfiding, NiMo/Al_2O_3 yielded the highest conversion of SRL to distillate, whereas NiW/SiO_2-Al_2O_3 yielded the lowest. Tin chloride powder was also evaluated by either adding it directly to the reaction mixture or impregnating it on an alumina support. The conversion for the supported catalyst was lower, but that for the powder was comparable to the NiMo/Al_2O_3 catalyst. No effect of pressure on the conversion was observed for the various catalysts; however, conversion increased linearly with temperature.

SYNTHOIL and H-coal liquids, like the COED oils, have been evaluated for their potential use in the production of aviation fuel [59, 121]. From the analyses, it is apparent that H-coal liquids could be used in the production of jet fuel; however, except through severe processing, SYNTHOIL did not yield significant quantities of fuel meeting all aviation-fuel specifications.

Only a few investigations have been devoted to synthesizing new catalysts for coal-liquid upgrading. In one investigation [148] numerous metallic chlorides and several commercial catalysts were evaluated for their effectiveness in hydrogenating SRC-I. The metal oxide catalysts were all inactive at promoting the hydrogenation and solubilization of SRC-I in benzene at 573 K and 13.78 MPa. Under the same conditions, metallic chloride catalysts were found to yield benzene-soluble products with higher H/C and lower N/C ratios. Of these chloride catalysts, only $ZnCl_2$ and $SnCl_2$ increased the yield of soluble products. Increasing catalyst loading increased the H/C and aliphatic/aromatic ratios and decreased the N/C ratio of the products. These catalysts were also effective in hydrogenating the cyclohexane insoluble residue of the SRC-I feed but were ineffective for heteroatom removal. The H/C and aliphatic/aromatic ratios reached a maximum at 623 K. Again, this can be explained by the aromatic-saturation model. Hydrogenation and solubilization increased sharply with pressure; however, in the absence of a catalyst, no dependence was observed.

In another study directed at catalyst development, Berg and McCandless [11–17] evaluated numerous catalysts for hydrotreating SRC-I, SYNTHOIL, and COED oils. Many commercially available catalysts were screened to evaluate the effects of various metals, metal loadings, and catalyst-support properties on catalyst activity. Several of these are listed in Table 6-23. Although their investigations were quite extensive and cannot be elaborated on here, some of their major findings are as follows: (1) NiW catalysts were inferior to CoMo and NiMo catalysts; (2) NiMo catalysts had higher denitrogenation and lower desulfurization activities than CoMo catalysts; (3) denitrogenation activity improved with larger pore diameters and less silica in the support; and (4) sulfur removal was diffusion limited.

6.4.4 *Summary*

As can be seen, a considerable amount of work on hydrotreating coal liquids has been conducted in an attempt to reduce the heteroatom content and increase the H/C ratio of the liquids. Numerous feedstocks have been investigated, including those from the more advanced coal liquefaction processes. Catalytic studies, although numerous, have been primarily concerned with evaluating commercially available catalysts for their effectiveness in heteroatom removal and hydrogenation activity. All of these catalysts were initially developed for use in processing petroleum, but because of the vast differences in the properties of coal- and petroleum-derived liquids, these catalysts may not be optimum for upgrading coal-derived liquids. Of the catalysts evaluated, those containing nickel and molybdenum on alumina supports with large pore diameters appear to yield acceptable products. More research is needed to understand the various reaction mechanisms involved in hydrotreating coal liquids and to evaluate catalyst properties so that more efficient catalysts can be developed. Catalyst development is mostly required in the areas which will affect processing conditions and process economics the most.

Although most of the above investigations have employed coal liquids from a variety of liquefaction processes and coals, only a few have evaluated the effect of the type of coal on the extent of upgrading required to obtain a specific product. Because of this, no firm conclusions can be drawn; however, the results indicate that the ease of upgrading decreases as the rank of the starting coal is increased. That is, liquids derived from coals such as Wyodak (subbituminous coal) are easier to upgrade than those derived from such coals as Illinois #6 and Pittsburgh seam (bituminous coals). Further investigations in this area may also provide useful information for the design of liquefaction reactors.

Hydrotreating reactors for upgrading coal liquids should be quite similar to those currently used in petroleum refining. For the more fluid products, packed beds operating in either trickle flow or concurrent upflow can be employed. For heavier products such as SRC-I residual, ebullating-bed reactors may be advantageous because the increased turbulence may help to reduce local hot spots and plugging in the reactors, and to extend catalyst life. Having two or more reactors in series containing different catalysts is a possibility for these feedstocks. The severity of operation will depend on the nature of the feedstock and the desired product specification. Because of high concentrations of polynuclear aromatics, coal liquids will undoubtedly have to undergo more severe processing than their petroleum counterparts using currently available catalysts. The kinetic models presented in this section should be adequate for the preliminary design of these reactors; however, some refinement is required before detailed designs can be developed.

6.5 Catalytic Hydrocracking

With the production of low-cost hydrogen from catalytic reformers and the shift in demand to gasoline, catalytic hydrocracking has become an important refining process. Hydrocrackers are primarily used to process refractory feeds such as cycle oils and coker distillates, which are resistant to catalytic cracking but are relatively easy to process under hydrocracker conditions. Recently, it has also been possible to process residual fuel oils and reduced crudes in the catalytic hydrocracker. However, a somewhat different technology is required because of the presence of asphaltenes, ash, and metals. These refinements, therefore, make catalytic hydrocracking a viable technique for upgrading coal liquids.

6.5.1 *Chemistry*

In catalytic hydrocracking, the reaction mechanism consists of catalytic cracking superimposed upon catalytic hydrogenation, the two reactions complementing one another. The catalytic cracking reactions which are endothermic result in the production of olefins. These olefins are subsequently hydrogenated, and the heat produced from the reaction is used to provide that required for the cracking process. The resulting products from a catalytic hydrocracker are essentially saturated components with high concentrations of isoparaffins and naphthenes. Typical hydrocracking reactions are shown below:

$$n\text{-}C_7H_{16} \xrightarrow{\text{heat}} n\text{-}C_3H_8 + CH_2{=}CH{-}CH_2{-}CH_3 \xrightarrow{H_2} n\text{-}C_4H_{10} + \text{heat}, \quad (6\text{-}39)$$

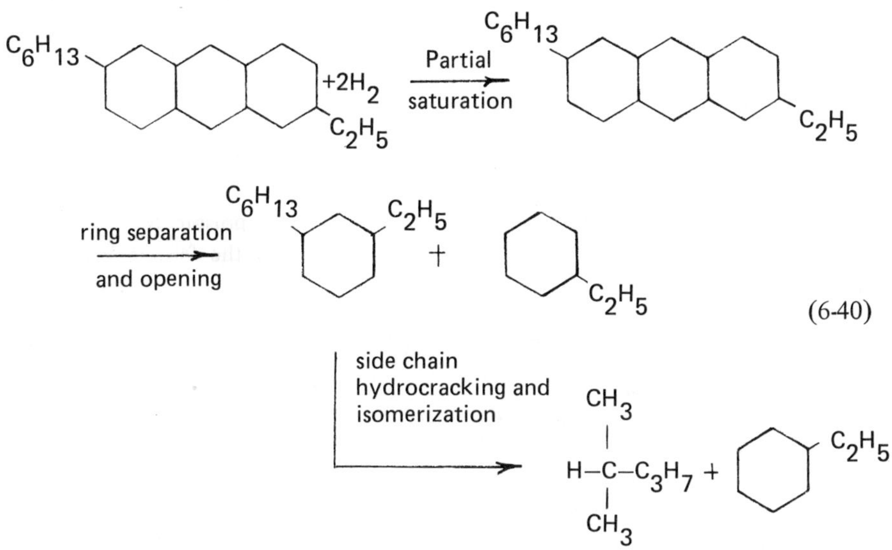

(6-40)

As shown above, isomerization is another type of reaction which occurs during the hydrocracking process. The olefinic molecule produced as a result of dealkylation isomerizes and rapidly hydrogenates, thus preventing the occurrence of the reverse reaction and thereby maintaining the desired high-octane isoparaffin.

Typical operating conditions for a catalytic hydrocracker range from 478 to 700 K and 0.69 to 13.8 MPa. Operating severity and hydrogen consumption are dependent upon the nature of the feedstock, the required product distribution, and the hydrocracking catalyst employed.

6.5.2 *Catalysts*

Hydrocracking catalysts currently employed consist of metallics such as platinum, palladium, nickel, tungsten oxide, uniformly distributed over a crystalline silica-alumina support. The metals composing the catalyst are responsible for its hydrogenation activity, while the silica-alumina support provides the cracking activity. As the activity of the catalyst decreases with age, the reactor operating temperature is usually raised in order to maintain a desired reaction rate. With this increase in temperature comes a change in selectivity and a higher gas yield. Numerous hydrocracking catalysts are commercially available, with an actual composition that is usually tailored to a specific process, feed material, and product distribution.

Hydrocracking catalysts are susceptible to poisoning by metallic salts and organic nitrogen, oxygen, and sulfur compounds in the feedstock. Hydrotreating prior to hydrocracking is often employed to remove these poisons. In addition to the above undesirable components, the water content of the feed must also be reduced to less than 25 ppm; otherwise the steam produced in the reactor can cause the crystalline structure of most hydrocracking catalysts to collapse and the metals to agglomerate. For certain catalysts, low concentrations of hydrogen sulfide have been shown to inhibit the saturation of aromatic components, resulting in a decrease in hydrogen consumption and an increase in the octane rating of the products.

6.5.3 *Catalytic Hydrocracking of Coal Liquids*

Several investigations on the catalytic hydrocracking of coal liquids have been reported. The majority of these studies either employed the whole product oil or the heavier mid-distillate and gas-oil fractions as feedstocks. Table 6-27 summarizes several of these investigations.

The catalytic hydrocracking of a low-temperature carbonization tar and a coal liquid produced by hydrogenation in an entrained-bed reactor was investigated by Qader and Hill [105-107]. Both feedstocks were derived from a high-volatile bituminous coal. These studies employed a dual functional sulfided catalyst containing 6% Ni and 19% W supported on silica-alumina (212 m^2/g), a conventional cracking catalyst composed of 89% silica and 16% alumina (200 m^2/g), and a refining catalyst containing 6% Ni and 19% W on alumina (200 m^2/g). For the low-temperature tar, the rate of gasoline formation was high up to 10.34 MPa and then decreased over the range 10.34–17.23 MPa. At pressures higher than 17.23 MPa, gasoline formation once again increased.

The aromatic content of the products increased up to 17.23 MPa and then rapidly decreased. This can be attributed to an initial suppression of coke formation followed by subsequent cracking of naphthenes at higher pressures. Figure 6-15 shows the naphtha yield as a function of temperature and pressure for both coal liquids investigated. As shown, at high pressures the naphtha yield increases linearly with temperature, whereas at lower pressures the rate of naphtha production diminishes as the temperature increases. These trends indicate that higher pressures are required to hydrogenate the aromatics and heterocyclics in the coal liquids. At the higher pressures, the coal liquids were found to behave similarly to a petroleum gas oil. Less gas products were formed during the hydrocracking of coal liquids than with the gas oil; however, the yield of +473 K oil was higher. As shown in Figure

Table 6-27
Summary of Hydrocracking Studies on Coal-Derived Liquids

Investigators	Reactor data	Pressure, MPa	Temp., K	Residence time	Feedstocks
Peters [98]	Fixed bed	10.34–17.23	727–811	0.3–0.7 LHSV	COED mid-distillate
		17.23	727	1.23 LHSV	COED gas oil
Peters [100]	Fixed bed	17.23	739–366	0.35–1.45 LHSV	H-coal mid-distillate
		17.23	722–809	0.69–1.68 LHSV	H-coal gas oil
Peters [101]	Fixed bed	10.34–17.23	766–783	0.62–1.47 LHSV	SYNTHOIL
Qader & Hill [105–107]	Fixed bed downflow, 0.75-in. i.d.	10.34–13.78	623–773	0.5–2 LHSV	Low-temperature carbonization tar oil
Gallagher et al. [43]	Packed bed, 2-in. i.d.	13.78	607–644	1–1.05 WHSV	COED oil (561K–700K) fraction)
Sullivan et al. [140–143]	Fixed bed	17.23	671–689	0.09–0.2 LHSV	SRC-I, SRC-II
deRosset et al. [29]	Fixed bed	13.78–20.68	648–723	0.25–1 hr^{-1}	SYNTHOIL, H-coal, SRC
Gagarin et al. [42]	Fixed bed	4.05	613–673	0.5–2 LHSV	Coal-derived distillate (453–593 K)

6-16, isomerization reactions increased with temperature but were not as prevalent as with the gas oil, and therefore the iso-to-normal ratio was lower. Because of their higher aromatic content, the coal liquids had higher proportions of light to heavy naphtha than the petroleum gas oil, as shown in Figure 6-17. For these coal liquids, the use of the NiW/Al_2O_3 catalyst resulted in higher yields and better-quality

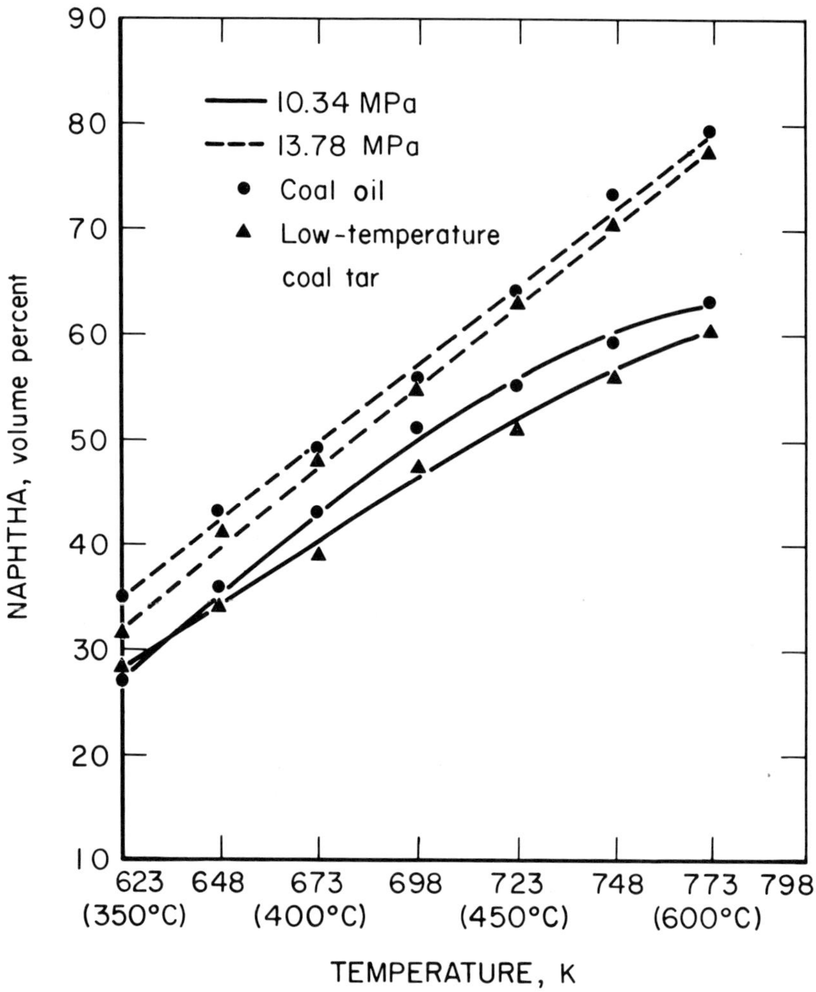

Figure 6-15. Influence of temperature and pressure on naphtha yield during the hydrocracking of coal liquids [107].

diesel oil than the other catalysts. The dual-functional NiW/SiO_2-Al_2O_3 catalyst had a higher cracking activity and was more effective in producing gasoline, while the NiW/Al_2O_3 catalyst had the highest desulfurization and hydrogenation activity. Activity for isomerization decreased in the following order: NiW/SiO_2-Al_2O_3, cracking catalyst, NiW/Al_2O_3.

For the reactions at 13.8 MPa, the data were well correlated by the first-order kinetics. From an Arrhenius plot for the first-order rate constant such as the one shown in Figure 6-18, the rate expression for coal oil is given by

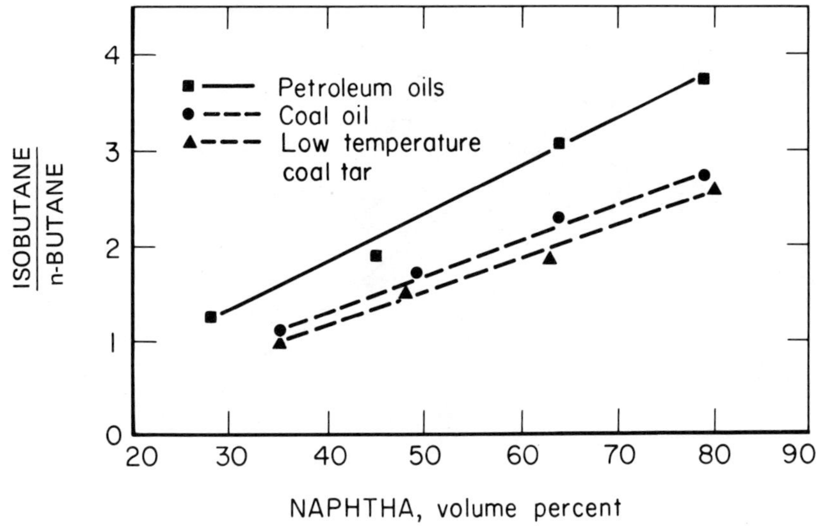

Figure 6-16. Relationship between the formation of butanes and naphthas [107].

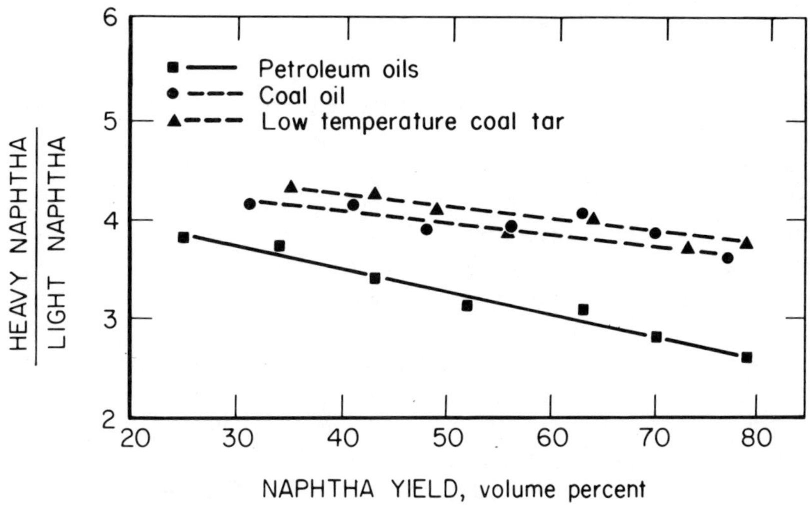

Figure 6-17. Ratio of heavy to light naphtha at different levels of conversion [107].

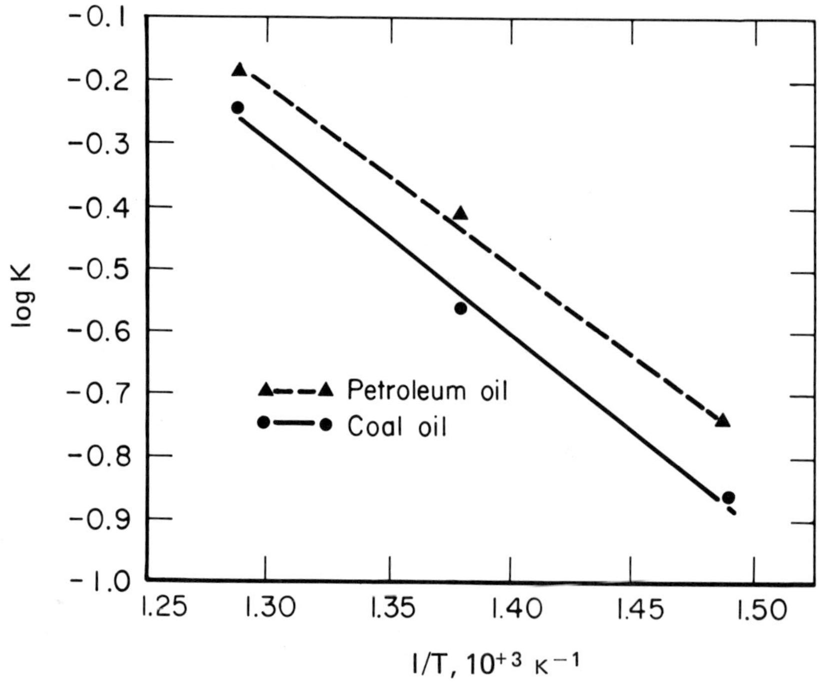

Figure 6-18. Arrhenius plot for hydrocracking of coal oils and petroleum gas oil [107].

$$k = 2.33 \exp [- 64.88/RT] \text{ sec}^{-1}. \tag{6-41}$$

The results shown in Figure 6-18 also indicate that coal liquids and petroleum gas oil exhibit similar kinetic behavior. Cracking of C–C, C–O, C–S, and C–N bonds were suggested to be the controlling reactions. Preliminary tests showed that hydrotreating prior to hydrocracking was beneficial to catalyst life because it reduced coke formation and the level of nitrogen (a catalyst poison) in the hydrocracker feed.

The hydrocracking of COED, SYNTHOIL, SRC-I, and H-coal whole oils and distillate fractions has been the subject of several studies. Commercial hydrocracking catalysts consisting of Group VI and VIII metals supported on silica-alumina were employed for the conversion of SYNTHOIL, SRC-I, and H-coal liquids [30, 31]. For SYNTHOIL whole oil, the conversion to −616 K distillate was low (15–36%) even at the most severe conditions, whereas the conversion of the mid-distillate was higher, ranging from 50 to 67%. Nitrogen removal was also higher from the mid-distillate (97%) than from the whole oil (80%), indicating the difficulty in converting the nitrogen in the heptane-insoluble fraction of the whole oil.

The conversion of hydrotreated SRC-I was also low, ranging from 2 to 11% with hydrogen consumption of 45 to 93 sm^3/m^3. No catalyst deactivation was observed, but the data showed that the high-molecular-weight asphaltenes interfered with conversion. For the production of fuel oil from H-coal distillate, hydrogen consumption of approximately 60 sm^3/m^3 was required; however, about 534 sm^3/m^3 was required for complete conversion to gasoline with a RON of 92.

Single- and two-stage hydrocracking of H-coal gas oil, distillate, and atmospheric-still bottoms were investigated using commercial catalysts (UOP-DCA, UOP-DCB, UOP-HCA) in a downflow fixed-bed reactor [146]. For the first-stage hydrocracking of gas oil over the UOP-DCA catalyst, nitrogen conversion, as would be expected, increased for lower LHSVs with hydrogen consumptions of 1.12 to 1.92 wt% of feed. Catalyst deactivation in this stage was equated to a temperature increase of 5 K over 827 hours of processing. In the second-stage reactor, UOP-HCA catalyst was employed. Both pressure and space velocity had drastic effects on the temperature required for 100% conversion. The C_5^+ gasoline yield varied inversely with the aromatic content of the feed. A high-nitrogen-containing H-coal distillate was also processed in a series-flow (two reactors) second-stage hydrocracking process [146]. UOP-DCB and UOP-HCA were used in the first and second stages. The results showed that feeds containing high levels of nitrogen could be processed at low severities in both stages. The two-reactor, series-flow second stage, when preceded by a single-reactor first stage, resulted in improved catalyst stability and temperature control. Single-stage hydrocracking of H-coal atmospheric-still bottoms to produce fuel oil was also examined using UOP-DCA catalyst; however, complete conversion was obtained only at the most severe conditions.

Mid-distillate and gas-oil fractions of H-coal, SYNTHOIL, and COED liquids were processed using HT-400E catalyst [98, 100, 101]. This catalyst was employed because of its neutral support and its high activity for the removal of and resistance to, deactivation by, nitrogen, oxygen, and sulfur. The neutral support was desired because the high levels of heteroatoms in the feed would cause the Lewis acid sites found in silica-containing catalysts to deactivate, and also because Lewis acid sites yield C_4–C_9 paraffin products with low iso-to-normal ratios. The conversion of mid-distillate to −477 K product was as high as 90%, while the conversion of gas oil to −617 K product reached 80%. Heteroatom removal was high in all cases.

For the three mid-distillate feedstocks and the H-coal gas oil, the data were modeled, assuming first-order kinetics, by the following equation:

$$\ln(1-x) = -k \exp[-E/RT] \; P_{H_2}^{a} \, \mathrm{LHSV}^{b} \, . \qquad (6\text{-}42)$$

The values of a and b for the COED, H-coal, and SYNTHOIL mid-distillate and the H-coal gas oil are (1.25, −0.5), (1, −1), (1, −0.5), and (1, −1) respectively. The activation energies in all cases were 159 kJ/mol. Because of limited data, no model for the COED gas oil could be established.

In other studies, Sullivan [140] hydrocracked a 450–727 K fraction of hydrotreated SRC-I-creosote mixture over ICR-106 catalyst. The aromatic fraction of the feed was found to decrease as a result of ring saturation. Wiser [154, 158] hydrocracked a 523–703 K fraction of SYNTHOIL in a fluidized bed with NiW/Al_2O_3-SiO_2 catalyst. Under the assumption of complete mixing, the data obtained did not follow pseudo first-order kinetics as they did with a petroleum cycle oil.

In an investigation directed at evaluating hydrocracking catalysts, Berg and McCandless [11, 12] showed that Harshaw Ni-4401 (NiW/SiO_2-Al_2O_3) catalyst performed best in a fixed-bed reactor. Initial catalyst screening was conducted in a batch reactor at 723 K, 13.78 MPa (cold) and 30-min reaction time. Catalysts containing iron, chromium, and tungsten produced heavy black products, whereas catalysts composed of nickel, nickel-tungsten, nickel-molybdenum, and cobalt-molybdenum yielded light-brown products with over 80% sulfur removal. Catalyst presulfiding proved beneficial. A catalyst was synthesized containing 1–2% CoO, 0.4% NiO, 18.2% MoO_3 on a high-area Norton 6176 support that yielded a product with a high volume percentage boiling below 644 K. The investigators concluded that an understanding of the effect of catalyst-support properties, especially acid-base features, is required before new catalysts can be developed.

Ralston et al. [108] hydrocracked solvent-refined lignite over $NiSO_4$-MoO_3 supported on TiO_2, ZrO_2, TiO_2-SiO_2, and three types of aluminas at 723 K and 17.23 MPa in a batch autoclave. NiMo on fibrous Al_2O_3 was found to have the highest activity of the catalysts evaluated. Presulfiding of the catalysts increased the liquid yield from 52 to 64%. No correlations between acidic properties of the supports and activity or selectivity were reported.

Based on model compound studies, Wu and Haynes [161] suggested that non-acidic or low-acidity catalysts may offer the advantage of selectivity towards cracking of inner rings of partially saturated aromatics. In preliminary studies, chromia-alumina catalysts were found to offer this advantage and were also resistant to poisons. In contrast, the results of Berg and McCandless [11] showed that chromium-containing catalysts did not function well with coal liquids. Further investigations are therefore necessary in these areas.

The hydrocracking reactions of a coal-derived distillate (453–593 K) over a zeolite catalyst have been modeled using first-order kinetics to describe the overall hydrocracking process [42]. The reaction scheme proposed is

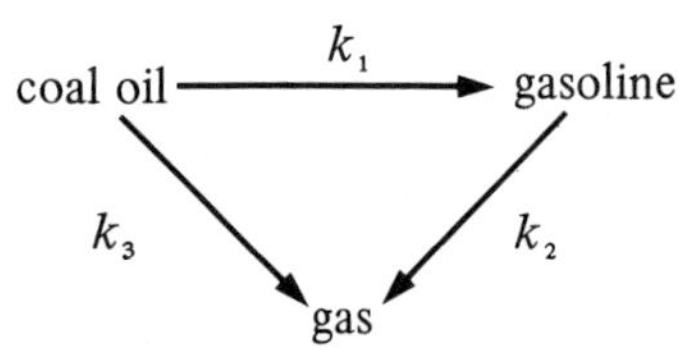

For first-order kinetics

$$\frac{dC_{\text{oil}}}{dt} = -(k_1 + k_3)\, C_{\text{oil}} = -k_{13}\, C_{\text{oil}}, \tag{6-43}$$

$$\frac{dC_{\text{prod}}}{dt} = k_1 C_{\text{oil}} - k_2 C_{\text{prod}}, \tag{6-44}$$

$$\frac{dC_{\text{gas}}}{dt} = k_2 C_{\text{prod}} + k_3 C_{\text{oil}}. \tag{6-45}$$

The values of the kinetic constants are presented in Table 6-28. Since the formation of gas directly from coal oil is less than from the gasoline ($k_2 \gg k_3$), the activation energies presented in Table 6-28 neglect the former. The low values reported for the activation energies could be due to lumping several complex reactions into one simple reaction. A polynomial equation was used to optimize the hydrocracking process with the constraints of 10% gas yield, 75% conversion of raw materials, and maximum yield of gasoline. Their results showed a multiplicity of values for temperature (K) and space velocity (sec^{-1}) and are expressed by

$$T = 591 + 1.98 \times 10^5 \, V_0. \tag{6-46}$$

For the prescribed constraints, the gasoline yield was 40%. The final choice of optimum conditions depends on economic feasibility.

Table 6-28

Kinetic Rate Constants and Activation Energies for the Hydrocracking Model of Gagarin et al. [42]

Temp., K	k_1	k_2	k_3	k_{13}
613	1.08	2.78	0.83	1.17
633	1.53	3.33	2.33	1.75
653	2.08	8.33	1.89	2.28
673	2.50	8.33	2.36	2.72

Temp., K	Activation energies, kJ/mol Reaction (6-44)	Reaction (6-45)
613–633	70.56	106.68
633–673	38.64	21.00

6.5.4 *Summary*

Catalytic hydrocracking offers considerable versatility in the upgrading of coal liquids because of its ability to handle feedstocks having vastly different properties. Naphthas, mid-distillates, gas oils, and residuals from coal liquids have been investigated with varying degrees of success. Many of the problems encountered in these investigations are probably due to the fact that the original process units were designed for petroleum feedstocks, whereas coal liquids, because of their different compositions, offer vastly different constraints.

Conventional hydrocracking catalysts such as nickel and tungsten on silica-alumina supports should be adequate during the initial stages of coal-liquid upgrading; however, further research directed at developing new catalysts for efficiently converting the high-molecular-weight polyaromatic components will contribute greatly to improving the economics of coal-liquid refining. Because of the high heteroatom content of coal liquids, catalytic hydrocrackers will have to be preceded by a catalytic hydrotreating unit in order to lower the heteroatom concentration and metal content to acceptable levels to prevent catalyst poisoning.

The overall hydrocracking process will most likely employ packed-bed reactors in two separate stages with intermediate separation. The first stage will be the hydrotreating or pretreatment section; the second stage will be the hydrocracking portion. The number of reactors employed in the hydrocracking section will depend on the origin, and hence the nature, of the feedstock. Temperatures similar to those used in petroleum refining should be adequate; however, higher pressures ($\geqslant$ 13.78 MPa) will be required because of the presence of polyaromatic compounds. A preliminary reactor design could be obtained using the kinetic models presented in this section for particular feedstocks, although further refinement may be necessary in specific cases.

6.6 Catalytic Cracking

Catalytic cracking has become one of the most widely used processes in the refining of crude oil. The process has been used primarily on distilled gas oil and other heavy charge stocks to increase the yield of motor-fuel blending stocks and other lighter products over that produced by thermal cracking. Residual petroleum feedstocks are also currently being processed in catalytic crackers. As a result of this apparent flexibility in feedstocks, catalytic cracking may become an important process in the refining of coal liquids.

6.6.1 *Chemistry*

The overall reaction mechanism of the catalytic cracking process primarily involves four types of reactions. The first of these is the thermal decomposition of paraffins to form olefinic components. It is believed that the resulting olefins extract protons from the cracking catalyst to form large carbonium ions. These large ions subsequently decompose to produce olefins and small carbonium ions, which then extract protons from larger paraffinic molecules. This results in the formation of other large carbonium ions, and a chain reaction occurs.

The second type of reaction is classified as a primary reaction on the catalyst surface and involves the cracking of paraffinic, olefinic, naphthenic, and aromatic molecules. The cracking of paraffins result in a high yield of C_3 and C_4 gases, the rates of reaction and product distribution being dependent upon the size and structure of the initial molecule. The effect of the catalyst on the reaction rates increases with the number of carbon atoms but is not significant for molecules with less than six carbon atoms. Paraffins preferentially crack where the resulting molecule contains three or four carbon atoms. Tertiary carbon atoms are easily cracked, while quarternary carbon atoms are the most resistant; if both occur in the same molecule, they tend to nullify each other. *n*-Paraffins generally crack either at the gamma carbon-carbon bond or towards the center of the molecule, but long chains will crack in several places.

The catalytic cracking of olefins involve carbon-carbon bond scission, isomerization, polymerization, saturation, aromatization, and coking. The overall reaction rates for olefin cracking are higher than those for paraffins. The isomerization of olefinic molecules and subsequent saturation and aromatization are the reactions primarily responsible for the high octane rating of catalytically cracked products.

The dehydrogenation of naphthenes to aromatics is the most important cracking reaction of naphthenic molecules. At temperatures above 800 K, the cracking of carbon-carbon bonds both within the ring and in peripheral groups (especially if these groups possess more than three carbon atoms) occurs; however, below 800 K the reaction is not significant. The reaction products are generally molecules containing three or four carbon atoms.

For aromatic components, the removal of peripheral alkyl groups from the ring structure is the predominant reaction. For long-chain alkyl groups, the carbon-carbon bond between the ring and the alkyl group is the one that is broken, whereas for groups with three or less carbon atoms, the cracking reactions are not very important. The cracking of aromatic rings seldom occurs.

Secondary reactions between primary products are the third type of reaction which occurs in catalytic cracking. These reactions involve the isomerization of paraffinic molecules to branched structures and the formation of aromatics from olefinic reaction products.

The fourth type of reaction involves the removal of polymerizable products from the reaction by adsorption on the catalyst surface. This process permits these components to decompose and deposit coke on the catalyst surface. As a result, coke deposition is restricted primarily to the catalyst, and recycling of these components is reduced. Typical operating conditions for a catalytic cracking process are:

Temperature	744–1033 K (dependent on feedstock)
Pressure	0.07–0.21 MPa
Weight hourly space velocity	1–3 wt/hr/wt
Catalyst-to-oil ratio	8–12

6.6.2 *Catalysts*

Catalytic cracking processes employ microspherical or powdered catalysts of either natural or synthetic origin. These catalysts consist of acid-treated aluminosilicates, amorphous synthetic silica-aluminas, or zeolites, which are crystalline synthetic silica-aluminas. Since their introduction, zeolites have been the most widely used cracking catalyst because of both design and operational advantages. Some of these advantages are: (1) simplification of design and operation with short-contact-time reactors, (2) higher conversions without severe cracking, (3) quick separation of products and catalyst to prevent excessive cracking, (4) higher catalyst activities, (5) reduction in coke production, and (6) improved selectivity.

Metals such as iron, nickel, copper, and vanadium, which are present in the feedstock, deposit and accumulate on the cracking catalyst. This reduces catalyst activity, alters selectivity (i.e., decreases gasoline and C_4 yields while increasing hydrogen yield), and increases the formation of coke. Metals also reduce the amount of coke removed from the catalyst in the regenerator per unit of air supplied by catalyzing carbon to carbon dioxide rather than to carbon monoxide. Metal deposition on the catalyst can be controlled or eliminated by proper selection of feedstocks, increasing catalyst makeup rate, demetallization of the feedstock, or treatment after the accumulation of metals.

6.6.3 *Catalytic Cracking of Coal-Derived Liquids*

Investigations into the catalytic cracking of coal liquids are few. Most of the reported studies have processed either liquids produced by the more promising coal-liquefaction processes, or blends of coal liquids and petroleum fuels. Table 6-29 summarizes some of these investigations.

Wiser [159] investigated the fluid catalytic cracking of hydrogenated coal liquids derived from a laboratory-scale coal-liquefaction facility. A conventional zeolite catalyst (Mobil Durabead) and a newly developed, strongly acidic, large-pore molecular sieve formed by the crosslinking of silicate layers were employed. Because of its high naphthenic content, resulting from hydrogenation, the hydrotreated coal liquid was an excellent catalytic-cracker feedstock. Details of the investigation, however, were not reported. Wiser [160] also investigated the fluid catalytic cracking of a coal liquid blended with a typical petroleum-derived catalytic-cracker feedstock. Details are provided in Table 6-29.

The effect of hydrotreatment on the catalytic cracking of COED oil was examined by White et al. [152]. Product yields, as shown in Figure 6-19, varied

Table 6-29

Summary of Catalytic Cracking Investigations

Investigators	Reactor	Catalyst	Pressure, MPa	Temp., K	Feedstock
Wiser [159, 160]	Fluid catalytic reactor	Amorphous & crystalline zeolite alumina-silicate catalyst, Mobil Durabead zeolite catalyst	0.10	658–740	Blends of petroleum-derived catalytic cracker feedstock and coal liquids derived from subbituminous and high and medium volatile bituminous coals
White et al. [152]	–	–	–	755	Hydrotreated COED oils derived from IL #6 and Utah coals and raw COED oil from IL #6 coal
Gallagher et al. [44, 45]	Fluid catalytic reactor	–	–	772	Hydrotreated COED gas oil (+613 K)
Sullivan [142]	Fixed fluidized-bed reactor	Equilibrium catalyst from an FCC unit	–	797	Hydrotreated blend of SRC-I and creosote oil (450–727 K)
deRosset et al. [30] Tan & deRosset [145]	Riser catalytic reactor	Equilibrium zeolite catalyst from an FCC unit	–	–	Raw and hydrotreated H-coal gas oil and H-coal and EDS distillates

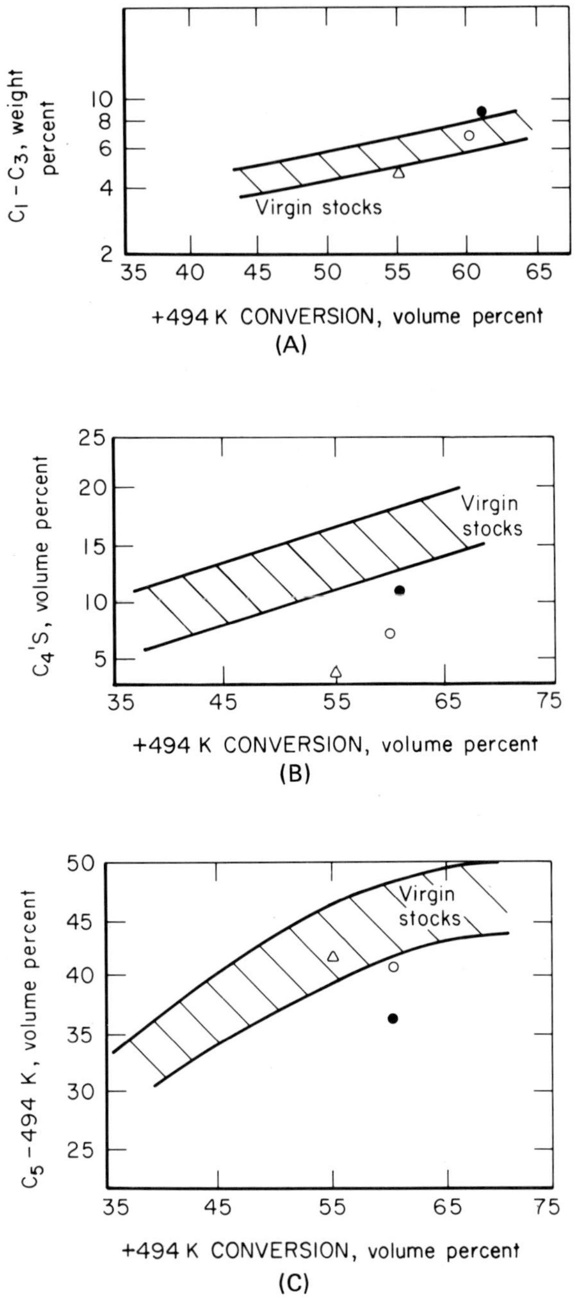

Figure 6-19. Yields of products from catalytic cracking of COED oils as compared to virgin petroleum feeds [152]. △ Furnace oil (Illinois coal), ○ furnace oil (Utah coal), ● gas oil (Utah coal).

over a range and were dependent on the conversion and feedstock properties. For the hydrotreated feedstocks, the production of C_1 to C_3 components [Figure 6-19(a)] was comparable to that of virgin petroleum stock, whereas the C_4 yield [Figure 6-19(b)] was lower, probably because of a lower paraffinic content. Gasoline production [Figure 6-18(c)] from the furnace-oil fractions of the two coal liquids were comparable to that of the virgin petroleum stock, whereas the gas-oil fraction was somewhat lower. For the raw coal liquid, approximately 80% of the feed was converted to coke. Hydrotreatment prior to catalytic cracking reduced coke production by a factor of 3 at a constant conversion.

In a preliminary investigation of the fluid catalytic cracking of hydrotreated COED gas oil (+631 K), it was reported that if the hydrogen produced by the catalytic cracker were recovered, over 30% of the hydrogen requirement for the hydrotreater could be supplied by the fluid-catalytic-cracker unit [44, 45]. Comparing the results for the catalytic cracking of the gas-oil fraction with those obtained using hydrocracking, it was found that the FCC unit produced about 5% less gasoline; however, the distillate yield was about 9% higher. The hydrogen requirement for the catalytic cracking process was, including hydrotreatment, considerably lower than that of the hydrocracker.

Sullivan [142] reported on the catalytic cracking of a hydrotreated blend of SRC-I and creosote oil. Conversions ranged from 48 to 62% with a coke yield of 1% of feed. At low-severity conditions, gasoline selectivity was excellent and gas yield was low.

deRosset and coworkers [30, 145] employed a once-through riser catalytic cracker to investigate the catalytic cracking of raw and hydrotreated H-coal gas oil and H-coal and EDS distillate materials. An equilibrium zeolite catalyst from a commercial fluid-catalytic-cracking unit was employed. For the raw H-coal gas oil, conversions ranged from 31 to 43 vol%, and research octane numbers (RON) ranged from 97.4 to 101.2, the higher values being obtained at high catalyst-to-oil ratios and temperatures. For the hydrotreated H-coal gas oil, conversions increased to 63–66% with RONs of 95.5 to 98.8. This feedstock contained more naphthenes and less aromatics and heteroatoms than the raw gas oil, which improved its performance. Carbon deposition was also reduced. Although the aromatic concentration was still somewhat high, multiring aromatics were partially saturated and readily cracked to lower-boiling aromatics. At +468 K fraction of severely hydrotreated H-coal gas oil which contained the highest concentration of hydrogen and the lowest aromatic and nitrogen content was also processed. Conversion and gasoline yield were again increased, with a further reduction in coke production. The above results, together with data from petroleum feedstocks, are graphically displayed in Figure 6-20. Preliminary results for the H-coal and EDS distillates indicated that gasoline yields were low while carbon yields and product RONs were high. Conversions were 31 and 46% respectively. These investigators concluded that hydrotreating of coal-derived catalytic-cracker feedstocks is necessary to improve product yields and quality. Extremely high saturation of multiring aromatics is not required to improve cracking; and in reality, the final gasoline product could have a lower RON if all multiring aromatics were saturated.

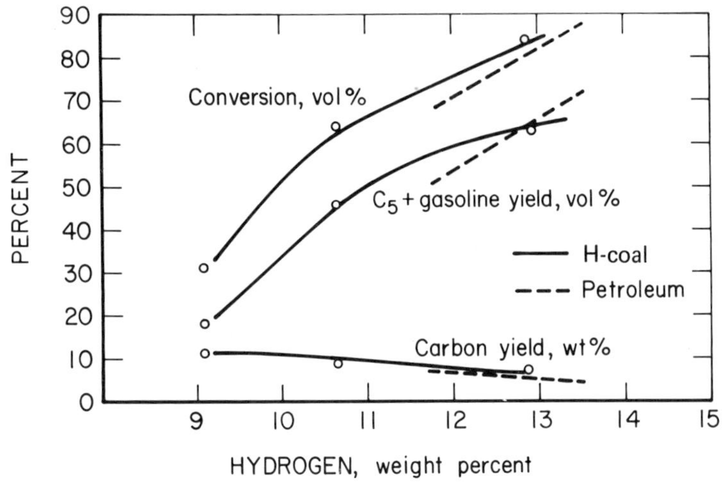

Figure 6-20. Comparison of FCC product distribution for H-coal distillate with petroleum feedstocks [30].

6.6.4 *Summary*

The above studies on catalytic cracking of coal liquids are not extensive, but they do demonstrate the capabilities and potential of this technique for upgrading coal liquids. As has been described earlier, hydrotreating of the feedstock is required in order to improve product yields and minimize coke deposition on the catalyst surface. Because of the potential for excessive coke deposition, the riser catalytic crackers and, to a somewhat lesser degree, the fluid bed reactors should be employed. The riser reactors provide more flexibility for varying catalyst-oil contact times, which is a critical factor in coke deposition. Attention should also be directed to recovering the hydrogen produced in the cracking process. If the amount produced is substantial enough to warrant a separation unit, a large saving in hydrogen cost may be realized, which could significantly improve the economics for upgrading coal liquids to useful fuels and chemicals.

6.7 Catalytic Reforming

With the increasing demand for high-octane unleaded gasoline, catalytic reforming has become an extremely important process in refining because of its ability to substantially improve the octane rating of fuels. With the advent of coal liquefaction, reformer feedstocks will gradually be shifted from petroleum-based to coal-derived fuels. Due to the difference in the chemical constitution of coal liquids and petroleum, different operating conditions and/or catalyst modifications may be required for optimizing the reforming process for coal liquids. The operation of the catalytic reformer may also have a substantial impact on the overall economics of coal liquefaction because of its hydrogen production capabilities. These aspects will be discussed further in this section.

6.7.1 *Chemistry*

In petroleum refining, the typical catalytic reformer feedstock consists of paraffinic, naphthenic, aromatic, and olefinic components, of which the first two constitute 75% or more of the feed. The boiling range for the reformer feedstock is generally between 366 and 477 K. Materials boiling below 366 K are primarily low-molecular-weight paraffins, which tend to crack to lighter components under reformer conditions, whereas those boiling above 477 K have a tendency to hydrocrack and produce significant amounts of coke, which deactivate reforming catalysts.

Within the catalytic reformer, numerous complex chemical reactions are known to occur which produce both desirable and undesirable products. The major reactions are presented below:

1. *Dehydrogenation of naphthenes.*

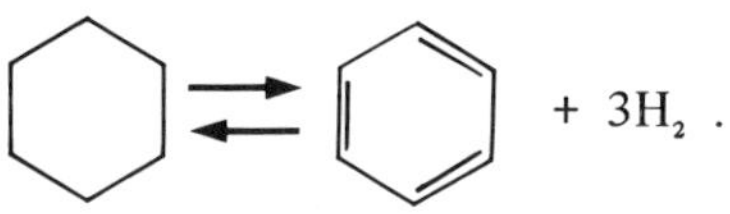

(6-47)

2. *Dehydroisomerization of naphthenes.*

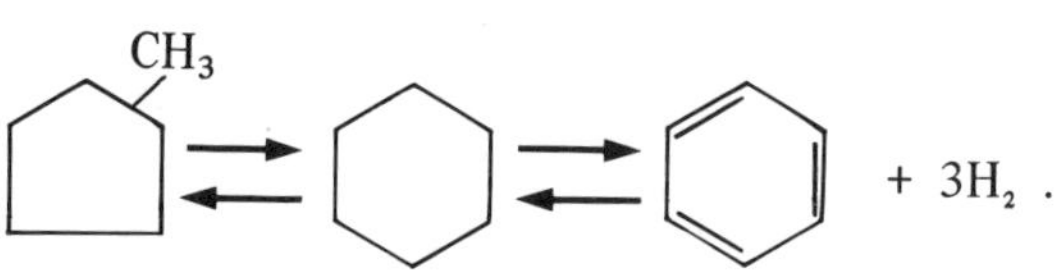

(6-48)

3. *Dehydrocyclization of paraffins.*

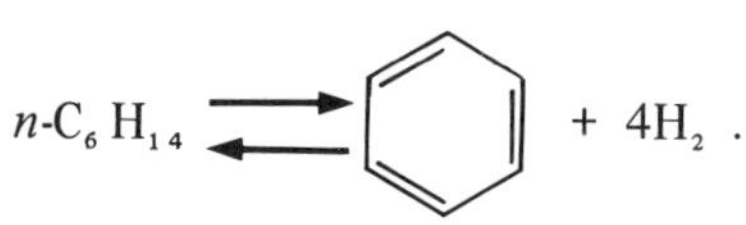

(6-49)

4. *Isomerization of paraffins.*

$$CH_3-(CH_2)_4-CH_3 \leftrightharpoons CH_3-\underset{\substack{|\\ CH_3}}{CH}-(CH_2)_2-CH_3 \ . \qquad (6\text{-}50)$$

5. *Hydrogenation and dehydrocyclization or isomerization of olefins.*

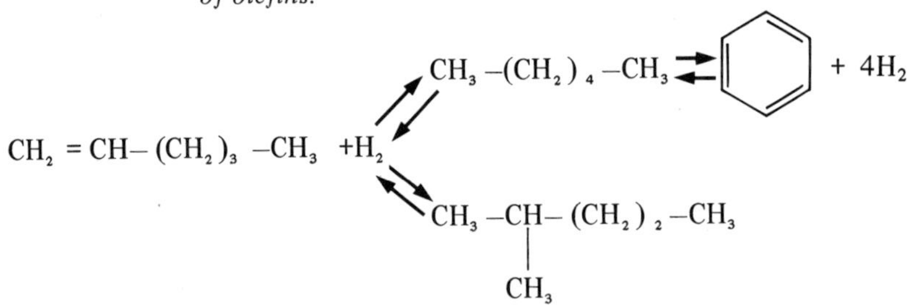

(6-51)

(6-52)

6. *Hydrocracking of paraffins.*

$$C_{10}H_{22} + H_2 \rightleftharpoons C_5H_{12} + CH_3{-}CH_2{-}\underset{\displaystyle CH_3}{\underset{|}{CH}}{-}CH_3 \; . \qquad (6\text{-}53)$$

7. *Dealkylation of naphthenes and aromatics.*

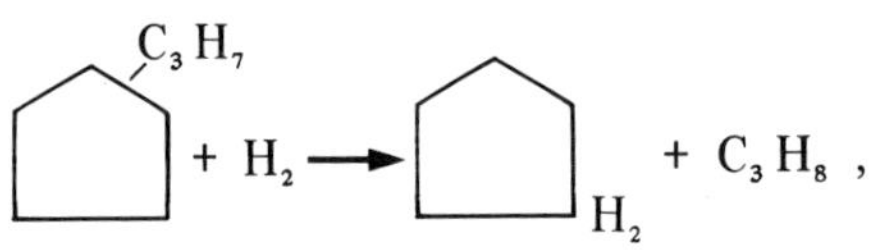

(6-54)

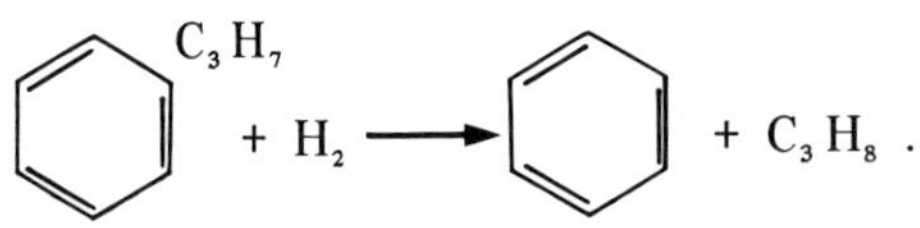

(6-55)

The desired chemical reactions are those which result in the formation of aromatic and isoparaffinic components which have high octane numbers. The reactions (6-47) through (6-50) therefore represent the most desirable reactions. Together with this is a large production of molecular hydrogen from the reactions (6-47), (6-48), and (6-49), which can be used in other hydrogen-consuming processes. Hydrogenation of olefins is undesirable because they consume molecular hydrogen produced in the reformer. However, if the reaction is accompanied by isomerization (6-52) or is followed by dehydrocyclization (6-51), the net reaction results in desirable products.

The hydrocracking of paraffins, Reaction (6-53), is undesirable when compared to the isomerization and dehydrogenation reactions. The former reactions lead to desirable isomers, but the resulting excess coke production, net hydrogen consumption, and lower liquid yields cause these reactions to be classified as undesirable. The extent of hydrocracking is dependent upon the concentration of paraffinic components in the reformer feedstock, whereas the amount of isomers produced in any molecular-weight range is independent of the feedstock. To suppress the formation of coke, a reasonably high partial pressure of hydrogen is required in the reformer:

Temperature	700–824 K
Pressure	0.3–5.2 MPa
LHSV	0.7–5 vol/vol hr
H_2 recycle rate	267–445 sm^3/m^3

6.7.2 *Catalysts*

The reforming catalysts currently employed are mostly Group VIII metals of the periodic table supported on silica or silica-alumina. Platinum is one of the important metallic components of the catalyst and is believed to be responsible for its hydrogenation and dehydrogenation reactivity. Rhenium is often added to the catalyst for stability, as it allows operation at lower pressures. Prior to use, the catalyst must be chlorinated to provide acid sites for reaction. It is the acidity of the catalyst which is believed to be responsible for isomerization, cyclization, and hydrocracking activity of the catalyst. Deactivated catalysts can be restored by oxidizing the carbon and rechlorinating the catalyst. Generally, the catalyst can be regenerated a few times before it must be replaced.

Prior to contacting the catalyst with the reformer feedstock, the feed must be pretreated to remove catalyst poisons. This is accomplished by hydrotreating the feedstock with a conventional hydrotreating catalyst such as cobalt molybdenum on Al_2O_3. By doing so, the concentrations of organic sulfur and nitrogen components together with certain metals are reduced to levels which will not readily deactivate reforming catalysts.

6.7.3 *Reforming of Coal Liquids*

Only a few investigations have been reported on the reforming of coal liquids, and these have evaluated coal liquids from the more advanced coal liquefaction processes. In a major study, Peters [98–101] evaluated the reforming of the naphtha fraction (IBP–450 K) from COED, H-coal, SYNTHOIL, and SRC-II coal liquids. The straight-run naphthas were hydrotreated prior to reforming. Several reforming runs were also made using hydrotreated naphthas, which were subsequently hydrocracked, as feedstocks. The reforming was performed in a single-pass fixed-bed reactor at 786–798 K, 1.72 MPa, and LHSVs of 2–2.25. American Cyanamid Aeroform PHF-4, 0.16-cm extrudate catalyst was employed and contained 0.3% platinum and 0.6% chloride on an alumina support. This catalyst was selected over the newer bimetallic reforming catalysts because it is much less susceptible to poisoning, and at severe conditions the catalyst increases the aromatics content by selectively cracking away the paraffinic components in the feed. The resulting products should be suitable as feedstocks to a hydrodealkylator.

The hydrotreated COED and H-coal straight-run naphtha reformer feeds contained approximately 50% and 41% naphthenes and 11% and 14% C_6–C_9 paraffins, respectively. Initial catalyst activity was high for both feedstocks but decreased rapidly, causing a decrease in the conversion of naphthenes to aromatics. Naphthene conversion and hydrogen production were 72–98% and 266 sm^3/m^3 for the COED naphtha, and 67–92% and 107–480 sm^3/m^3 for the H-coal feed. SRC-II hydrotreated straight-run naphtha possessed the highest concentration of naphthenes (71%) of the three hydrotreated straight-run naphthas investigated; its paraffin content was comparable. This feedstock was also processed at similar conditions but at a higher pressure (3.38 MPa). Naphthene conversions were approximately 93%, with a hydrogen production of 2.7–2.9 g/(100 g feed). It is interesting to note that the yield of C_1–C_5 and C_6–C_9 paraffins was higher than with either the COED or H-coal feedstock. Paraffin conversion was lower for the SRC-II feed, which indicates that cyclization reactions were minimal.

COED, H-coal, and SYNTHOIL hydrotreated and hydrocracked straight-run naphthas were all processed at 791–793 K, 3.4 MPa, and an LHSV of 2. The COED and H-coal feedstocks possessed approximately the same naphthene (36 versus 37%) and paraffin (14 versus 11%) concentrations, while the SYNTHOIL feedstock contained considerably more naphthenes (49%) and less paraffins (3%). Catalyst deactivation was not a problem for the COED and H-coal feedstocks as reported earlier; however, for the SYNTHOIL feed, some deactivation was observed. For the COED and H-coal feedstocks, this was attributed to a twofold increase in pressure over the previous case, although the runs were not of sufficient duration to confirm this conclusion. Naphthene conversions for the three feedstocks ranged from 89 to 95% with hydrogen productions of 53–249 sm^3/m^3.

In another major study, deRosset and coworkers [30, 144, 146] investigated the reforming of H-coal, EDS, and SRC-II hydrotreated naphthas and H-coal hydrocracked naphtha using a commercial platforming catalyst. For the two H-coal reformer feeds, higher reaction temperatures were found to increase the RON of the products from 99.2 to 102.6. At all process conditions, dehydrogenation of naphthenes was the primary reaction; however, at the highest severity, dehydrocyclization of paraffins became significant. The feedstocks contained approximately 62% naphthenes and 15% paraffins. Hydrogen yields ranged from 2.5 to 3.4 wt% of feed. Hydrotreated and hydrocracked EDS naphthas were processed under similar conditions to the H-coal feedstocks. For both cases, the RON of the products increased while the C_{5+} yield decreased, with hydrogen production ranging from 2.1 to 3.6 wt% of feed. Again, naphthene dehydrogenation was significant, whereas the paraffins remained unchanged. The naphthene and paraffin concentrations were very similar to those of the H-coal naphthas. Results from the reforming of SRC-II naphtha were not reported; supposedly they were very similar to both the H-coal and EDS results.

6.7.4 *Summary*

From the preliminary investigations presented above, no major technological problems are evident with the use of current reforming technology for coal-derived feedstocks. Because of their high heteroatom content, especially nitrogen, the coal-derived reformer feedstock will require hydrotreatment to remove catalyst poisons prior to reforming. As shown above, the naphthene content of coal liquids is much higher than that of petroleum and may cause minor design problems. Because naphthene dehydrogenation reactions are rapid and endothermic, they will occur at the beginning of the reactor and require heat. Supplying this heat to a localized area to sustain the reactions may require minor design modifications to the reactor and auxiliary equipment. Such reactions can be slowed by adjusting the operating conditions of the reactor; however, the economics of alternatives should be evaluated.

6.8 Conclusions

Upgrading coal liquids through individual processes provides preliminary information on the applicability of such processes as well as potential problem areas. Investigations which process coal liquids in integrated pilot-scale refineries will provide the final data for optimization of operating conditions and product distributions for specific feedstocks from which economic evaluations can be made.

The high nitrogen and aromatic content of coal liquids will have a profound effect on the selection of upgrading processes necessary to produce finished products [36]. The consensus is that severe hydrotreating must be the first step in the upgrading process [71, 88]. This is required in order to reduce the high concentration of heteroatoms to levels which will not adversely affect further processing and in order to hydrogenate the polyaromatics to saturated components. Hydrotreating can be performed either on the raw oil or on individual fractions. If coal liquids are to be blended with petroleum fuels in the early years of the synthetic fuels industry, these liquids will have to be compatible with the desired petroleum components. Figure 6-1, which presents a plot of the aromaticity of various coal liquids and petroleum fuels versus their carbon-to-hydrogen ratio, illustrates the difference between petroleum and coal liquid and explains why compatibility is a problem. This figure can be used to estimate the hydrogen requirements for upgrading coal liquids to products which are compatible with petroleum fuels.

Following the hydrotreating process, the heavy distillates can be further processed in either a hydrocracker or a catalytic cracker. Oblad [88] had indicated that catalytic cracking is a more viable alternative because hydrocracking results in lower liquid yields and higher gas make with high hydrogen consumption. Mid-distillates can also be hydrocracked to yield desirable gasoline-range products. The naphtha fraction produced during upgrading must first be hydrotreated and then fed to a catalytic reformer to improve the quality of the products. The hydrogen

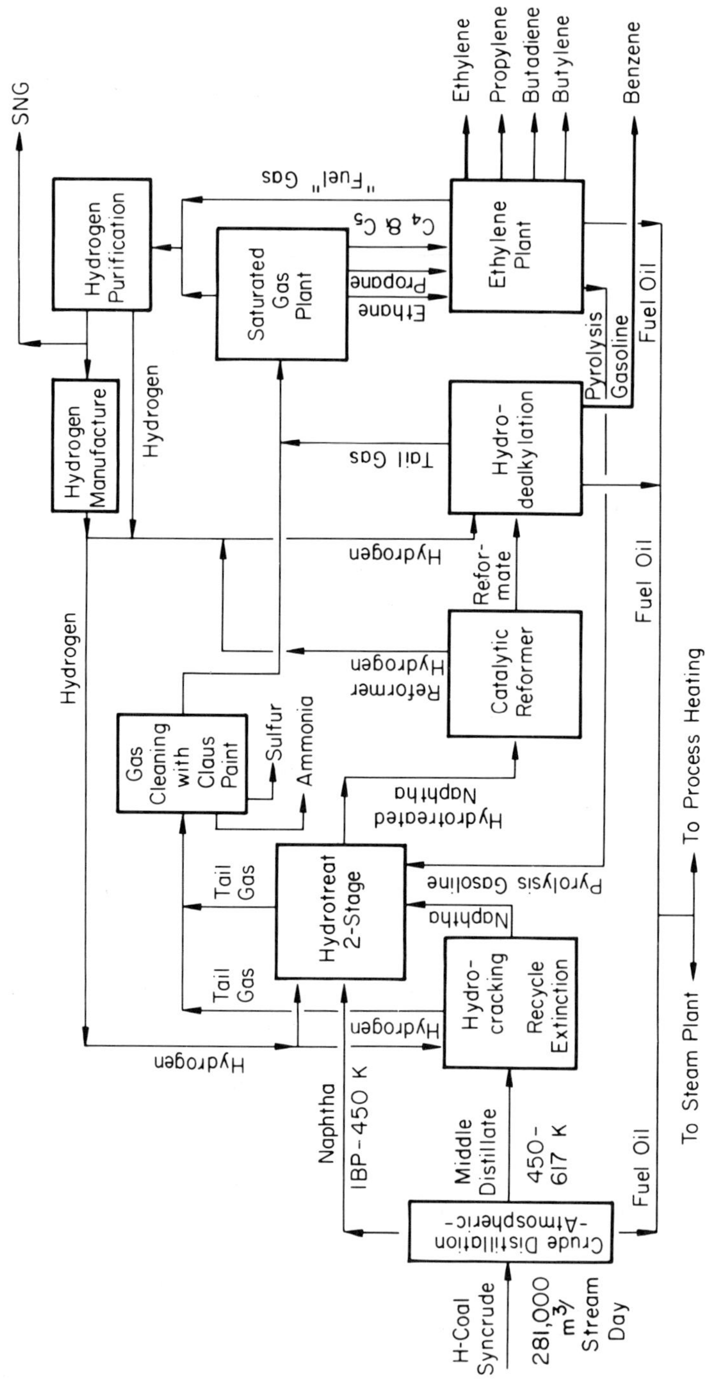

Figure 6-21. Chemical refinery for the production of benzene and olefins [102].

produced in the reformer will probably meet the hydrogen demand of the hydrotreating and hydrocracking operations [49, 71]. As mentioned earlier, some investigators [44, 45] believe a portion of the total hydrogen requirement for upgrading coal liquids can also be obtained from catalytic cracking.

It is believed by some that the most economical uses for coal liquids will be those requiring the least extensive upgrading for direct use or blending with other products [49]. The major questions confronting the development of upgrading coal liquids are: (1) how coal liquids will fit into typical petroleum refining situations, (2) what product slate will best fit the feedstocks, and (3) what plant modifications will be required. As mentioned previously, these questions can only be answered when the coal liquids are evaluated in integrated pilot-scale refineries.

Recently, several reports have been published on the economic evaluation of coal-liquid upgrading. O'Hara et al. [91, 92] presented an economic evaluation of a total facility including the liquefaction process. Their evaluation was based on products from the SRC-II process, and are therefore of current interest.

Based on data previously obtained for processing COED, H-coal, SYNTHOIL, and SRC-II coal liquids, Peters [102] concluded that the H-coal liquids were the most desirable for producing petrochemical feedstocks, giving an overall conversion of 34% versus 15, 27, and 23% for the COED, SYNTHOIL and SRC-II liquids. From his results, a preliminary process flow diagram was constructed (Figure 6-21), and an economic analysis of a refining facility for the production of ethylene and benzene was carried out. Although the basis of the analysis is questionable (i.e. a price for H-coal liquid of $16 per barrel), the results showed the technical and economic feasibility of upgrading coal liquids to petrochemical feedstocks.

Finally, Dabkowski et al. [28] evaluated the economics of a 566-m^3/day hydrotreater with auxiliary waste-water treater and sulfur plant for upgrading SRC-I light organic liquid, recycle solvent, and residual and H-coal distillate and fuel oil. An ebullating-bed reactor was used for the residual feeds, whereas multistage reactors with multiple quench zones were used for the distillates. The analyses were based on the previously reported data.

In conclusion, this chapter has illustrated that for the most part, coal liquids can be upgraded in conventional petroleum refineries. Additional process and catalyst development work is required for a few of the processes so that the cost of production can be decreased. Processing of coal liquids in an integrated pilot plant will be necessary to develop the final optimum operating conditions and product distribution for specific liquids. A reliable economic analysis must await such investigations.

6.9 Nomenclature

6.9.1 *Latin Letters*

a	Decay constant
c	Constant
C	Concentration
C_H	Concentration of heavy oil, g/(g feed)
C_L	Concentration of light oil, g/(g feed)
$C_{\mathrm{N}0}$	Initial nitrogen concentration
C_{S}	Concentration of sulfur, g/(g feed)
$C_{\mathrm{S}0}$	Initial concentration of sulfur, g/(g feed)
E	Activation energy, kJ/mol
E_{ij}	Activation energy for the ith component in the jth reaction
f_A	Fraction of carbon atoms which are aromatic
F	Feed rate
ΔH_{rx}	Heat of reaction
k	Reaction rate constant
k'	Reaction rate constant
k_d	Deactivation rate constant
k_f	Forward reaction rate constant
k_{H}	Rate constant for pyridine hydrogenation
k_i	First-order rate constant for fraction i [Equation (6-29)]; second-order rate constant for species i [Equation (6-31)]
k_{ij}	Rate constant for jth reaction involving species i
k_k	Rate constant for highest-boiling fraction
k_n	Rate constant for piperidine denitrogenation
k_r	Reverse-reaction rate constant
k^o	Frequency factor
k_1, k_2, k_3, k_{13}	First-order rate constants
K	Adsorption coefficient
K_{A}	Adsorption coefficient for ammonia
K'_{A}	Adsorption coefficient for ammonia on the hydrogenation sites of a catalyst
K'_{B}	Adsorption coefficient for butane
K_{H_2}	Adsorption coefficient for hydrogen
K_{HS}	Adsorption coefficient for hydrogen sulfide
K_i	Adsorption coefficient for species i
K_{N}	Adsorption coefficient for piperidine

K_S, K_S'	Adsorption coefficient for solvent
K_T	Adsorption coefficient for thiophene
LHSV	Liquid hourly space velocity
n	Reaction order
N	Constant
N_{fi}	Nitrogen concentration
N_p	Total nitrogen concentration
N_{pi}	Nitrogen concentration in the ith fraction of the product
P	Pressure
P_A	Partial pressure of ammonia
P_B	Partial pressure of butene
P_{H_2}	Partial pressure of hydrogen
P_{HS}	Partial pressure of hydrogen sulfide
P_N	Partial pressure of piperidine
P_P	Partial Pressure of pyridine
P_S	Solvent partial pressure
P_T	Partial pressure of thiophene
r	Reaction rate
R	Universal gas constant
r_i	Ratio of k_i/k_k
r_{ij}	Reaction rate for the jth reaction involving species i
r_T	Reaction rate for thiophene
SV	Space velocity
t	Time
T	Temperature
V_o	Space velocity
VHSV	Volume hourly space velocity
W	Mass fraction
WHSV	Weight hourly space velocity
x	Conversion

6.9.2 *Greek Letters*

α	Fraction of initial sulfur reacting rapidly
θ_v	Volumetric hourly space velocity

6.9.3 *Subscripts*

A	Asphaltenes
f	Forward reaction
H	Heavy oil
i	Boiling-range fraction of heteroatom species

L	Light oil
N	Nitrogen
O	Oil
r	Reverse reaction
R	Resins
S	Sulfur
0	Initial conditions
1	Fast reaction
2	Slow reaction

REFERENCES

1. Aboul Gheit, A. K. and Abdou, I. K., *J. Inst. Petroleum,* **59** (568), 188 (1973).
2. Aboul Gheit, A. K., *Can. J. Chem.,* **53**, 2575 (1975).
3. Ahmed, M. M., M. S. Thesis, Oklahoma State University, 1975.
4. Ahmed, M. M., and Crynes, B. L., *Coal Processing Technology,* Vol. 3, American Institute of Chemical Engineers, 1977, p. 23.
5. Alekseeva, K. A. and Moldavskii, B. L., *Khim. Tekhnol. Topliv. Masel.,* **4** (1), 43 (1959).
6. Anderson, S. S., Badawy, M. L., Qader, S. A., and Hill, G. R., *ACS Div. of Fuel Chem. Prepr.,* **12** (3), 181 (1968).
7. Antoine, A. C. and Gallagher, J. P., *Coal Processing Technology,* Vol. 3, American Institute of Chemical Engineering, 1977, p. 107.
8. Badilla-Ohlbaum, R., Pratt, K. C., and Trimm, D. L., *Fuel,* **58**, 311 (1979).
9. Bartsch, R. and Tanelian, C., *J. Catal.,* **35**, 353 (1974).
10. Bendoraitis, J. G., Cabal, A. V., Callen, R. B., Stein, T. R., and Voltz, S. E., EPRI Report No. EPRI 361-1, 1976.
11. Berg, L. and McCandless, F. P., U. S. ERDA Report No. FE-2034-4, 1976.
12. Berg, L. and McCandless, F. P., U. S. ERDA Report No. FE-2034-5, 1976.
13. Berg, L. and McCandless, F. P., U. S. ERDA Report No. FE-2034-9, 1977.
14. Berg, L. and McCandless, F. P., U. S. ERDA Report No. FE-2034-10, 1978.
15. Berg, L. and McCandless, F. P., U. S. DOE Report No. FE-2034-11, 1978.
16. Berg, L. and McCandless, F. P., U. S. DOE Report No. FE-2034-12, 1978.
17. Berg, L. and McCandless, F. P., U. S. DOE Report No. FE-2034-13, 1978.
18. Bhinde, M. S., Ph.D. Thesis, University of Delaware, 1979.
19. Bobyshev, V. I., Dyakova, M. K., and Lozovoi, A. V., *Sh. Prkl. Khim.,* **13**, 942 (1940).
20. Brahmachari, B. B., Saha, M. R., Choudhury, P. B., and Ganguli, A. K., *Proceedings of the Symposium on Chemicals and Oil from Coal,* Dhanbad, India, 1969.
21. Brinkman, D. W., Whisman, M. L., and Bowden, J. L., U. S. DOE Report No. BETC/RI-78/23, 1979.

22. Cabal, A. V., Voltz, S. E., and Stein, T. R., *Ind. Eng. Chem. Prod. Res. Dev.,* **16** (1), 58 (1977).
23. Callen, R. B., Simpson, C. A., Bendoraitis, J. G., and Voltz, S. E., *Ind. Eng. Chem. Prod. Res. Dev.,* **15** (4), 222 (1976).
24. Cawley, C. M. and Hall, C. C., *J. Soc. Chem. Ind.,* **62,** 116 (1943).
25. Cox, K. E. and Berg, L., *Chem. Eng. Progr.,* **58** (12), 54 (1962).
26. Cusamano, J. A., Dalla Betta, R. A., and Levy, R. B., U. S. ERDA Report No. FE-2017-1, 1976.
27. Cusamano, J. A., Dalla Betta, R. A., and Levy, R. B., *Catalysis in Coal Conversion,* Academic Press, New York, 1978.
28. Dabkowski, M. J., Heck, R. H., Perrella, A. V., Schreiner, M., and Stein, T. R., EPRI Report No. EPRI AF-710, 1978.
29. deRosset, A. J., Tan, G., and Gatsis, J. G., *ACS Div. Pet. Chem. Prepr.,* **22** (3), 962 (1977).
30. deRosset, A. J., Tan, G., and Helfman, L., *Hydrocarbon Processing,* **58** (5), 152 (1979).
31. deRosset, A. J., Tan, G., Gatsis, J. G., Shaffner, J. P., and Swenson, R. F., U.S. ERDA Report No. FE-2010-09, 1977.
32. Desikan, P. and Amberg, C. H., *Can. J. Chem.,* **41,** 1966 (1963).
33. Desikan, P. and Amberg, C. H., *Can. J. Chem.,* **42,** 843 (1964).
34. Doelman, J. and Vlaghter, J. C., *Proceedings of the Sixth World Petroleum Congress,* Sec. III, The Hague, Netherlands, 1973, p. 247.
35. Donath, E. E., "Hydrogenation of Coal and Tar," in *Chemistry of Coal Utilization,* H. H. Lowry, Ed., John Wiley and Sons, New York, 1963.
36. Dooley, J. E., Lanning, W. C., and Thompson, C. J., *ACS Div. of Pet. Chem. Prepr.,* **22** (3), 904 (1977).
37. Eisen, F. S., U.S. Navy Contract NOO14-74-C-0568, Report No. AD-A007 923, 1975.
38. Eisen, F. S., "Gas Turbine Engine Fuel and Synthetic Crude," in *Coal Processing Technology,* Volume 3, American Institute of Chemical Engineers, 1977, p. 115.
39. Fant, B. T., U.S. DOE Report No. FE 2353-2, Vol. 1, 1978.
40. Finseth, D., Hough, M., Quiser, J. A., and Retcofsky, H. L., *ACS Div. of Pet. Chem. Prepr.,* **24** (4), 979 (1979).
41. Flinn, R. A., Larson, O. A., and Beuther, H., *Hydrocarbon Processing and Petroleum Refiner.,* **42** (9), 129 (1953).
42. Gagarin, S. G., Zamanova, L. P., Krichko, A. A., and Yulin, M. K., *Khimiya Tverdogo Topliva,* **10** (1), 94 (1976).
43. Gallagher, J. P., Collins, T. A., Nelson, T. J., Pederson, M. J., Robison, M. G., and Wisinski, L. J., NASA Report No. CR-135112, 1976.
44. Gallagher, J. P., Humes, W. H., and Siemssen, J. O., AIChE 71st Annual Meeting, Paper 44f, 1978.
45. Gallagher, J. P., Humes, W. H., and Siemssen, J. O., *Chem. Eng. Progr.,* **75** (6), 56 (1979).
46. Givens, E. N., Collura, M. A., Alexander, W., Greskovich, E. J., Engleman, C. D., Wetherington, J. B., Clump, C. W., and Levy, E. K., U.S. DOE Report No. FE-2003-27, 1977.
47. Givens, E. N., Collura, M. A., Skinner, R. W., and Greskovich, E. J., *Hydrocarbon Processing,* **57** (11), 195 (1978).

48. Givens, E. N. and Venuto, P. B., *ACS Div. of Pet. Chem. Prepr.,* **15** (4), A138 (1970).

49. Goelzer, A. R., Rubero, P. A., and Ellington, R. T., "Coal Liquefaction from a Refiner's Viewpoint," Fifth Annual International Conference on Commercialization of Coal Gasification, Liquefaction, and Conversion to Electricity, University of Pittsburgh, 1978.

50. Goudriaan, F., Gierman, H., and Vlughter, J. C., *J. Inst. Petroleum,* **59** (565), 40 (1973).

51. Hall, C. C. and Cawley, C. M., *J. Soc. Chem. Med.,* **58,** 7 (1939).

52. Hartung, G. K., Jewell, D. M., Larson, O. A., and Flinn, R. A., *ACS Div. of petroleum Chemistry Prepr.,* **5** (3), 27 (1960).

53. Houalla, M., Nag, N. K., Sapre, A. V., Broderick, D. H., and Gates, B. C., *AIChE J.,* **24** (6), 1015 (1978).

54. Hoog, H., *Rec. Trav. Chim.,* **69**, 1289 (1950).

55. Jacobs, H. E., Jones, J. F., and Eddinger, R. T., *ACS Div. of Fuel Chem. Prepr.,* **14** (4), 42 (1970).

56. Jacobs, H. E., Jones, J. F., and Eddinger, R. T., *Ind. Eng. Chem. Proc. Des. Dev.,* **10** (4), 558 (1971).

57. Jaffe, S. B., *Ind. Eng. Chem. Proc. Des. Dev.,* **13** (1), 34 (1974).

58. Johns, J. J., Jones, J. F., and McMunn, B. D., *ACS Div. of Fuel Chem. Prepr.,* **16** (1) 26 (1972).

59. Kalfadelis, C. F., Air Force Aeropropulsion Laboratory Report No. AFAPL-TR-75-10, Vol. II, 1976.

60. Kang, C. C. and Gendler, J., EPA Report No. EPA-600/7-78-159, 1978.

61. Karn, F. S., Brown, F. R., and Sharkey, A. G., *ACS Div. of Fuel Chem. Prepr.,* **19** (5), 2 (1974).

62. Kolboe, S. and Amberg, C. H., *Can. J. Chem.,* **44,** 2623 (1966).

63. Korosi, A., Woebeke, H. N., and Virk, P. A., *ACS Div. of Fuel Chem. Prepr.,* **21** (6), 190 (1976).

64. Krishna, M. G. and Ehsan, M., *Erdöl und Kohle* **29** (6), 261 (1979).

65. Krishnamurthy, S., Shah, Y. T., and Stiegel, G. J., *Ind. Eng. Chem. Proc. Des. Dev.,* **18** (2), 254 (1979).

66. Krishnamurthy, S., Shah, Y. T., and Stiegel, G. J., *Ind. Eng. Chem. Proc. Des. Dev.,* **18** (3), 466 (1979).

67. Krishnamurthy, S., Shah, Y. T., and Stiegel, G. J., Paper presented at the American Chemical Society–Chemical Society of Japan Joint Meeting, Honolulu, HI, 1979.

68. Landa, S. and Mrnkova, A., *Coll. Czech. Chem. Commun.,* **31** 2202 (1966).

69. Landa, S., Mrnkova, A., and Bertova, N., *Sci. Papers Inst. Chem. Techn. Prague,* **D16,** 159 (1969).

70. Landa, S., Kafka, Z., Galik, V., and Safar, M., *Coll. Czech. Chem. Commun.,* **34,** 3967 (1967).

71. Lanning, W. C., U.S. ERDA Report No. BERC/IC-76/2, 1976.

72. Lanning, W. C., U.S. DOE Report No. BETC/IC-78/1, 1978.

73. Lanning, W. C., Green, J. B., and Dooley, J. E., U.S. DOE Report No. BRTC/RI-78/10, 1978.

74. Lee, H. C. and Butt, J. B., *J. Catal.,* **49,** 320 (1977).

75. Lin, Y. Y., Anderson, L. L., and Wiser, H. H., *ACS Div. of Fuel Chem. Prepr.,* **19** (5), 2 (1974).

76. Low, J. F. Y., Klabunde, K. J., Woolsey, N. F., Batlisberger, R. J., Govele, M., and Steinberg, V. I., *ACS Div. of Fuel Chem. Prepr.*, **20** (1), 122 (1975).
77. Low, J. Y. F., Klabunde, K. J., Woolsey, N. F., Baltisberger, R. J., Govele, M., and Steinberg, V. I., *Energy Sources,* **13** (1), 19 (1976).
78. Madkour, M. M., Mahoud, B. H., Abdou, I. K., and Vlughter, J. C., *J. Indian Chem. Soc.,* **46**, 720 (1969).
79. McIlvried, H. G., paper presented at the ACS Div. of Pet. Chem., Houston, TX, 1970.
80. Mirza, A., Masood, M. A., Mallinkayunan, M. M., and Vardyeswaran, R., *Proceedings of the Symposium on Chemicals and Oil from Coal,* Dhanbad, India, 1969.
81. Moldavskii, B. L. and Prokopchuk, N. J., *Zh. Prikl. Khim.*, **5**, 619 (1932).
82. Moldavskii, B. L. and Livshits, S. E., *Zh. Obshch. Khim.,* **3**, 603 (1933).
83. Moldavskii, B. L. and Kumare, Z. I., *Zh. Obshch. Khim.,* **4**, 298 (1934).
84. Morooka, S. and Hamrin, C. E., *Chem. Eng. Sci.,* **32**, 125 (1977).
85. Morparia, N. P. and Sarkar, S., *Proceedings of the Symposium on Chemicals and Oil from Coal,* Dhanbad, India, 1969.
86. Nai, M., Sawahiraki, K., and Kabe, T., paper presented at the American Chemical Society–Chemical Society of Japan Joint Meeting, Honolulu, HI, 1979.
87. Nowack, C. J., Solash, J., and Delfosse, R. J., *Coal Processing Technology,* Vol. 3, American Institute of Chemical Engineers, 1977, p. 122.
88. Oblad, A. G., *Catal. Rev.,* **14** (1), 83 (1976).
89. Obolentsev, R. D. and Mashkina, A. V., *Doklady Akad. Nauk SSSR,* **119**, 1187 (1958).
90. Obolentsev, R. D. and Mashkina, A. V., *Heftenprod. Akad. Nauk SSSR,* **2**, 228 (1959).
91. O'Hara, J. B., Klumpe, H. W., Bela, A., and Jentz, N. E., AIChE 70th Annual Meeting, Paper 22d, 1977.
92. O'Hara, J. B., Klumpe, H. W., Bela, A., and Jentz, N. E., *Chem. Eng. Prog.,* **74** (8), 49 (1978).
93. O'Rear, D. J., Sullivan, R. F., and Strangeland, B. D., *ACS Div. of Fuel Chem. Prepr.,* **25** (1), 78 (1980).
94. Owens, P. J. and Amberg, C. H., *Adv. Chem. Ser.,* **33**, 18 (1961).
95. Owens, P. J. and Amberg, C. H., *Can. J. Chem.,* **40**, 941 (1962).
96. Owens, P. J. and Amberg, C. H., *Can. J. Chem.,* **40**, 947 (1962).
97. Pease, R. N. and Keighton, W. B., *Ind. Eng. Chem.,* **25**, 1012 (1933).
98. Peters, B. C., U.S. ERDA Report No. FE-1534-44, 1977.
99. Peters, B. C., U.S. ERDA Report No. FE-1534-46, 1977.
100. Peters, B. C., U.S. ERDA Report No. FE-1534-48, 1977.
101. Peters, B. C., U.S. ERDA Report No. FE-1534-49, 1977.
102. Peters, B. C., U.S. ERDA Report No. FE-1534-50, 1977.
103. Polozov, V. F., *Khim. Tverd. Topliva,* **6**, 78 (1932).
104. Potts, J. D., Hastings, K. E., and Unger, H., "Expanded Bed Hydroprocessing of Solvent Refined Coal (SRC-I) Extract," Fifth International Conference on Coal Gasification, Liquefaction, and Conversion to Electricity, University of Pittsburgh, 1978.
105. Qader, S. A. and Hill, G. R., *Ind. Eng. Chem. Proc. Des. Dev.,* **8** (4), 450 (1969).

106. Qader, S. A. and Hill, G. R., *Ind. Eng. Chem. Proc. Des. Dev.,* **8** (4), 462 (1969).

107. Qader, S. A. and Hill, G. R., *ACS Div. Fuel Chem. Prepr.,* **13** (1), 87 (1969).

108. Ralston, D., Govek, M., Graf, C., Jones, O. S., Tanabe, K., Klabunde, K. J., Woolsey, N. F., Baltisberger, R. J., and Sternberg, V. I., *Fuel Processing Technology,* **1**, 143 (1977/78).

109. Reynolds, T. W., NASA Report No. NASA TMX-3551 (1971).

110. Roberti, G., *Ann. Chim. Applicata,* **21**, 217 (1931).

111. Roberti, G., *Ann. Chim. Applicata,* **22**, 3 (1932).

112. Rollmann, L. D., *ACS Div. Fuel Chem. Prepr.,* **21**, 59 (1976).

113. Rollmann, L. D., *J. Catal.,* **46**, 243 (1977).

114. Ruberto, R. G. and Cronauer, D. C., *ACS Div. Petroleum Chem. Prepr.,* **23** (1), 264 (February 1978).

115. Satchell, D. P., Ph.D. Thesis, Oklahoma State University, 1974.

116. Satchell, D. P., U.S. ERDA Report No. FE-496-T7, 1974.

117. Satterfield, C. N. and Roberts, G. W., *AIChE J.,* **14**, 159 (1968).

118. Satterfield, C. N., Modell, M., and Mayer, J. R., *AIChE J.,* **21** (6), 1100 (1975).

119. Satterfield, C. N. and Cocchetto, J. R., *AIChE J.,* **21** (6), 1107 (1975).

120. Satterfield, C. N., Modell, M., Hites, R. A., and Dederck, C. J., *Ind. Eng. Chem. Proc. Des. Dev.,* **17** (2), 141 (1978).

121. Shaw, H., Kalfadelis, C. D., and Jahnig, C. E., Air Force Aero-Propulsion Laboratory Report No. AFAPL-TR-75-10, 1975.

122. Shih, S. S., Katzer, J. R., Kwart, H., and Stiles, A. B., *ACS Div. Pet. Chem. Prepr.,* **22** (3), 919 (1977).

123. Shih, S., Reiff, E., Zawadski, R., and Katzer, J. R., *ACS Div. Fuel Chem. Prepr.,* **23** (1), 99 (1978).

124. Sivasubramanian, R., Ph.D. Thesis, Oklahoma State University, 1977.

125. Sivasubramanian, R. and Crynes, B. L., *Ind. Eng. Chem. Prod. Res. Dev.,* **18** (3), 179 (1979).

126. Sonnemans, J., Goudriaan, F., and Mars, P., Fifth International Congress on Catalysis, Paper 76, Palm Beach, FL, 1972.

127. Sonnemans, J. and Mars, P., *J. Catal.,* **31**, 209 (1973).

128. Sonnemans, J., VanDenBerg, G. H., and Mars, P., *J. Catal.,* **31**, 220 (1973).

129. Sonnemans, J. and Mars, P., *J. Catal.,* **34**, 215 (1974).

130. Sonnemans, J., Neyens, W. J., and Mars, P., *J. Catal.,* **34**, 230 (1974).

131. Sooter, M. C., U.S. ERDA Report No. FE-496-T6, 1974.

132. Sooter, M. C. and Crynes, B. L., *Ind. Eng. Chem. Prod. Res. Dev.,* **19** (3), 199 (1975).

133. Stein, T. R., Bendoraitis, J. G., Cabal, A. V., Callen, R. B., Dabkowski, M. J., Heck, R. H., Ireland, H. R., and Simpson, C. A., EPRI Report No. EPRI AF-444, 1977.

134. Stein, T. R., Cabal, A. V., Callen, R. B., Dabkowski, M. J., Heck, R. H., Simpson, C. A., and Shih, S. S., EPRI Report No. EPRI AF-873, 1978.

135. Stein, T. R., Heck, R. H., Callen, R. B., and Voltz, S. E., AIChE 71st Annual Meeting, Paper 44e, 1978.

136. Stein, T. R., Voltz, S. E., and Callen, R. B., *Ind. Eng. Chem. Prod. Res. Dev.,* **16** (1), 61 (1977).

137. Stengler, W., Welker, J., and Leibnitz, E., *Freiberger Forschungsh.* **329A**, 51 (1964).
138. Stern, E. W., *J. Catal.*, **57**, 390 (1979).
139. Storch, H. H., "Hydrogenation of Coal and Tar," in *Chemistry of Coal Utilization*, H. H. Lowry, Ed., John Wiley and Sons, New York 1945.
140. Sullivan, R. F., U.S. DOE Report No. FE-2315-31, 1978.
141. Sullivan, R. F., U.S. DOE Report No. FE-2315-34, 1979.
142. Sullivan, R. F., U.S. DOE Report No. FE-2315-37, 1979.
143. Sullivan, R. F., Rudy, C. E., and Chem., H. C., U.S. DOE Report No. FE-2315-28, 1978.
144. Tan, G. and deRosset, U.S. DOE Report No. FE-2566-12, 1978.
145. Tan, G. and deRosset, A. J., U.S. DOE Report No. FE-2566-20, 1978.
146. Tan, G. and deRosset, A. J., U.S. DOE Report No. FE-2566-23, 1978.
147. Tan, G. and deRosset, A. J., U.S. DOE Report No. FE-2566-25, 1978.
148. Tanner, K. I. and Bell, A. T., Lawrence Berkeley Laboratory Report No. LBL-6807, 1978.
149. Uehara, K. and Fu, Y. C., *Petrotech.*, **2** (6), 514 (1979).
150. Veluswamy, L., Shabtai, J., and Oblad, A. G., *ACS Div. Fuel Chem. Prepr.*, **16** (2), 1 (1972).
151. Wan, K. T. and Crynes, B. L., *ACS Div. Fuel Chem. Prepr.* **16** (2), 1 (1972).
152. White, P. J., Jones, J. F., and Eddinger, R. T., *Hydrocarbon Processing*, **47** (12), 97 (1968).
153. Wilson, E. L. and Angelo, J. B., U.S. Patent 3,617,513, 1971.
154. Wiser, W. H., U.S. ERDA Report No. FE-2006-1, 1975.
155. Wiser, W. H., U.S. ERDA Report No. FE-2006-2, 1976.
156. Wiser, W. H., U.S. ERDA Report No. FE-2006-5, 1977.
157. Wiser, W. H., U.S. ERDA Report No. FE-2006-6, 1977.
158. Wiser, W. H., U.S. DOE Report No. FE-2006-11, 1978.
159. Wiser, W. H., U.S. DOE Report No. FE-2006-12, 1978.
160. Wiser, W. H., U.S. DOE Report No. FE-2006-13, 1978.
161. Wu, W. and Haynes, H. W., *ACS Div. Pet. Chem. Prepr.*, **20**, 466 (1975).
162. Wu, W. R. K. and Storch, H. H., *U.S. Bureau of Mines Bulletin* 633, 1968.
163. Zawadski, R., Shi, S. S., Katzer, J. R., and Kwart, H., submitted for publication to J. Catalysis (1979).

Author Index

Numbers in parentheses indicate the numbers of the references when these are cited in the text without the name of the author. Numbers set in *italics* designate those page numbers on which the complete literature citations are given.

Subject Index